SAT Subject Test:

MATHEMATICS LEVEL 2

Tenth Edition

RELATED KAPLAN TITLES FOR COLLEGE-BOUND STUDENTS

AP Biology

AP Calculus AB & BC

AP Chemistry

AP English Language & Composition

AP English Literature & Composition

AP Environmental Science

AP European History

AP Human Geography

AP Macroeconomics/Microeconomics

AP Physics B & C

AP Psychology

AP Statistics

AP U.S. Government & Politics

AP U.S. History

AP World History

ACT Strategies, Practice, and Review

ACT Premier

8 Practice Tests for the ACT

SAT Strategies, Practice, and Review

SAT Premier

SAT Total Prep

8 Practice Tests for the SAT

Evidence-Based Reading, Writing, and Essay Workbook for the SAT

Math Workbook for the SAT

SAT Subject Test: Biology E/M

SAT Subject Test: Chemistry

SAT Subject Test: Literature

SAT Subject Test: Mathematics Level 1

SAT Subject Test: Physics

SAT Subject Test: U.S. History

SAT® Subject Test:

MATHEMATICS LEVEL 2

Tenth Edition

New York

Published by Kaplan Publishing, a division of Kaplan, Inc.
750 Third Avenue
New York, NY 10017

Printed in the United States of America

10 9 8 7 6 5 4 3 2 1

ISBN 13: 978-1-5062-0923-4

Kaplan Publishing books are available at special quantity discounts to use for sales promotions, employee premiums, or educational purposes. For more information or to purchase books, please call the Simon & Schuster special sales department at 866-506-1949.

Table of Contents

PART THREE: Practice Tests

AVAILABLE ONLINE

FOR ANY TEST CHANGES OR LATE-BREAKING DEVELOPMENTS

kaptest.com/publishing

The material in this book is up-to-date at the time of publication. However, the College Board and Educational Testing Service (ETS) may have instituted changes in the test after this book was published. Be sure to read carefully the materials you receive when you register for the test. If there are any important late-breaking developments—or any changes or corrections to the Kaplan test preparation materials in this book—we will post that information online at **kaptest.com/publishing.**

FEEDBACK AND COMMENTS

kaplansurveys.com/books

We'd love to hear your comments and suggestions about this book. We invite you to fill out our online survey form at **kaplansurveys.com/books.** Your feedback is extremely helpful as we continue to develop high-quality resources to meet your needs.

Part One

The Basics

Chapter 1: **Getting Ready for the SAT Subject Test: Mathematics**

- Understand the SAT subject tests
- Content of SAT Subject Test: Mathematics
- Finding Your Level
- Level of Difficulty and Scoring

You're serious about going to the college of your choice. You wouldn't have opened this book otherwise. You've made a wise choice, because this book can help you to achieve your goal. It'll show you how to score your best on the SAT Subject Test: Mathematics. But before turning to the math content, let's look at the SAT subject tests generally.

UNDERSTAND THE SAT SUBJECT TESTS

The following background information about the SAT subject test is important to keep in mind as you get ready to prep for the SAT Subject Test: Mathematics Level 2.

What Are the SAT Subject Tests?

Known until 1994 as the College Board Achievement Tests and until 2004 as the SAT IIs, the SAT Subject Tests focus on specific disciplines: English, U.S. History, World History, Mathematics, Physics, Chemistry, Biology, and many foreign languages. Each test lasts one hour and consists entirely of multiple-choice questions. On any one test date, you can take one, two, or three subject tests.

How Do the SAT Subject Tests Differ from the SAT?

The SAT is largely a test of verbal and math skills. True, you need to know some vocabulary and some formulas for the SAT, but it's designed to measure how well you read and think rather than how much you remember. The SAT subject tests are very different. They're designed to measure what you know about specific disciplines. Sure, critical reading and thinking skills play a part on these tests, but their main purpose is to determine exactly what you know about math, history, chemistry, and so on.

How Do Colleges Use the SAT Subject Tests?

DUAL ROLE

Colleges use your SAT subject test scores in both admissions and placement decisions.

Many people will tell you that the SAT measures only your ability to perform on standardized exams—that it measures neither your reading and thinking skills nor your level of knowledge. Maybe they're right. But these people don't work for colleges. Those schools that require the SAT feel that it is an important indicator of your ability to succeed in college. Specifically, they use your scores in one or both of two ways: to help them make admissions and/or placement decisions.

Like the SAT, the SAT subject tests provide schools with a standard measure of academic performance, which they use to compare you to applicants from different high schools and different educational backgrounds. This information helps them to decide if you're ready to handle their curriculum.

SAT subject test scores may also be used to decide what course of study is appropriate for you once you've been admitted. A high score on an SAT Subject Test: Mathematics Level 2 may mean that you'll be exempted from an introductory math course.

Which SAT Subject Tests Should I Take?

CALL YOUR COLLEGES

Many colleges require you to take certain SAT subject tests. Check with all of the schools you're interested in applying to before deciding which tests to take.

The simple answer is: those that you'll do well on. High scores, after all, can only help your chances for admission. Unfortunately, many colleges demand that you take particular tests, usually including one of the Mathematics tests. Some schools will give you some choice in the matter, especially if they want you to take a total of three tests. Before you register to take any tests, therefore, check with the colleges you're interested in to find out exactly which tests they require. Don't rely on high school guidance counselors or admissions handbooks for this information. They might not give you accurate or current information.

When Are the SAT Subject Tests Administered?

Most of the SAT subject tests are administered six times a year: in October, November, December, January, May, and June. A few of the tests are offered less frequently. Due to admissions deadlines, many colleges insist that you take SAT subject tests no later than December or January of your senior year in high school. You may even have to take them sooner if you're interested in applying for "early admission" to a school. Those schools that use scores for placement decisions only may allow you to take SAT subject tests as late as May or June of your senior year. You should check with colleges to find out which test dates are most appropriate for you.

How Do I Register for the SAT Subject Tests?

The College Board administers the SAT Subject Tests, so you must sign up for the tests with them. The easiest way to register is online. Visit the College Board's website at www.collegeboard.org for registration information. If you register online, you immediately get to choose your test date and test center and you have 24-hour access to print your admission ticket. You'll need access to a credit card to complete online registration.

If you would prefer to register by mail, you must obtain a copy of the *Student Registration Guide for the SAT and SAT Subject Tests*. This publication contains all of the necessary information, including current test dates and fees. It can be obtained at any high school guidance office or directly from the College Board. If you have previously registered for an SAT or SAT Subject Test, you can reregister by telephone for an additional fee ($15 at the time of this printing). If you choose this option, you should still read the College Board publications carefully before you make any decisions.

CONTACT THE TEST MAKERS

Want to register for the SAT subject tests or get more info? You can get copies of the *Student Registration Guide for the SAT and the SAT Subject Tests* from the College Board. If you have a credit card, you can also register for the SAT subject test online. You can register by phone *only* if you have registered for an SAT or SAT subject test in the past (fee = $15).

College Board SAT Program
Domestic: 866-756-7346
International: 212-713-7789
www.collegeboard.org

How Are the SAT Subject Tests Scored?

The SAT subject tests are scored on a 200–800 scale.

What's a "Good" Score?

That's tricky. The obvious answer is: the score that the colleges of your choice demand. Keep in mind, though, that SAT subject test scores are just one piece of information that colleges will use to evaluate you. The decision to accept or reject you will be based on many criteria, including your high school transcript, your SAT scores, your recommendations, your personal statement, your interview (where applicable), your extracurricular activities, and the like. So, failure to achieve the necessary score doesn't automatically mean that your chances of getting in have been damaged. If you really want a numerical benchmark, a score of 600 is considered very solid.

What Should I Bring to the SAT Subject Test?

It's a good idea to get your test materials together the day before the test. You'll need an admission ticket; a form of identification (check the *Registration Guide* or College Board website to find out what is and what is not permissible); a few sharpened No. 2 pencils; a good eraser; and an approved calculator. Also, make sure that you know how to get to the test center.

CONTENT OF THE SAT SUBJECT TEST: MATHEMATICS

There's a lot of overlap between what's tested on Level 1 and what's tested on Level 2. But there's also a lot that's tested on Level 2 only and even some math that's tested on Level 1 only.

CONTENT AT A GLANCE

Level 1 covers two years of algebra and one year of geometry. Level 2 covers two years of algebra, one year of geometry, and one year of trigonometry and/or precalculus. There is no calculus on either test.

Level 1 is meant to cover the math you'd get in two years of algebra and one year of geometry. Level 2 is meant to cover that much math plus what you'd get in a year of trigonometry and/or precalculus. There is no calculus on either test.

In order to make room for more questions on more advanced topics, Level 2 has fewer questions on the more basic topics. In fact, it has no plane geometry questions at all. While we've included the official breakdown here, please visit the College Board's website for additional information regarding how the two tests differ in topic area.

Approximate Percentage of Content Coverage by Topic

Topic	Approx. % Level 1	Approx. % Level 2
Number and Operations	10–14%	10–14%
Algebra and Functions	38–42%	48–52%
Geometry and Measurement	38–42%	28–32%
Data Analysis, Statistics, and Probability	8–12%	8–12%

Level 2 is weighted toward the more advanced topics, but it still tests your understanding of the basics. For example, Level 2 has *no* plane geometry questions. But to do a lot of the more advanced Level 2 questions—solid geometry, coordinate geometry, trigonometry—you have to know all about plane geometry.

FIRM UP THE FOUNDATIONS

Don't review math haphazardly. Start with the fundamentals and work your way up to more advanced and esoteric topics.

The topics listed in the chart are not equally difficult. However, they do overlap. Think about how you learned these subjects. You didn't start with trigonometry or functions, did you? Of course not. Math is cumulative. Advanced subjects are built upon basic subjects. Firm up the foundations, and work your way up to more advanced topics.

The emphasis in Level 1 is on the foundations, while in Level 2 it's more on the advanced topics. But because the more advanced topics are built upon the basics, it can be said that for Level 2 you need to know everything that's tested on Level 1, plus a lot more.

FINDING YOUR LEVEL

The first thing to do to get ready for SAT Subject Test: Mathematics Level 2 is to be sure you are taking the right test. The information you need to make that decision, besides the differences in content, concerns level of difficulty, scoring, and reputation.

YOU DON'T NEED TO BE PERFECT

On Level 2 you can leave several questions unanswered, or even get them wrong, and still get an 800.

Level of Difficulty and Scoring

After content, the second and third factors to consider in deciding which test to take are *level of difficulty* and *scoring*. Level 2 questions are considerably more difficult than Level 1 questions. Some Level 2 questions are more difficult because they test more advanced topics. But even the Level 2 questions on basic math are generally more difficult than their counterparts on Level 1. This big difference in level of difficulty, however, is partially offset by differences in the score conversion tables. On Level 1, you would probably need to answer every question correctly to get an 800. On Level 2, however, you can get six or seven questions wrong and still get an 800. On Level 1, you would need a raw score of more than 20 (out of 50) to get a 500, but on Level 2, you can get a 500 with a raw score as low as about 11.

You don't need as many right answers to achieve a particular score on Level 2, so don't assume that you'll get a higher score by taking Level 1. If you've had a year of trigonometry and/or precalculus, you might actually find it easier to reach a particular score goal by taking Level 2.

TEST YOUR BEST

If you have the background to take Level 2, don't jump to the conclusion that you'll get a higher score by taking Level 1 instead.

Reputation

The final factor to consider is *reputation.* Admissions people know how much more math you have to know to get a good score on Level 2 than on Level 1. Your purpose is to demonstrate how much you've learned in high school. If you've learned enough math to take Level 2, then show it off!

Chapter 2: **SAT Subject Test Mastery**

- Use the structure of the test to your advantage
- Approaching SAT subject test questions
- Work strategically
- Stress management
- The final countdown

Now that you know a little about the SAT subject tests, it's time to let you in on a few basic test-taking skills and strategies that can improve your performance on them. You should practice these skills and strategies as you prepare for these tests.

USE THE STRUCTURE OF THE TEST TO YOUR ADVANTAGE

The SAT subject tests are different from the tests that you're used to taking. On your high school tests, you probably go through the questions in order. You probably spend more time on hard questions than on easy ones, since hard questions are generally worth more points. And you often show your work since your teachers tell you that how you approach questions is as important as getting the right answers.

None of this applies to the SAT subject tests. You can benefit from moving around within the tests, hard questions are worth the same as easy ones, and it doesn't matter how you calculate the answers—only what your answers are.

Take Advantage of the So-Called "Guessing Penalty"

You might have heard it said that the SAT subject tests have a "guessing penalty." That's a misnomer. It's really a *wrong-answer penalty.* If you guess wrong, you get a small penalty. If you guess right, you get full credit.

The fact is, if you can eliminate one or more answer choices as definitely wrong, you'll turn the odds in your favor and actually come out ahead by guessing. The fractional points that you lose are meant to offset the points you might get "accidentally" by guessing the correct answer. With practice, however, you'll see that it's often easy to eliminate *several* answer choices on some of the questions.

The Answer Grid Has No Heart

HIT THE SPOT

A common cause of major SAT subject test disasters is filling in all of the right answers—in the wrong spots. Every time you skip a question, circle it in your test booklet and be double sure that you skip it on the answer grid as well.

It sounds simple, but it's extremely important: Don't make mistakes filling out your answer grid. When time is short, it's easy to get confused going back and forth between your test booklet and your grid. If you know the answers but misgrid, you won't get the points. Here's how to avoid mistakes.

Always circle the questions you skip. Put a big circle in your test booklet around any question numbers that you skip. When you go back, these questions will be easy to locate. Also, if you accidentally skip a box on the grid, you'll be able to check your grid against your booklet to see where you went wrong.

Always circle the answers you choose. Circling your answers in the test booklet makes it easier to check your grid against your booklet.

Grid five or more answers at once. Don't transfer your answers to the grid after every question. Transfer them after every five questions. That way, you won't keep breaking your concentration to mark the grid. You'll save time and gain accuracy.

The SAT Subject Tests Are Highly Predictable

DON'T GET LOST

Learn SAT subject test directions as you prepare for the tests. You'll have more time to spend answering the questions on Test Day.

Because the format and directions of the SAT subject tests remain unchanged from test to test, you can learn the tests' setups in advance. On Test Day, the various question types on the tests shouldn't be new to you.

One of the easiest things you can do to help your performance on the SAT subject tests is to understand the directions before taking the test. Since the instructions are always the same, there's no reason to waste a lot of time on Test Day reading them. Learn them beforehand as you work through this book and the College Board publications.

Order of Difficulty

THERE'S A PATTERN

SAT Mathematics questions are generally arranged in order of difficulty—basic questions first, harder questions last.

SAT Subject Test: Mathematics questions are arranged in order of difficulty. The questions generally get harder as you work through different parts of a test. This pattern can work to your benefit. Try to be aware of where you are in a test. Be careful though. A few hard questions may appear early, a few easy ones late.

When working on more basic problems, you can generally trust your first impulse—the obvious answer is likely to be correct. As you get to the end of a test section, you need to be a bit more suspicious. Now the answers probably won't come as quickly and easily—if they do, look again because the obvious answers may be wrong.

Watch out for answers that just "look right." They may be distractors—wrong answer choices deliberately meant to entice you.

Move Around

You're allowed to skip around on the SAT subject tests. High scorers know this fact. They move through the tests efficiently. They don't dwell on any one question, even a hard one, until they've tried every question at least once.

When you run into questions that look tough, circle them in your test booklet and skip them for the time being. Go back and try again after you've answered the easier ones if you've got time. After a second look, troublesome questions can turn out to be remarkably simple.

If you've started to answer a question but get confused, quit and go on to the next question. Persistence might pay off in high school, but it usually hurts your SAT subject test scores. Don't spend so much time answering one hard question that you use up three or four questions' worth of time. That'll cost you points, especially if you don't even get the hard question right.

LEAP AHEAD

You should do the questions in the order that's best for you. Don't pass up the opportunity to score easy points by wasting time on hard questions. Skip hard questions until you've gone through every question once. Come back to them later.

APPROACHING SAT SUBJECT TEST QUESTIONS

Apart from knowing the setup of the SAT subject tests that you'll be taking, you've got to have a system for attacking the questions. You wouldn't travel around an unfamiliar city without a map, and you shouldn't approach any SAT subject test without a plan. What follows is the best method for approaching the questions systematically.

Think First

Think about the questions before you look at the answers. The test makers love to put distracters among the answer choices. Distracters are answers that look like they're correct, but aren't. If you jump right into the answer choices without thinking first about what you're looking for, you're much more likely to fall for one of these traps.

THINK FIRST

Always try to think of the answer to a question before you shop among the answer choices. If you've got some idea of what you're looking for, you'll be less likely to be fooled by "trap" choices.

Be a Good Guesser

You already know that the "guessing penalty" can work in your favor. Don't simply skip questions that you can't answer. Spend some time with them in order to see whether you can eliminate any of the answer choices. If you can, it pays for you to guess.

GUESSING RULE

Don't guess unless you can eliminate at least one answer choice. Don't leave a question blank unless you have absolutely no idea about it.

Pace Yourself

The SAT subject tests give you a lot of questions in a short period of time. To get through the tests, you can't spend too much time on any single question. Keep moving through

the tests at a good speed. If you run into a hard question, circle it in your test booklet, skip it, and come back to it later if you have time.

> **SPEED LIMIT**
>
> Work quickly on easier questions to leave more time for harder questions. But don't work so quickly that you lose points by making careless errors. And it's okay to leave some questions blank if you have to—you can still get a high score.

Don't spend the same amount of time on every question. Ideally, you should be able to work through the easier questions at a brisk, steady clip, and use a little more time on the harder questions. One caution: Don't rush through basic questions just to save time for the harder ones. The basic questions are points in your pocket. All questions are worth the same number of points. Therefore, don't worry about answering the more difficult questions—work methodically through the easier questions and rack up the points! Remember, you don't earn any extra credit for answering hard questions.

WORK STRATEGICALLY

As you'll see throughout this book, there is often more than one way to solve a problem. Be on the lookout for the quicker route to the answer. Making such choices requires awareness of the options and lots of practice.

- You can set up and solve an equation with or without the aid of a calculator (the subject of chapter 3).
- You can Pick Numbers, for example, when answer choices are algebraic expressions. We'll look at several examples of this approach in the "Mathematics Level 2 Review" chapters.
- You can Backsolve, which is often the quickest route to an answer when the question describes an equation and the answer choices are simple numbers. To Backsolve, generally start with choice (B) or (D). That gives you a 2-in-5 chance that a single calculation will give you the answer (as long as you can tell whether a larger or smaller number is desirable). If you try (B), you'll know the right answer (1) if (B) is correct, or (2) if it's too high—making (A) correct. If that calculation isn't enough, only one more is needed. Try the middle number of the remaining three, (D). If it's right, it's your answer. If it's too high (C) is correct, and if it's too low (E) is.
- You may be able to "eyeball" for the answer. Look at the figure provided; mark it up if that helps. Often, when no figure is provided, just drawing one will make the answer apparent.

Locate quick points. Some questions can be done more quickly than others because they require less work or because choices can be eliminated more easily. If you start to run out of time, look for these quicker questions.

Set a Target Score. Naturally, you want the best score you can earn to maximize your college options. But choose a realistic target score that is above the average for the school or schools you want to attend. By keeping an eye on which questions you are sure of and which you guessed on, you can monitor your progress toward this goal. Of course, you shouldn't stop practicing (or taking the test) when you reach your target score—but you can be more relaxed and confident.

When you take an SAT subject test, you have one clear objective in mind: to score as many points as you can. It's that simple. The rest of this book will show you how to do that on the SAT Subject Test: Mathematics 2.

STRESS MANAGEMENT

The countdown has begun. Your date with THE TEST is looming on the horizon. Anxiety is on the rise. The butterflies in your stomach have gone ballistic. Perhaps you feel as if the last thing you ate has turned into a lead ball. Your thinking is getting cloudy. Maybe you think you won't be ready. Maybe you already know your stuff, but you're going into panic mode anyway. Worst of all, you're not sure of what to do about it.

Don't freak! It is possible to tame that anxiety and stress—before and during the test. We'll show you how. You won't believe how quickly and easily you can deal with that killer anxiety.

Identify the Sources of Stress

In the space provided, jot down anything you identify as a source of your test-related stress. The idea is to pin down that free-floating anxiety so that you can take control of it. Here are some common examples to get you started:

- I always freeze up on tests.
- I'm nervous about trig (or functions, or geometry, etc.).
- I need a good/great score to go to Acme College.
- My older brother/sister/best friend/girl- or boyfriend did really well. I must at least match their scores.
- My parents, who are paying for school, will be really disappointed if I don't test well.
- I'm afraid of losing my focus and concentration.
- I'm afraid I'm not spending enough time preparing.
- I study like crazy, but nothing seems to stick in my mind.
- I always run out of time and get panicky.
- I feel as though thinking is becoming like wading through thick mud.

THINK GOOD THOUGHTS

Create a set of positive but brief affirmations and mentally repeat them to yourself just before you fall asleep at night. (That's when your mind is very open to suggestion.) You'll find yourself feeling a lot more positive in the morning.

Periodically repeating your affirmations during the day makes them more effective.

Sources of Stress

______________________ ______________________

______________________ ______________________

______________________ ______________________

______________________ ______________________

Take a few minutes to think about the things you've just written down. Then rewrite them in order like this: List the statements you most associate with your stress and anxiety first and put the least disturbing items last. Chances are, the top of the list is a fairly accurate description of exactly how you react to test anxiety, both physically and mentally. The later items usually describe your fears (disappointing Mom and Dad, looking bad, etc.). As you write the list, you're forming a hierarchy of items so you can deal first with the anxiety provokers that bug you most. Very often, taking care of the major items from the top of the list goes a long way toward relieving overall testing anxiety. You probably won't have to bother with the stuff you placed last.

VERY SUPERSTITIOUS

Stress expert Stephen Sideroff, PhD, tells of a client who always stressed out before, during, and even after taking tests. Yet, she always got outstanding scores. It became obvious that she was thinking superstitiously—subconsciously believing that the great scores were a result of her worrying. She didn't trust herself and believed that if she didn't worry she wouldn't study hard enough. Sideroff convinced her to take a risk and work on relaxing before her next test. She did, and her test results were still as good as ever—which broke her cycle of superstitious thinking.

Make the Most of Your Prep Time

Lack of control is one of the prime causes of stress. A ton of research shows that if you don't have a sense of control over what's happening in your life, you can easily end up feeling helpless and hopeless. So, just having concrete things to do and to think about—taking control—will help reduce your stress.

Strengths and Weaknesses

Take one minute to list the areas of the test that you are good at. They can be general ("algebra") or specific ("quadratic equations"). Put down as many as you can think of and, if possible, time yourself. Write for the entire time; don't stop writing until you've reached the one-minute stopping point.

Strong Test Subjects

______________________ ______________________

______________________ ______________________

______________________ ______________________

______________________ ______________________

Next, take one minute to list areas of the test you're not so good at, just plain bad at, have failed at, or keep failing at. Again, keep it to one minute and continue writing until you reach the cutoff. Don't be afraid to identify and write down your weak spots! In all probability, as you do both lists, you'll find you are strong in some areas and not so strong in others. Taking stock of your assets and liabilities lets you know the areas you don't have to worry about and the ones that will demand extra attention and effort.

Weak Test Subjects

_______________ _______________

_______________ _______________

_______________ _______________

_______________ _______________

Facing your weak spots gives you some distinct advantages. It helps a lot to find out where you need to spend extra effort. Increased exposure to tough material makes it more familiar and less intimidating. (After all, we mostly fear what we don't know and are probably afraid to face.) You'll feel better about yourself because you're dealing directly with areas of the test that bring on your anxiety. You can't help feeling more confident when you know you're actively strengthening your chances of earning a higher overall test score.

Now, go back to the "good" list and expand it for two minutes. Take the general items on that first list and make them more specific; take the specific items and expand them into more general conclusions. Naturally, if anything new comes to mind, jot it down. Focus all of your attention and effort on your strengths. Don't underestimate yourself or your abilities. Give yourself full credit. At the same time, don't list strengths you don't really have; you'll only be fooling yourself.

Expanding from general to specific might go as follows. If you listed "algebra" as a broad topic you feel strong in, you would then narrow your focus to include areas of this subject about which you are particularly knowledgeable. Your areas of strength might include multiplying polynomials, working with exponents, factoring, solving simultaneous equations, and so forth.

Whatever you know comfortably goes on your "good" list. Okay. You've got the picture. Now, get ready, check your starting time, and start writing down items on your expanded "good" list.

Strong Test Subjects: An Expanded List

_______________ _______________

_______________ _______________

_______________ _______________

_______________ _______________

After you've stopped, check your time. Did you find yourself going beyond the two minutes allotted? Did you write down more things than you thought you knew? Is it possible you know more than you've given yourself credit for? Could that mean you've found a number of areas in which you feel strong?

You just took an active step toward helping yourself. Notice any increased feelings of confidence? Enjoy them.

Here's another way to think about your writing exercise. Every area of strength and confidence you can identify is much like having a reserve of solid gold at Fort Knox. You'll be able to draw on your reserves as you need them. You can use your reserves to solve difficult questions, maintain confidence, and keep test stress and anxiety at a distance. The encouraging thing is that every time you recognize another area of strength, succeed at coming up with a solution, or get a good score on a test, you increase your reserves. And there is absolutely no limit to how much self-confidence you can have or how good you can feel about yourself.

Imagine Yourself Succeeding

This next little group of exercises is both physical and mental. It's a natural follow-up to what you've just accomplished with your lists.

OCEAN DUMPING

Visualize a beautiful beach with white sand, blue skies, sparkling water, a warm sun, and seagulls. See yourself walking on the beach, carrying a small plastic pail. Stop at a good spot and put your worries and whatever may be bugging you into the pail. Drop it at the water's edge and watch it drift out to sea. When the pail is out of sight, walk on.

First, get yourself into a comfortable sitting position in a quiet setting. Wear loose clothes. If you wear glasses, take them off. Then, close your eyes and breathe in a deep, satisfying breath of air. Really fill your lungs until your rib cage is fully expanded and you can't take in any more. Then, exhale the air completely. Imagine you're blowing out a candle with your last little puff of air. Do this two or three more times, filling your lungs to their maximum and emptying them totally. Keep your eyes closed, comfortably but not tightly. Let your body sink deeper into the chair as you become even more comfortable.

With your eyes shut you can notice something very interesting. You're no longer dealing with the worrisome stuff going on in the world outside of you. Now you can concentrate on what happens *inside* you. The more you recognize your own physical reactions to stress and anxiety, the more you can do about them. You might not realize it, but you've begun to regain a sense of being in control.

Let images begin to form on the "viewing screens" on the back of your eyelids. You're experiencing visualizations from the place in your mind that makes pictures. Allow the images to come easily and naturally; don't force them. Imagine yourself in a relaxing situation. It might be in a special place you've visited before or one you've read about. It can be a fictional location that you create in your imagination, but a real-life memory of a place or situation you know is usually better. Make it as detailed as possible and notice as much as you can.

Stay focused on the images as you sink farther back into your chair. Breathe easily and naturally. You might have the sensations of any stress or tension draining from your muscles and flowing downward, out your feet and away from you.

Take a moment to check how you're feeling. Notice how comfortable you've become. Imagine how much easier it would be if you could take the test feeling this relaxed and in this state of ease. You've coupled the images of your special place with sensations of comfort and relaxation. You've also found a way to become relaxed simply by visualizing your own safe, special place.

Now, close your eyes and start remembering a real-life situation in which you did well on a test. If you can't come up with one, remember a situation in which you did something (academic or otherwise) that you were really proud of—a genuine accomplishment. Make the memory as detailed as possible. Think about the sights, the sounds, the smells, even the tastes associated with this remembered experience. Remember how confident you felt as you accomplished your goal. Now start thinking about the upcoming test. Keep your thoughts and feelings in line with that successful experience. Don't make comparisons between them. Just imagine taking the upcoming test with the same feelings of confidence and relaxed control.

This exercise is a great way to bring the test down to earth. You should practice this exercise often, especially when the prospect of taking the exam starts to bum you out. The more you practice it, the more effective the exercise will be for you.

COUNSELING

Don't forget that your school probably has counseling available. If you can't conquer test stress on your own, make an appointment at the counseling center. That's what counselors are there for.

Control Physical Stress

Exercise Your Frustrations Away

Whether it is jogging, walking, biking, mild aerobics, pushups, or a pickup basketball game, physical exercise is a very effective way to stimulate both your mind and body and to improve your ability to think and concentrate. A surprising number of students get out of the habit of regular exercise, ironically because they're spending so much time prepping for exams. Also, sedentary people—this is a medical fact—get less oxygen to the blood and hence to the head than active people. You can live fine with a little less oxygen; you just can't think as well.

Any big test is a bit like a race. Thinking clearly at the end is just as important as having a quick mind early on. If you can't sustain your energy level in the last sections of the exam, there's too good a chance you could blow it. You need a fit body that can weather the demands any big exam puts on you. Along with a good diet and adequate sleep, exercise is an important part of keeping yourself in fighting shape and thinking clearly for the long haul.

There's another thing that happens when students don't make exercise an integral part of their test preparation. Like any organism in nature, you operate best if all your "energy systems" are in balance. Studying uses a lot of energy, but it's all mental. When you take a study break, do something active instead of raiding the fridge or vegging out in front of the TV. Take a 5- to 10-minute activity break for every 50 or 60 minutes that you study. The physical exertion gets your body into the act, which helps to keep your mind and body in sync. Then, when you finish studying for the night and hit the sack,

PLAY THE MUSIC

If you want to play music, keep it low and in the background. Music with a regular, mathematical rhythm—reggae, for example—aids the learning process. A recording of ocean waves is also soothing.

CYBERSTRESS

If you spend a lot of time in cyberspace anyway, do a search for the phrase *stress management*. There's a ton of stress advice on the Net, including material specifically for students.

NUTRITION AND STRESS: THE DOS AND DON'TS

Do eat:

- Fruits and vegetables (raw is best, or just lightly steamed or nuked)
- Low-fat protein such as fish, skinless poultry, beans, and legumes (like lentils)
- Whole grains such as brown rice, whole-wheat bread, and pastas (no bleached flour)

Don't eat:

- Refined sugar; sweet, high-fat snacks (Simple carbohydrates like sugar make stress worse, and fatty foods lower your immunity.)
- Salty foods (They can deplete potassium, which you need for nerve functions.)

you won't lie there, tense and unable to sleep because your head is overtired and your body wants to pump iron or run a marathon.

One warning about exercise, however: It's not a good idea to exercise vigorously right before you go to bed. This could easily cause sleep-onset problems. For the same reason, it's also not a good idea to study right up to bedtime. Make time for a "buffer period" before you go to bed: For 30 to 60 minutes, just take a hot shower, meditate, or simply veg out.

Take a Deep Breath

Conscious attention to breathing is an excellent way to manage test stress (or any stress, for that matter). The majority of people who get into trouble during tests take shallow breaths. They breathe using only their upper chests and shoulder muscles, and they may even hold their breath for long periods of time. Conversely, the test taker who by accident or design keeps breathing normally and rhythmically is likely to be more relaxed and in better control during the entire test experience.

So, now is the time to get into the habit of relaxed breathing. Do the next exercise to learn to breathe in a natural, easy rhythm. By the way, this is another technique you can use during the test to collect your thoughts and ward off excess stress. The entire exercise should take no more than three to five minutes.

With your eyes still closed, breathe in slowly and *deeply* through your nose. Hold the breath for a bit, and then release it through your mouth. The key is to breathe slowly and deeply by using your diaphragm (the big band of muscle that spans your body just above your waist) to draw air in and out naturally and effortlessly. Breathing with your diaphragm encourages relaxation and helps minimize tension. Try it and notice how relaxed and comfortable you feel.

THE FINAL COUNTDOWN

Quick Tips for the Days Just Before the Exam

- The best test takers do less and less as the test approaches. Taper off your study schedule and take it easy on yourself. You want to be relaxed and ready on the day of the test. Give yourself time off, especially the evening before the exam. By then, if you've studied well, everything you need to know is firmly stored in your memory banks.
- Positive self-talk can be extremely liberating and invigorating, especially as the test looms closer. Tell yourself things such as "I choose to take this test," rather than "I have to"; "I will do well," rather than "I hope things go well"; "I can," rather than "I cannot." Be aware of negative, self-defeating thoughts and images and immediately counter any you become aware of. Replace them with affirming statements that encourage your self-esteem and confidence. Create and practice visualizations that build on your positive statements.
- Get your act together sooner rather than later. Have everything (including choice of clothing) laid out days in advance. Most important, know where the test will be held and the easiest, quickest way to get there. You will gain great peace of mind if you know that all the little details—gas in the car, directions, etc.—are firmly in your control before the day of the test.
- Experience the test site a few days in advance. This is very helpful if you are especially anxious. If at all possible, find out what room your part of the alphabet is assigned to and try to sit there (by yourself) for a while. Better yet, bring some practice material and do at least a section or two, if not an entire practice test, in that room. In this situation, familiarity doesn't breed contempt; it generates comfort and confidence.
- Forego any practice on the day before the test. It's in your best interest to marshal your physical and psychological resources for 24 hours or so. Even race horses are kept in the paddock and treated like royalty the day before a race. Keep the upcoming test out of your consciousness; go to a movie, take a pleasant hike, or just relax. Don't eat junk food or tons of sugar. And—of course—get plenty of rest the night before. Just don't go to bed too early. It's hard to fall asleep earlier than you're used to, and you don't want to lie there thinking about the test.

DRESS FOR SUCCESS

On the day of the test, wear loose layers. That way, you'll be prepared no matter what the temperature of the room is. (An uncomfortable temperature will just distract you from the job at hand.)

And, if you have an item of clothing that you tend to feel "lucky" or confident in—a shirt, a pair of jeans, whatever—wear it. A little totem couldn't hurt.

PACK YOUR BAG

Gather your test materials the day before the test. You'll need the following:

- Admission ticket
- Proper form of ID
- Some sharpened No. 2 pencils
- Good eraser
- An approved graphing or scientific calculator
- Spare calculator batteries

Handling Stress During the Test

The biggest stress monster will be the test itself. Fear not; there are methods of quelling your stress during the test.

- Keep moving forward instead of getting bogged down in a difficult question. You don't have to get everything right to achieve a fine score. The best test takers skip difficult material temporarily in search of the easier stuff. They mark the ones that require extra time and thought. This strategy buys time and builds confidence so you can handle the tough stuff later.
- Don't be thrown if other test takers seem to be working more furiously than you are. Continue to spend your time patiently thinking through your answers; it's going to lead to better results. Don't mistake the other people's sheer activity as signs of progress and higher scores.
- *Keep breathing!* Weak test takers tend to forget to breathe properly as the test proceeds. They start holding their breath without realizing it, or they breathe erratically or arrhythmically. Improper breathing interferes with clear thinking.
- Some quick isometrics during the test—especially if concentration is wandering or energy is waning—can help. Try this: Put your palms together and press intensely for a few seconds. Concentrate on the tension you feel through your palms, wrists, forearms, and up into your biceps and shoulders. Then, quickly release the pressure. Feel the difference as you let go. Focus on the warm relaxation that floods through the muscles. Now you're ready to return to the task.
- Here's another isometric that will relieve tension in both your neck and eye muscles. Slowly rotate your head from side to side, turning your head and eyes to look as far back over each shoulder as you can. Feel the muscles stretch on one side of your neck as they contract on the other. Repeat five times in each direction.

WHAT ARE "SIGNS OF A WINNER," ALEX?

Here's some advice from a Kaplan instructor who won big on *Jeopardy!*™ In the green room before the show, he noticed that the contestants who were quiet and "within themselves" were the ones who did great on the show. The contestants who did not perform as well were the ones who were fact-cramming, talking a lot, and generally being manic before the show. Lesson: Spend the final hours leading up to the test getting sleep, meditating, and generally relaxing.

With what you've just learned here, you're armed and ready to do battle with the test. This book and your studies will give you the information you'll need to answer the questions. It's all firmly planted in your mind. You also know how to deal with any excess tension that might come along when you're studying for and taking the exam. You've experienced everything you need to tame your test anxiety and stress. You're going to get a great score.

Chapter 3: **The Calculator**

- Have the right calculator
- Use your calculator strategically

In this chapter we'll take an important look at the calculator. We'll discuss what kind to use and how and when to use it during the SAT Subject Test: Mathematics 2.

HAVE THE RIGHT CALCULATOR

First of all, get a graphing or scientific calculator. Don't try to take a Math subject Test without one. It's not like the SAT, for which a calculator is permitted but not really necessary. On the SAT Subject Test: Mathematics 2, a calculator is essential.

Know what kind of calculator you should use. Almost any kind of calculator is allowed, even a programmable or graphing calculator. According to the College Board, the majority of students bring a graphing calculator to the Math 2 test; therefore, the tests are developed with the expectation that most students are using graphing calculators. Be sure your calculator performs the following functions:

- Sine, cosine, and tangent
- Arcsine, arccosine, and arctangent
- Squares, cubes, and other powers
- Square roots and cube roots
- Base-10 logarithms

No Laptops Allowed

Excluded are laptops or other computers, tablets, cell phones, or smartphones; machines that print, make noise, or need to be plugged in; or calculators with a typewriter keypad or an angled readout screen. Check the College Board website for further specifics.

Advantages of a graphing or scientific calculator. In addition to that list of necessary functions, there are other advantages to using a scientific or graphing calculator.

Scientific calculators with a two-line display let users see what they are typing before performing the operation, so they can catch input errors. Most scientific calculators will also have parentheses keys; these calculators will handle PEMDAS—order of operations.

DON'T FORGET PEMDAS

Parentheses
Exponents
Multiplication
Division
Addition
Subtraction

Most scientific calculators offer statistical operations: calculating permutations (${}_nP_r$), combinations (${}_nC_r$), and factorials (!). Newer calculators also have fraction and mixed number capabilities. Many have two very useful keys—the ANS key and the ENTRY key. The ANS key is a substitute for typing in the last calculated answer—saving time and avoiding errors. The ENTRY key will reenter the last string of calculator keystrokes. This is handy for repeated calculations or to correct an incorrect keystroke sequence.

In addition to all those features, graphing calculators have graphing features, including the ability to:

- display graphs
- find the intersection of simultaneous functions
- find the roots of functions
- display the associated table for a graphed function
- find the equation for a table of values (regressions)
- determine maximums and minimums of functions

Here are a few cautions when using calculators:

- Keep in mind that the graphing calculator graphs functions only. It is imperative that you understand this if you try to graph a circle or an ellipse.
- Don't round any values until the final step in a problem. Rounding too soon could result in an incorrect solution.
- Always double-check the keystrokes that you enter. It is easy to make a mistake. The home screen of a graphing calculator will allow you to see what was input even after the function is executed.

CALCULATORS: WHEN AND HOW

Just as important as knowing when and how to use your calculator is knowing when and how not to use your calculator.

Get to know your calculator. Getting the right calculator is only the first step. You must then get used to working with it. Going into the test with an unfamiliar calculator is not much better than going in with no calculator at all.

Practice using your calculator on testlike questions. It's not enough to know how to use your calculator. You need to know how to use it effectively on SAT Subject Test: Mathematics questions. Whenever you work with this book, whenever you take a practice test, practice with the very calculator you will use on Test Day. With some experience you will learn when to use and when *not* to use, how to use and how *not* to use your calculator on test questions.

Make sure your calculator is in good working order. You'll feel a lot more confident if you put new batteries in your calculator the night before the test and then check it out to see that it's working properly. You should also take spare batteries with you to the test.

USE YOUR CALCULATOR STRATEGICALLY

Here are some tips for using your calculator strategically during the test.

Don't Use Your Calculator Too Often

You will not need your calculator for every question. Top scorers use their calculators for 40–50 percent of the questions on Level 1 and for 55–65 percent of the questions on Level 2. This is still a math test, not a calculator test. Success depends more on your problem-solving skills than on your calculator skills. The calculator is just one tool. Use it wisely and sparingly.

According to the test makers, your calculator will be of little or no use on about 35–45 percent of the questions on the test. Here's an example where the calculator is no help:

1. If $x \neq \pm 1$, then $\frac{x+1}{x-1} - \frac{x-1}{x+1} =$

(A) $\frac{2x}{x-1}$ (B) $\frac{2x}{x^2+1}$ (C) $\frac{2x}{x^2-1}$ (D) $\frac{4x}{x^2+1}$ (E) $\frac{4x}{x^2-1}$

To answer this question, you need to be adept at algebraic manipulation. Your calculator won't help here. To subtract fractions, even algebraic fractions like these, you need a common denominator, which in this case is the product of the denominators $x - 1$ and $x + 1$.

$$\frac{x+1}{x-1} - \frac{x-1}{x+1} = \frac{(x+1)(x+1)}{(x-1)(x+1)} - \frac{(x-1)(x-1)}{(x-1)(x+1)}$$

$$= \frac{x^2+2x+1}{x^2-1} - \frac{x^2-2x+1}{x^2-1}$$

$$= \frac{x^2+2x+1-x^2+2x-1}{x^2-1}$$

$$= \frac{4x}{x^2-1}$$

The answer is (E).

Here's a question for which you have a choice. You can answer it just as quickly and easily with or without a calculator:

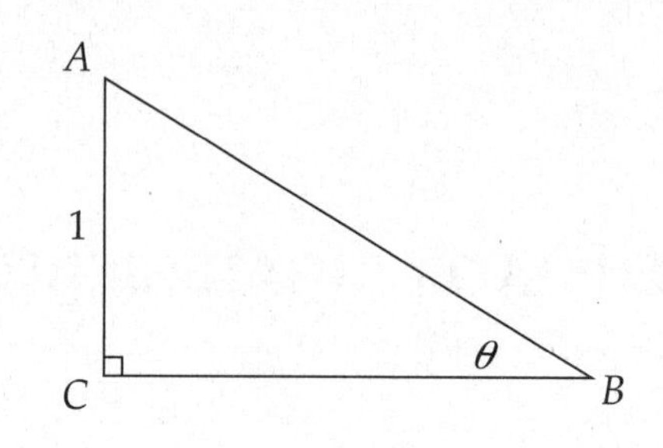

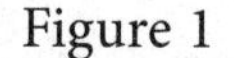
Figure 1

2. In Figure 1, if $\sin\theta = 0.5$, what is the length of BC ?

(A) 1.25
(B) 1.41
(C) 1.50
(D) 1.73
(E) 2.00

You could use your calculator to find out what angle has a sine of 0.5, but you might just know that it's 30°. This is a 30-60-90 triangle, and BC is the longer leg, which is equal to the shorter leg times $\sqrt{3}$. Since $AC = 1$, then $BC = \sqrt{3}$. You could use your calculator to find the square root of 3, but you should probably know that it's about 1.73. The answer is (D).

Don't Use Your Calculator Too Early

When you do use your calculator, it will usually be at a later stage in solving a problem. Never start punching calculator keys before you've given the problem a little thought. Know what you're doing, and where you're going, before you start calculating.

DON'T GET PUNCHY

Don't start punching calculator keys until you've given the problem some thought.

Here's a question that really requires a calculator, but not until late in the problem solving:

3. If $0° < x < 90°$, and $\tan^2 x - \tan x = 6$, which of the following could be the value of x in degrees (rounded to the nearest degree)?

(A) 63 (B) 67 (C) 72 (D) 77 (E) 81

What you have here is essentially a quadratic equation in which the unknown is tan *x*. For simplicity's sake, substitute *y* for tan *x*:

$$\tan^2 x - \tan x = 6$$
$$y^2 - y = 6$$

Now solve for *y*:

$$y^2 - y = 6$$
$$y^2 - y - 6 = 0$$
$$(y-3)(y+2) = 0$$
$$y = 3 \text{ or } -2$$

Now you know that tan x = 3 or –2. Since x is a positive acute angle, the tangent is positive, and tan x = 3. Now's the time to use your calculator. You want to know what acute angle has a tangent of 3. In other words, what you're looking for is the arctangent of 3. Nobody expects you to know the value of arctan(3) off the top of your head. Use your calculator:

$$\tan x = 3$$
$$x = \arctan(3) \approx 71.57$$

The answer is (C).

In the next example, a calculator is not required, but knowledge of certain operations could save you time.

4. Given the table below, which function represents the values in the table?

x	y
1	–1
3	5
5	11
7	17
9	23

(A) $y = 2x - 3$ (B) $x = 3y - 4$ (C) $x = 2y - 3$ (D) $y = 3x - 4$ (E) $y = x - 2$

One way to solve this problem is to substitute the values of x from the table and see which function results in the corresponding y-values. The correct answer, (D), will give an outcome of each y-value when the x-value to its left is plugged in for x. For example, 3(1) – 4 = 3 – 4 = –1, so the first pair satisfies the equation. Since 3(3) – 4 = 9 – 4 = 5, the second pair also satisfies the equation, and so on.

An alternative method would be to input the *x*- and *y*-values into two separate lists on a graphing calculator and perform a linear regression on the data. This approach may prove to be much less time consuming than trial and error; however, you need to be both familiar with and comfortable with the method for finding a linear regression on your particular calculator.

Here is a question in which the calculator is essential.

5. Solve for x:

   $7^x = 18.52$

   (A) 2.65 (B) 0.18 (C) 21.9 (D) 129.64 (E) 1.5

$$\log 7^x = \log 18.52$$

$$x(\log 7) = \log 18.52$$

$$x = \frac{\log 18.52}{\log 7}$$

$$x \approx 1.5, \text{ or choice (E)}$$

It is best, however, to use the calculator on the last step and enter both log functions in the same set of keystrokes. If you calculate the log of 18.52 alone, you will get 1.267640982. You could make careless errors and waste precious time as you try to retype this number into the calculator. Another approach to this problem could be to try each answer choice, using the power key on your keypad.

In the following problem, a graphing calculator would give you a decided advantage:

6. **What is a possible solution to the system?**

$$y = x^2 + 3x + 2$$

$$x = \frac{1}{2}y - 2$$

   (A) (–1,2) (B) (0,4) (C) (0,2) (D) (1,6) (E) (2,12)

This type of problem can be solved by the substitution method for simultaneous equations. An efficient alternative method is first to solve each equation for y and then graph the two functions on a graphing calculator. You can then use the Intersect function to solve the system, arriving at the points (1,6) and (–2,0). Choice (D) names one of these.

To solve the following volume problem, you can use the cube root, or $\sqrt[3]{x}$, key on the calculator. If your calculator does not have this key, remember that you can use the power key and enter the exponent of $\frac{1}{3}$.

7. What is the edge length of a cube whose volume is 4,096 cm^3?

   (A) 16 cm (B) 1,365.33 cm (C) 64 cm (D) 455.11 cm (E) None of these

$$V = s^3$$

$$4{,}096 = s^3$$

$$\sqrt[3]{4{,}096} = s$$

$$16 \text{ cm} = s$$

Answer choice (A) is correct.

In any of the given examples, keep in mind that the calculator is a tool for calculating. Your brain is the most important tool you will take into the test. Pay attention as you read further in the lessons to see how and when the calculator should be used. Keep your trusty calculator at your side but remember to practice restraint.

Part Two

Mathematics Level 2 Review

Chapter 4: **Algebra**

- Expressions and exponents
- Operations with polynomials and factoring
- The Golden Rule of equations
- Quadratic equations

According to the test makers' official breakdown, about 18 percent of Math 2 questions are algebra questions. But that's counting only the ones that are explicitly and primarily algebra questions. In fact, almost all questions involve algebra. Most functions questions and coordinate geometry questions are algebraic. Most plane geometry and solid geometry questions involve algebraic formulas, and many use algebraic expressions for lengths and angle measures. Most word problems are best solved algebraically. *Algebra is fundamental to Math 2. You can't get a good score without it.*

ALGEBRA FACTS AND FORMULAS IN THIS CHAPTER

- The Five Rules of Exponents
- Combining Like Terms
- Multiplying Monomials
- Multiplying Binomials—FOIL
- Classic Factorables
- Quadratic Formula
- Inequalities and Absolute Value

HOW TO USE THIS CHAPTER

Maybe you know all the algebra you need. Find out by taking the Algebra Diagnostic Test. The 10 Diagnostic questions are typical of those on the Mathematics subject tests. Check your answers using the answer key following the test. No matter how you score, don't worry! The answer key also shows where to find a detailed explanation for each question. The "Find Your Study Plan" section that follows the test will suggest next steps based on your performance on the Diagnostic.

Find Your Level

How you use this chapter really depends on how much time you have to prep. Find your level and pace below.

Standard Plan. No matter how well you do on the Diagnostic Test, go on and read everything in this chapter and do all the practice problems.

Shortcut: If you can answer at least seven of the ten questions on the Algebra Diagnostic Test correctly, then you already know the material in this chapter well enough to move on.

Panic Plan. The material in this chapter is vital. If you can't get most of the questions on the Algebra Diagnostic Test right, you should study this chapter—even if it's just two days before Test Day.

ALGEBRA DIAGNOSTIC TEST

DO YOUR FIGURING HERE

10 Questions (12 Minutes)

Directions: Solve the following problems. Fill in the oval corresponding to the best answer choice in the grid to the right of each question. (Answers are on page 39.)

1. If $a = b^3$, and $b = \frac{\sqrt{5}}{c}$, what is the value of a when $c = \frac{1}{3}$?

(A) 0.42
(B) 1.89
(C) 60.37
(D) 100.62
(E) 301.87

Ⓐ Ⓑ Ⓒ Ⓓ Ⓔ

2. For all $def \neq 0$, $\frac{4d^3e^4f^2}{(2de^3f)^2} =$

(A) $2de^2$
(B) $2de^{-2}$
(C) de^2
(D) de^{-2}
(E) $\frac{de^2}{2}$

Ⓐ Ⓑ Ⓒ Ⓓ Ⓔ

3. When $4g^3 - 3g^2 + g + k$ is divided by $g - 2$, the remainder is 27. What is the value of k?

(A) 3
(B) 5
(C) 8
(D) 12
(E) 25

Ⓐ Ⓑ Ⓒ Ⓓ Ⓔ

DO YOUR FIGURING HERE

4. For all $h \neq \pm 4$, $\dfrac{3h^2 - 9h - 12}{16 - h^2} =$

(A) $\dfrac{-3(h+1)}{(h-4)}$

(B) $\dfrac{3(h-1)}{(h-4)}$

(C) $\dfrac{-3(h+1)}{(h+4)}$

(D) $\dfrac{(h+1)}{3(h+4)}$

(E) $\dfrac{3(h-1)}{(h+4)}$

Ⓐ Ⓑ Ⓒ Ⓓ Ⓔ

5. If $\sqrt[3]{5j-7} = -\dfrac{1}{2}$, what is the value of j?

(A) 1.375
(B) 2.118
(C) 2.599
(D) 5.125
(E) 6.875

Ⓐ Ⓑ Ⓒ Ⓓ Ⓔ

6. If $\dfrac{7}{m-\sqrt{3}} = \dfrac{\sqrt{3}}{m} + \dfrac{4}{2m}$, what is the value of m?

(A) −3.464
(B) −1.978
(C) −0.918
(D) 1.978
(E) 3.464

Ⓐ Ⓑ Ⓒ Ⓓ Ⓔ

7. If $9^n = 27^{n+1}$, then $2^n =$

(A) $-\dfrac{10}{3}$

(B) $-\dfrac{8}{3}$

(C) $-\dfrac{3}{8}$

(D) $\dfrac{1}{8}$

(E) $\dfrac{3}{8}$

Ⓐ Ⓑ Ⓒ Ⓓ Ⓔ

DO YOUR FIGURING HERE

8. If $p=\dfrac{\sqrt{q}+z}{r^2+z}$, what is the value of z in terms of p, q, and r?

(A) $\dfrac{p-r^2\sqrt{q}}{p-1}$

(B) $\dfrac{pr^2-\sqrt{q}}{r^2}$

(C) $\sqrt{\dfrac{p+qr}{r^2}}$

(D) $\dfrac{r^2\sqrt{q}-p}{p+1}$

(E) $\dfrac{\sqrt{q}-pr^2}{p-1}$

Ⓐ Ⓑ Ⓒ Ⓓ Ⓔ

9. If $3s + 5t = 10$, and $2s - t = 7$, what is the value of $\frac{1}{2}s + 3t$?

(A) 0
(B) 1.5
(C) 2.8
(D) 3.4
(E) 3.5

Ⓐ Ⓑ Ⓒ Ⓓ Ⓔ

10. Which of the following is the solution set of $\left|\frac{2}{3}u - 5\right| > 8$?

(A) $\left\{u: -\dfrac{39}{2} < u < \dfrac{9}{2}\right\}$

(B) $\left\{u: -\dfrac{9}{2} < u < \dfrac{39}{2}\right\}$

(C) $\left\{u: u > \dfrac{39}{2} \text{ or } u < -\dfrac{9}{2}\right\}$

(D) $\left\{u: u > \dfrac{9}{2} \text{ or } u < -\dfrac{39}{2}\right\}$

(E) $\left\{u: u < -\dfrac{39}{2} \text{ or } u > -\dfrac{9}{2}\right\}$

Ⓐ Ⓑ Ⓒ Ⓓ Ⓔ

Find Your Study Plan

The answer key on the next page shows where in this chapter to find explanations for the questions you missed. Here's how you should proceed based on your Diagnostic Test score.

9–10: Superb! You really know your algebra. Unless you have lots of time and just love to read about algebra, you might consider skipping this chapter. You seem to know it all already. To make absolutely sure, you could look over the facts, formulas, and strategies in the margins of this chapter. And if you just want to do some more algebra problems, go to the Follow-Up Test at the end of this chapter.

7–8: You're quite good at algebra. Some of these are especially difficult questions. If you're taking a "shortcut" or you're on the Panic Plan, you don't really have time to study this chapter, and you don't really need to. You might want to look at those pages that address the questions you didn't get right. And if you just want to do some more algebra problems, go to the Follow-Up Test at the end of the chapter.

0–6: No matter which level test you're taking, or how pressed for time you are, you should continue to read this chapter and do the Follow-Up Test at the end. You need to brush up on your algebra before you can take full advantage of later chapters.

ALGEBRA TEST TOPICS

The questions in the Algebra Diagnostic Test are typical of the Mathematics subject tests. They cover a wide range of algebra topics, from simple expressions to quadratic equations to inequalities with absolute value signs. In this chapter, we'll use these questions to review the algebra you need to know. We will also use these questions to demonstrate effective problem-solving techniques, alternative methods, and test-taking strategies that apply to algebra questions.

Evaluating Expressions

The simplest algebra questions on Math 2 are like Example 1:

Example 1

If $a = b^3$, and $b = \frac{\sqrt{5}}{c}$, what is the value of a when $c = \frac{1}{3}$?

(A) 0.42 (B) 1.89 (C) 60.37 (D) 100.62 (E) 301.87

Answering a question like this is basically a matter of plugging in the numbers and cranking out the arithmetic. You are given an expression for *a* in terms of *b*, an expression for *b* in terms of *c*, and the value of *c*.

Plug $c = \frac{1}{3}$ into the second equation and find b:

$$b = \frac{\sqrt{5}}{c} = \frac{\sqrt{5}}{\frac{1}{3}} = \frac{\sqrt{5}}{1} \div \frac{1}{3} = \frac{\sqrt{5}}{1} \times \frac{3}{1} = 3\sqrt{5}$$

Now plug $b = 3\sqrt{5}$ into the first equation to find a:

$$a = b^3 = (3\sqrt{5})^3 = 3^3 \times (\sqrt{5})^3 = 27 \times 5\sqrt{5} \approx 301.87$$

The answer is (E).

Exponents—Key Operations

You can't be adept at algebra unless you're completely at ease with exponents. Here's what you need to know:

Multiplying powers with the same base: To multiply powers with the same base, keep the base and add the exponents:

$$x^3 \times x^4 = x^{3+4} = x^7$$

Dividing powers with the same base: To divide powers with the same base, keep the base and subtract the exponents:

$$y^{13} \div y^8 = y^{13-8} = y^5$$

Raising a power to an exponent: To raise a power to an exponent, keep the base and multiply the exponents:

$$(x^3)^4 = x^{3\times4} = x^{12}$$

Multiplying powers with the same exponent: To multiply powers with the same exponent, multiply the bases and keep the exponent:

$$(3^x)(4^x) = 12^x$$

Dividing powers with the same exponent: To divide powers with the same exponent, divide the bases and keep the exponent:

$$\frac{6^x}{2^x} = 3^x$$

On Test Day, you might encounter an algebra question, like Example 2, that specifically tests your understanding of the rules of exponents.

DIAGNOSTIC TEST ANSWER KEY

1. E
See "Evaluating Expressions."

2. D
See "Exponents—Key Operations."

3. B
See "Dividing Polynomials."

4. C
See "Factoring."

5. A
See "The Golden Rule of Equations."

6. B
See "Unknown in a Denominator."

7. D
See "Unknown in an Exponent."

8. E
See "In Terms Of."

9. B
See "Simultaneous Equations."

10. C
See "Inequalities and Absolute Value."

Example 2

For all $def \neq 0$, $\dfrac{4d^3e^4f^2}{(2de^3f)^2} =$

(A) $2de^2$ (B) $2de^{-2}$ (C) de^2 (D) de^{-2} (E) $\dfrac{de^2}{2}$

THE FIVE RULES OF EXPONENTS

1. $(x^m)(x^n) = x^{m+n}$
2. $\dfrac{x^m}{x^n} = x^{m-n}$
3. $(x^m)^n = x^{mn}$
4. $(x^n)(y^n) = (xy)^n$
5. $\dfrac{x^n}{y^n} = \left(\dfrac{x}{y}\right)^n$

There's nothing tricky about this question if you know how to work with exponents. The first step is to eliminate the parentheses by distributing the square in the denominator:

$$\frac{4d^3e^4f^2}{(2de^3f)^2} = \frac{4d^3e^4f^2}{(2)^2(d^1)^2(e^3)^2(f^1)^2} = \frac{4d^3e^4f^2}{4d^2e^6f^2}$$

The next step is to look for factors common to the numerator and denominator. The 4 on top and the 4 on the bottom cancel each other out, so it already looks like the answer's going to be (C) or (D).

The e^4 on top cancels with the e^6 on the bottom, leaving e^2 on bottom. You're actually subtracting the exponents: $4 - 6 = -2$, since e^{-2} is the same as $\frac{1}{e^2}$. The d^3 on top cancels with the d^2 on the bottom, leaving d on top. And the f^2 on top and the f^2 on bottom cancel each other out.

$$\frac{4d^3e^4f^2}{4d^2e^6f^2} = \frac{d}{e^2} = de^{-2}$$

The answer is (D).

Adding, Subtracting, and Multiplying Polynomials

Algebra is the basic language of the Mathematics subject tests, and you will want to be fluent in that language. You might not get a whole lot of questions that ask explicitly about such basic algebra procedures as combining like terms, multiplying binomials, or factoring algebraic expressions, but you will do all of those things in the course of working out the answers to more advanced questions. So it's essential that you be at ease with the mechanics of algebraic manipulations.

COMBINING LIKE TERMS

$ax + bx = (a + b)x$

$ax - bx = (a - b)x$

Combining like terms: To combine like terms, keep the variable part unchanged while adding or subtracting the coefficients:

$$2a + 3a = (2 + 3)a = 5a$$

Adding or subtracting polynomials: To add or subtract polynomials, combine like terms:

$$(3x^2 + 5x - 7) - (x^2 + 12) =$$
$$(3x^2 - x^2) + 5x + (-7 - 12) =$$
$$2x^2 + 5x - 19$$

Multiplying monomials: To multiply monomials, multiply the coefficients and the variables separately:

$$2x \times 3x = (2 \times 3)(x \times x) = 6x^2$$

MULTIPLYING MONOMIALS

$(ax)(bx) = (ab)x^2$

Multiplying binomials: To multiply binomials, use FOIL. To multiply $(x + 3)$ by $(x + 4)$, first multiply the **F**irst terms: $x \times x = x^2$. Next the **O**uter terms: $x \times 4 = 4x$. Then the **I**nner terms: $3 \times x = 3x$. And finally the **L**ast terms: $3 \times 4 = 12$. Then add and combine like terms:

$$x^2 + 4x + 3x + 12 = x^2 + 7x + 12$$

MULTIPLYING BINOMIALS—FOIL

$(a + b)(c + d) = ?$

First = ac

Outer = ad

Inner = bc

Last = bd

Product = $ac + ad + bc + bd$

Multiplying polynomials: To multiply polynomials with more than two terms, make sure you multiply each term in the first polynomial by each term in the second. (FOIL works only when you want to multiply two binomials.)

$$\begin{aligned}(x^2 + 3x + 4)(x + 5) &= x^2(x + 5) + 3x(x + 5) + 4(x + 5)\\ &= x^3 + 5x^2 + 3x^2 + 15x + 4x + 20\\ &= x^3 + 8x^2 + 19x + 20\end{aligned}$$

After multiplying two polynomials together, the number of terms in your expression before simplifying should equal the number of terms in one polynomial multiplied by the number of terms in the second. In the example above, you should have $3 \times 2 = 6$ terms in the product before you combine like terms.

Dividing Polynomials

To divide polynomials, you can use long division. For example, divide $2x^3 + 13x^2 + 11x - 16$ by $x + 5$:

$$x + 5\overline{)2x^3 + 13x^2 + 11x - 16}$$

The first term of the quotient is $2x^2$, because that's what will give you a $2x^3$ as a first term when you multiply it by $x + 5$:

$$\begin{array}{r} 2x^2 \\ x+5\overline{)2x^3+13x^2+11x-16} \\ \underline{2x^3+10x^2} \end{array}$$

Subtract and continue in the same way as when dividing numbers:

$$\begin{array}{r} 2x^2 + 3x - 4 \\ x+5\overline{)2x^3 + 13x^2 + 11x - 16} \\ \underline{2x^3 + 10x^2} \qquad\qquad \\ 3x^2 + 11x \qquad \\ \underline{3x^2 + 15x} \qquad \\ -4x - 16 \\ \underline{-4x - 20} \\ 4 \end{array}$$

The result is $2x^2 + 3x - 4$ with a remainder of 4.

Long division is the way to do Example 3:

Example 3

When $4g^3 - 3g^2 + g + k$ is divided by $g - 2$, the remainder is 27. What is the value of k?

(A) 3 (B) 5 (C) 8 (D) 12 (E) 25

To answer this question, start by cranking out the long division:

$$\begin{array}{r} 4g^2 + 5g + 11 \\ g-2\overline{)4g^3 - 3g^2 + g + k} \\ \underline{4g^3 - 8g^2} \qquad\qquad \\ 5g^2 + g \qquad \\ \underline{5g^2 - 10g} \qquad \\ 11g + k \\ \underline{11g - 22} \end{array}$$

The question says that the remainder is 27, so whatever k is, when you subtract −22 from it, you get 27:

$$\begin{aligned} k - (-22) &= 27 \\ k + 22 &= 27 \\ k &= 5 \end{aligned}$$

The answer is (B).

Factoring

Performing operations on polynomials is largely a matter of cranking it out. Once you know the rules, adding, subtracting, multiplying, and even dividing are simple tasks. Factoring algebraic expressions is a different matter.

To factor successfully, you have to do more thinking and less cranking. You have to try to figure out what expressions multiplied will give you the polynomial you're looking at. Sometimes that means having a good eye for the test makers' favorite factorables:

- Factor common to all terms
- Difference of squares
- Square of a binomial

Factor common to all terms: A factor common to all the terms of a polynomial can be factored out. This is essentially the distributive property in reverse. For example, all three terms in the polynomial $3x^3 + 12x^2 - 6x$ contain a factor of $3x$. Pulling out the common factor yields $3x(x^2 + 4x - 2)$.

Difference of squares: You will want to be especially keen at spotting polynomials in the form of the difference of squares. Whenever you have two identifiable squares with a minus sign between them, you can factor the expression like this:

$$a^2 - b^2 = (a + b)(a - b)$$

For example, $4x^2 - 9$ factors to $(2x + 3)(2x - 3)$.

Squares of binomials: Learn to recognize polynomials that are squares of binomials:

$$a^2 + 2ab + b^2 = (a + b)^2$$

$$a^2 - 2ab + b^2 = (a - b)^2$$

For example, $4x^2 + 12x + 9$ factors to $(2x + 3)^2$, and $a^2 - 10a + 25$ factors to $(a - 5)^2$.

Sometimes you'll want to factor a polynomial that's not in any of these classic factorable forms. When that happens, factoring becomes a kind of logic exercise, with some trial and error thrown in. To factor a quadratic expression, think about what binomials you could use FOIL on to get that quadratic expression. For example, to factor $x^2 - 5x + 6$, think about what **F**irst terms will produce x^2, what **L**ast terms will produce +6, and what **O**uter and **I**nner terms will produce $-5x$. Some common sense—and a little trial and error—will lead you to $(x - 2)(x - 3)$.

CLASSIC FACTORABLES

Factor common to all terms:

$ax + ay = a(x + y)$

Difference of squares:

$a^2 - b^2 = (a - b)(a + b)$

Square of binomial:

$a^2 + 2ab + b^2 = (a + b)^2$

Example 4 is a good instance of a Math 2 question that calls for factoring.

Example 4

For all $h \neq \pm 4$, $\dfrac{3h^2-9h-12}{16-h^2} =$

(A) $\dfrac{-3(h+1)}{(h-4)}$

(B) $\dfrac{3(h-1)}{(h-4)}$

(C) $\dfrac{-3(h+1)}{(h+4)}$

(D) $\dfrac{(h+1)}{3(h+4)}$

(E) $\dfrac{3(h-1)}{(h+4)}$

To reduce a fraction, you need to eliminate common factors from the top and the bottom. This is just as true for algebraic equations as it is for integers. The first step should be to try and simplify the numerator or denominator, which will help in factoring. Notice that 3 is a factor of every term in the numerator:

$$\frac{3h^2-9h-12}{16-h^2}=\frac{3(h^2-3h-4)}{16-h^2}$$

Next, factor both the numerator and denominator:

$$\frac{3(h^2-3h-4)}{16-h^2}=\frac{3(h-4)(h+1)}{(4-h)(4+h)}$$

Don't forget tricks such as the difference of squares, which makes factoring the denominator very easy. The key now is to notice that *h* – 4 and 4 – *h* are opposites of each other, so you can factor out –1 in the numerator and cancel them out:

$$\frac{3(h-4)(h+1)}{(4-h)(4+h)}=\frac{(-1)(3)(4-h)(h+1)}{(4-h)(4+h)}=\frac{-3(h+1)}{(4+h)}$$

This is equivalent to answer choice (C):

$$\frac{-3(h+1)}{(4+h)}=\frac{-3(h+1)}{(h+4)}$$

If you get stuck while factoring, it helps to realize that the top and bottom *must* have a factor in common—otherwise the test makers wouldn't be asking the question. If you can factor either one completely, it will provide a good start on the other expression.

Alternative method: Here's another way to answer this question. *Pick a number for* h *and see what happens.* One of the answer choices will give you the same value as the original fraction will, no matter what you plug in for *h*. Pick a number that's easy to work with—like 0.

When you plug $h = 0$ into the original expression, any term with an *h* drops out, and you end up with $-\frac{12}{16}$ or $-\frac{3}{4}$. Now plug $h = 0$ into each answer choice to see which ones equal $-\frac{3}{4}$.

When you get to (C), it works, but you can't stop there. It might just be a coincidence. When you pick numbers, *look at every answer choice.* (E) also works for $h = 0$. At least you know one of those is the correct answer, and you can decide between them by picking another value of *h*.

PICK NUMBERS.

When the answer choices are algebraic expressions, it often works to *pick numbers* for the unknowns, plug those numbers into the stem and see what you get, and then plug those same numbers into the answer choices to find matches.

Warning: When you pick numbers, you have to check *all* the answer choices. Sometimes more than one works with the number(s) you pick, in which case you have to pick numbers again.

The Golden Rule of Equations

You probably remember the basic procedure for solving algebraic equations: *Do the same thing to both sides.* You can do almost anything you want to one side of an equation as long as you preserve the equality by doing the same thing to the other side. Your aim in whatever you do to both sides is to get the variable (or expression) you're solving for all by itself on one side. Look at Example 5:

Example 5

If $\sqrt[3]{5j-7} = -\frac{1}{2}$, what is the value of j?

(A) 1.375 (B) 2.118 (C) 2.599 (D) 5.125 (E) 6.875

To solve this equation for *j*, do whatever is necessary to both sides of the equation to get *j* alone on one side. Layer by layer, peel away all those extra symbols and numbers around the *j*. First, get rid of that cube-root symbol. The way to undo a cube root is to cube both sides:

$$\sqrt[3]{5j-7} = -\frac{1}{2}$$

$$\left(\sqrt[3]{5j-7}\right)^3 = \left(-\frac{1}{2}\right)^3$$

$$5j-7 = -\frac{1}{8}$$

The rest is easy. Add 7 to both sides (expressing 7 as an improper fraction will make the addition easier) and then divide both sides by 5:

$$5j-7=-\frac{1}{8}$$

$$5j-\frac{56}{8}=-\frac{1}{8}$$

$$5j=\frac{55}{8}$$

$$j=\frac{55}{8}\div\frac{5}{1}=\frac{55}{8}\times\frac{1}{5}=\frac{55}{40}=1.375$$

The answer is (A).

The test makers have a couple of favorite equation types that you should be prepared to solve. Solving linear equations is usually pretty straightforward. Generally it's obvious what to do to isolate the unknown. But when the unknown is in a denominator or an exponent, it might not be as obvious how to proceed.

Unknown in a Denominator

The basic procedure for solving an equation is the same even when the unknown is in a denominator: Do the same thing to both sides. In this case, you multiply in order to undo division.

If you wanted to solve $1+\frac{1}{x}=2-\frac{1}{x}$, you would multiply both sides by x:

$$1+\frac{1}{x}=2-\frac{1}{x}$$

$$x\left(1+\frac{1}{x}\right)=x\left(2-\frac{1}{x}\right)$$

$$x+1=2x-1$$

Now you have an equation with no denominators, which is easy to solve:

$$x+1=2x-1$$

$$x-2x=-1-1$$

$$-x=-2$$

$$x=2$$

Another good way to solve an equation with the unknown in the denominator is to *cross multiply*. That's the best way to do Example 6.

Example 6

If $\frac{7}{m-\sqrt{3}}=\frac{\sqrt{3}}{m}+\frac{4}{2m}$, what is the value of m ?

(A) -3.464 (B) −1.978 (C) −0.918 (D) 1.978 (E) 3.464

Before you can cross multiply, you need to express the right side of the equation as a single fraction. That means giving the two fractions a common denominator and adding them. The common denominator is 2*m*.

$$\frac{7}{m-\sqrt{3}}=\frac{\sqrt{3}}{m}+\frac{4}{2m}$$

$$\frac{7}{m-\sqrt{3}}=\frac{2\sqrt{3}}{2m}+\frac{4}{2m}$$

$$\frac{7}{m-\sqrt{3}}=\frac{2\sqrt{3}+4}{2m}$$

Now you can cross multiply:

$$\frac{7}{m-\sqrt{3}}=\frac{2\sqrt{3}+4}{2m}$$

$$(7)(2m)=(m-\sqrt{3})(2\sqrt{3}+4)$$

$$14m=2\sqrt{3}m-6+4m-4\sqrt{3}$$

$$10m-2\sqrt{3}m=-6-4\sqrt{3}$$

$$(10-2\sqrt{3})m=-6-4\sqrt{3}$$

$$m=\frac{-6-4\sqrt{3}}{10-2\sqrt{3}}\approx-1.978$$

The answer is (B).

This problem could also be solved by backsolving, explained on the next page. Just try the different choices for *m*, and when one of them works, you have the answer. Very often the more complicated equations, such as the one in this problem, can be solved more easily and quickly by backsolving.

Unknown in an Exponent

The procedure for solving an equation when the unknown is in an exponent is a little different. What you want to do in this situation is to express one or both sides of the equation so that the two sides have the same base.

BACKSOLVE

One of the answer choices has to be the correct answer, so you can sometimes just try them out. Since the answer choices are generally in numerical order, start with (B) or (D). If it doesn't work, you might at least know whether you're looking for something bigger or something smaller. At most you only have to try the middle number of the remaining choices. *Note*: Often you don't have to try all the answer choices when backsolving.

Look at Example 7.

Example 7

If $9^n = 27^{n-1}$, then $2^n =$

(A) $-\frac{10}{3}$ (B) $-\frac{8}{3}$ (C) $-\frac{3}{8}$ (D) $\frac{1}{8}$ (E) $\frac{3}{8}$

In this case, the base on the left is 9, and the base on the right is 27. They're both powers of 3, so express both sides as powers of 3:

$$9^n = 27^{n+1}$$
$$(3^2)^n = (3^3)^{n+1}$$
$$3^{2n} = 3^{3n+3}$$

Now that both sides have the same base, you can simply set the exponent expressions equal and solve for n:

$$2n = 3n + 3$$
$$-n = 3$$
$$n = -3$$

Then plug in –3 for n in the final expression:

$$2^n = 2^{-3} = \frac{1}{2^3} = \frac{1}{8}$$

The answer is (D).

QUADRATIC FORMULA

If $ax^2 + bx + c = 0$, then:

$$x = \frac{-b \pm \sqrt{b^2 - 4ac}}{2a}$$

Note that Backsolving would work here if we were solving for n, but it doesn't work because of the added expression to evaluate. Don't depend on backsolving too much. Lots of math questions can't be backsolved at all, and many that can are more quickly solved by a more direct approach.

Quadratic Equations

To solve a quadratic equation, put it in the "$ax^2 + bx + c = 0$" form, factor the left side (if you can) and set each factor equal to 0 separately to get the two solutions. For example, to solve $x^2 + 12 = 7x$, first rewrite it as $x^2 - 7x + 12 = 0$. Then factor the left side.

$$x^2 - 7x + 12 = 0$$
$$(x - 3)(x - 4) = 0$$
$$x - 3 = 0 \text{ or } x - 4 = 0$$
$$x = 3 \text{ or } 4$$

Sometimes the left side may not be obviously factorable. You can always use the *quadratic formula*. Just plug in the coefficients a, b, and c from $ax^2 + bx + c = 0$ into the formula:

$$x = \frac{-b \pm \sqrt{b^2 - 4ac}}{2a}$$

To solve $x^2 + 4x + 2 = 0$, plug $a = 1$, $b = 4$, and $c = 2$ into the formula:

$$x = \frac{-4 \pm \sqrt{4^2 - 4 \cdot 1 \cdot 2}}{2 \cdot 1}$$

$$x = \frac{-4 \pm \sqrt{8}}{2} = -2 \pm \sqrt{2}$$

A third useful technique for solving quadratic equations is to graph them. Then, using your calculator, you can either approximate the x-intercepts (which are the solutions to the equation), or you can calculate them exactly if the answer choices are very close to each other.

"In Terms Of"

So far in this chapter, solving an equation has meant finding a numerical value for the unknown. When there's more than one variable and only one equation, it's impossible to get numerical solutions. Instead, what you do is solve for the unknown *in terms of* the other variables.

To solve an equation for one variable in terms of another means to isolate the one variable on one side of the equation, leaving an expression containing the other variable on the other side of the equation.

For example, to solve the equation $3x - 10y = -5x + 6y$ for x in terms of y, isolate x :

$$\begin{aligned} 3x - 10y &= -5x + 6y \\ 3x + 5x &= 6y + 10y \\ 8x &= 16y \\ x &= 2y \end{aligned}$$

Now look at the next example, which asks you to solve "in terms of."

Example 8

If $p = \frac{\sqrt{q} + z}{r^2 + z}$, what is the value of z in terms of p, q, and r?

(A) $\frac{p - r^2\sqrt{q}}{p - 1}$

(B) $\frac{pr^2 - \sqrt{q}}{r^2}$

(C) $\sqrt{\frac{p + qr}{r^2}}$

(D) $\dfrac{r^2\sqrt{q}-p}{p+1}$

(E) $\dfrac{\sqrt{q}-pr^2}{p-1}$

You want to get z on one side by itself. The first thing to do is to eliminate the denominator by multiplying both sides by $r^2 + z$:

$$p=\frac{\sqrt{q}+z}{r^2+z}$$

$$p(r^2+z)=\left(\frac{\sqrt{q}+z}{r^2+z}\right)(r^2+z)$$

$$pr^2+pz=\sqrt{q}+z$$

Next move all terms with z to one side and all terms without z to the other:

$$pr^2+pz=\sqrt{q}+z$$

$$pz-z=\sqrt{q}-pr^2$$

Now factor z out of the left side and divide both sides by the other factor to isolate z:

$$pz-z=\sqrt{q}-pr^2$$

$$z(p-1)=\sqrt{q}-pr^2$$

$$z=\frac{\sqrt{q}-pr^2}{p-1}$$

The answer is (E).

Simultaneous Equations

DON'T DO MORE WORK THAN YOU HAVE TO

You don't always have to find the value of each variable to answer a simultaneous equations question.

You can get numerical solutions for more than one unknown if you are given more than one equation. The test makers like simultaneous equations questions because they take a little thought to answer. Solving simultaneous equations almost always involves combining equations, but you have to figure out what's the best way to combine the equations.

You can solve for two variables only if you have two distinct equations. Two forms of the same equation will not be adequate. Combine the equations in such a way that one of the variables cancels out. For example, to solve the two equations $4x + 3y = 8$ and $x + y = 3$, multiply both sides of the second equation by –3 to get $-3x - 3y = -9$. Now add the two equations; the $3y$ and the $-3y$ cancel out, leaving $x = -1$. Plug that back into either one of the original equations, and you'll find that $y = 4$.

Example 9 is a simultaneous equations question.

Example 9

If $3s + 5t = 10$ and $2s - t = 7$, what is the value of $\frac{1}{2}s + 3t$?

(A) 0 (B) 1.5 (C) 2.8 (D) 3.4 (E) 3.5

If you just plow ahead without thinking, you might try to answer this question by solving for one variable at a time. That would work, but it would take a lot more time than this question needs. As usual, the key to this simultaneous equations question is to combine the equations, but combining the equations doesn't necessarily mean losing a variable. Look what happens here if you just subtract the equations as presented:

$$\begin{array}{r} 3s + 5t = 10 \\ -\ (2s - t = 7) \\ \hline s + 6t = 3 \end{array}$$

Suddenly, you're almost there. Just divide both sides by 2, and you get $\frac{1}{2}s + 3t = \frac{3}{2} = 1.5$. The answer is (B).

Absolute Value

To solve an equation that includes absolute value signs, think about the two different cases. For example, to solve the equation $|x - 12| = 3$, think of it as two equations:

$$x - 12 = 3 \text{ or } x - 12 = -3$$
$$x = 15 \text{ or } 9$$

Inequalities

To solve an inequality, do whatever is necessary to both sides to isolate the variable. Just remember that when you multiply or divide both sides by a negative number, you must reverse the sign. To solve $-5x + 7 < -3$, subtract 7 from both sides to get $-5x < -10$. Now divide both sides by –5, remembering to reverse the sign: $x > 2$.

Inequalities and Absolute Value

About the most complicated algebraic solving you'll have to do on the Math subject tests will involve inequalities and absolute value signs.

INEQUALITIES AND ABSOLUTE VALUE

For all $n > 0$,
if $|x| < n$, then
$-n < x < n$;
if $|x| > n$, then
$x < -n$ or $x > n$.

Look at Example 10.

Example 10

Which of the following is the solution set of $\left|\frac{2}{3}u - 5\right| > 8$?

(A) $\left\{u: -\frac{39}{2} < u < \frac{9}{2}\right\}$

(B) $\left\{u: -\frac{9}{2} < u < \frac{39}{2}\right\}$

(C) $\left\{u: u > \frac{39}{2} \text{ or } u < -\frac{9}{2}\right\}$

(D) $\left\{u: u > \frac{9}{2} \text{ or } u < -\frac{39}{2}\right\}$

(E) $\left\{u: u < -\frac{39}{2} \text{ or } u > -\frac{9}{2}\right\}$

What does it mean if $\left|\frac{2}{3}u - 5\right| > 8$? It means that if the expression between the absolute value bars is positive, it's greater than +8 or, if the expression between the bars is negative, it's less than –8. In other words, $\frac{2}{3}u - 5$ is outside the range of –8 to +8:

$\frac{2}{3}u - 5 > 8$	$\frac{2}{3}u - 5 < -8$
$\frac{2}{3}u > 13$	$\frac{2}{3}u < -3$
$u > \frac{39}{2}$	$u < -\frac{9}{2}$

The answer is (C).

In fact, there's a general rule that applies here: To solve an inequality in the form |whatever| < p, where $p > 0$, just put that "whatever" inside the range $-p$ to p:

$$|\text{whatever}| < p \text{ means } -p < \text{whatever} < p$$

For example, $|x - 5| < 14$ becomes $-14 < x - 5 < 14$.

And here's another general rule: To solve an inequality in the form $|\text{whatever}| > p$, where $p > 0$, just put that "whatever" outside the range $-p$ to p:

$$|\text{whatever}| > p \text{ means whatever} < -p \text{ OR whatever} > p$$

For example, $\left|\frac{3x+9}{2}\right| > 7$ becomes $\frac{3x+9}{2} < -7$ OR $\frac{3x+9}{2} > 7$.

Well, you've seen a lot of algebra in this chapter. You've seen ten of the test makers' favorite algebra situations. You've reviewed all the relevant Math 2 algebra facts and formulas. And you've learned some effective Kaplan test-taking strategies. Now it's time to take the Algebra Follow-Up Test to find out how much you've learned.

THINGS TO REMEMBER:

The Rules of Exponents

1. $(x^m)(x^n) = x^{m+n}$
2. $\dfrac{x^m}{x^n} = x^{m-n}$
3. $(x^m)^n = x^{mn}$
4. $(x^n)(y^n) = (xy)^n$
5. $\dfrac{x^n}{y^n} = \left(\dfrac{x}{y}\right)^n$

Combining Like Terms

$ax + bx = (a + b)x$

$ax - bx = (a - b)x$

Multiplying Monomials

$(ax)(bx) = (ab)x^2$

Multiplying Binomials—FOIL

$(a + b)(c + d) = ?$
First = ac
Outer = ad
Inner = bc
Last = bd
Product = $ac + ad + bc + bd$

Classic Factorables

Factor common to all terms:
$ax + ay = a(x + y)$

Difference of squares:
$a^2 - b^2 = (a - b)(a + b)$

Square of binomial:
$a^2 + 2ab + b^2 = (a + b)^2$

Quadratic Formula

If $ax^2 + bx + c = 0$, then

$$x = \frac{-b \pm \sqrt{b^2 - 4ac}}{2a}.$$

Inequalities and Absolute Value

For all $n > 0$,
if $|x| < n$, then
$-n < x < n$;

if $|x| > n$, then
$x < -n$ or $x > n$.

ALGEBRA FOLLOW-UP TEST

DO YOUR FIGURING HERE

10 Questions (12 Minutes)

Directions: Solve the following problems. Fill in the oval corresponding to the best answer choice in the grid to the right of each question. (Answers and explanations begin on page 59.)

1. If $a = 6 - b^3$ and $b = -2$, what is the value of a?

   (A) –8 (B) –2 (C) 2 (D) 14 (E) 18

   Ⓐ Ⓑ Ⓒ Ⓓ Ⓔ

2. For all z, $3^z + 3^z + 3^z =$

   (A) $3^z + 3$ (B) 3^{z+1} (C) 3^{z+3} (D) 3^{3z} (E) 9^z

   Ⓐ Ⓑ Ⓒ Ⓓ Ⓔ

3. For all $c \neq \pm\frac{1}{5}$, $\frac{5c^2 - 24c - 5}{1 - 25c^2} =$

   (A) $\frac{5c+1}{5+c}$

   (B) $\frac{5+c}{5c+1}$

   (C) $-\frac{5c-1}{5-c}$

   (D) $-\frac{5-c}{5c-1}$

   (E) $\frac{5-c}{5c-1}$

   Ⓐ Ⓑ Ⓒ Ⓓ Ⓔ

DO YOUR FIGURING HERE

4. When $2f^3 + 3f^2 - 1$ is divided by $f + 2$, the remainder is

(A) −9 (B) −5 (C) −1 (D) 2 (E) 3

Ⓐ Ⓑ Ⓒ Ⓓ Ⓔ

5. If $\sqrt[5]{\dfrac{g-1}{4}} = \dfrac{1}{3}$, then $g =$

(A) 0.984 (B) 0.996 (C) 1.004 (D) 1.016 (E) 1.029

Ⓐ Ⓑ Ⓒ Ⓓ Ⓔ

6. If $\dfrac{37}{4\sqrt{j-19}} = \dfrac{37}{17}$, then $j =$

(A) 64 (B) 72.25 (C) 81 (D) 90.25 (E) 144

Ⓐ Ⓑ Ⓒ Ⓓ Ⓔ

7. If $5^{k^2}(25^{2k})(625) = 25\sqrt{5}$ and $k < -1$, what is the value of k?

(A) −3.581
(B) −3.162
(C) −2.613
(D) −1.581
(E) −0.419

Ⓐ Ⓑ Ⓒ Ⓓ Ⓔ

8. If $n \neq 3p$, and $s = \dfrac{n^2 + p}{n - 3p}$, what is the value of p in terms of n and s?

(A) $\dfrac{n^2 - ns}{1 - 3s}$

(B) $\dfrac{ns + n^2}{3s + 1}$

(C) $-\dfrac{ns - n^2}{3s + 1}$

(D) $\dfrac{n^2 - ns}{3s + 1}$

(E) $-\dfrac{n^2 - ns}{3s + 1}$

Ⓐ Ⓑ Ⓒ Ⓓ Ⓔ

9. Which of the following can be a solution to the pair of equations $2q - rt = 21$ and $q + rt = 36$?

(A) $q = 18$ and $t = 2$
(B) $q = 19$ and $t = 1$
(C) $q = 20$ and $t = -1$
(D) $q = 21$ and $t = -2$
(E) $q = 23$ and $t = -3$

Ⓐ Ⓑ Ⓒ Ⓓ Ⓔ

10. How many integers are in the solution set of $|2x + 6| < \dfrac{19}{2}$?

(A) None
(B) Two
(C) Nine
(D) Fourteen
(E) Infinitely many

Ⓐ Ⓑ Ⓒ Ⓓ Ⓔ

DO YOUR FIGURING HERE

FOLLOW-UP TEST—ANSWERS AND EXPLANATIONS

1. D

Plug $b = -2$ into the first equation:

$$a = 6 - b^3 = 6 - (-2)^3 = 6 - (-8) = 6 + 8 = 14$$

2. B

The sum of three identical quantities is 3 times one of those quantities, so the sum of the three terms 3^z is 3 times 3^z:

$$3^z + 3^z + 3^z = 3(3^z) = (3^1)(3^z) = 3^{z+1}$$

3. E

Factor the top and the bottom and cancel the factors they have in common:

$$\frac{5c^2-24c-5}{1-25c^2}$$

$$=\frac{(5c+1)(c-5)}{(1+5c)(1-5c)}$$

$$=\frac{(1+5c)(c-5)}{(1+5c)(1-5c)}$$

$$=\frac{c-5}{1-5c}$$

Multiplying by $\frac{-1}{-1}$, you get

$$=\frac{5-c}{5c-1}$$

4. B

Use long division. Watch out: The expression that goes under the division sign needs a place-holding $0f$ term:

$$\begin{array}{r} 2f^2-f+2 \\ f+2\overline{)2f^3+3f^2-0f-1} \\ \underline{2f^3+4f^2} \\ -f^2-0f \\ \underline{-f^2-2f} \\ 2f-1 \\ \underline{2f+4} \\ -5 \end{array}$$

The remainder is –5.

5. D

To undo the fifth-root symbol, raise both sides to the fifth power:

$$\sqrt[5]{\frac{g-1}{4}}=\frac{1}{3}$$

$$\left(\sqrt[5]{\frac{g-1}{4}}\right)^5=\left(\frac{1}{3}\right)^5$$

$$\frac{g-1}{4}=\frac{1}{243}$$

Now cross multiply:

$$\frac{g-1}{4}=\frac{1}{243}$$

$$(g-1)(243)=(1)(4)$$

$$243g-243=4$$

$$243g=247$$

$$g=\frac{247}{243}\approx 1.016$$

6. C

Your first impulse might be to cross multiply, but doing so would actually complicate the question unnecessarily. Notice that the fractions on both sides have the same numerator, 37. So the numerator is irrelevant. If the two fractions are equal and they have the same numerator, then they must have the same denominator, so just write an equation reflecting as much:

$$\frac{37}{4\sqrt{j}-19}=\frac{37}{17}$$
$$4\sqrt{j}-19=17$$
$$4\sqrt{j}=36$$
$$\sqrt{j}=9$$
$$(\sqrt{j})^2=9^2$$
$$j=81$$

7. A

Express everything as powers of 5:

$$5^{k^2}(25^{2k})(625)=25\sqrt{5}$$
$$(5^{k^2})[(5^2)^{2k}](5^4)=5^{\frac{5}{2}}$$

The left side of the equation is the product of powers with the same base, so just add the exponents:

$$(5^{k^2})[(5^2)^{2k}](5^4)=5^{\frac{5}{2}}$$
$$(5^{k^2})(5^{4k})(5^4)=5^{\frac{5}{2}}$$
$$5^{k^2+4k+4}=5^{\frac{5}{2}}$$

Now the two sides of the equation are powers with the same base, so you can just set the exponents equal to each other:

$$5^{k^2+4k+4}=5^{\frac{5}{2}}$$
$$k^2+4k+4=\frac{5}{2}$$
$$(k+2)^2=\frac{5}{2}$$
$$k+2=\pm\sqrt{\frac{5}{2}}$$
$$k=\pm\sqrt{\frac{5}{2}}-2$$

Evaluate both the positive and negative square roots:

$k=\sqrt{\frac{5}{2}}-2$	$k=-\sqrt{\frac{5}{2}}-2$
$k\approx-0.419$	$k\approx-3.581$

Since $k < -1$ in this problem, use –3.581 for your answer.

8. E

First multiply both sides by $n - 3p$ to clear the denominator:

$$s=\frac{n^2+p}{n-3p}$$
$$s(n-3p)=n^2+p$$
$$ns-3ps=n^2+p$$

Now move all terms with p to the left and all terms without p to the right:

$$ns - 3ps = n^2 + p$$
$$-3ps - p = n^2 - ns$$

Now factor the left side and divide to isolate p.

$$-3ps - p = n^2 - ns$$
$$p(-3s-1) = n^2 - ns$$
$$p = \frac{n^2 - ns}{-3s-1} = -\frac{n^2 - ns}{3s+1}$$

Remember that this type of problem can also be solved by Picking Numbers.

9. B

With only two equations, you won't be able to get numerical solutions for three unknowns. But apparently you can get far enough to rule out four of the five answer choices. How? Look for a way to combine the equations that leads somewhere useful. Notice that the first equation contains $-rt$ and the second equation contains $+rt$, so if you add the equations as they are, you'll lose those terms:

$$\begin{aligned} 2q - rt &= 21 \\ +\ (q + rt &= 36) \\ \hline 3q &= 57 \\ q &= 19 \end{aligned}$$

There's not enough information to get numerical solutions for r or t, but you do know that $q = 19$, so the correct answer is the only choice that has $q = 19$.

10. C

If the absolute value of something is less than $\frac{19}{2}$, then that something is between $-\frac{19}{2}$ and $\frac{19}{2}$:

$\lvert 2x + 6\rvert < \frac{19}{2}$	
$2x+6<\frac{19}{2}$	$2x+6>-\frac{19}{2}$
$2x+\frac{12}{2}<\frac{19}{2}$	$2x+\frac{12}{2}>-\frac{19}{2}$
$2x<\frac{7}{2}$	$2x>-\frac{31}{2}$
$x<\frac{7}{4}$	$x>-\frac{31}{4}$
$x<1\frac{3}{4}$	$x>-7\frac{3}{4}$

Thus, $-7\frac{3}{4} < x < 1\frac{3}{4}$. There are nine integers in that range: –7, –6, –5, –4, –3, –2, –1, 0, and 1.

Chapter 5: **Plane Geometry**

- Line segments
- Triangles
- Quadrilaterals and polygons
- Circles

Officially, there are no plane geometry questions on Mathematics Level 2. But because plane geometry is fundamental to solid geometry, coordinate geometry, and trigonometry, the material in this chapter is relevant to about 40 percent of the Math 2 test.

PLANE GEOMETRY FACTS AND FORMULAS IN THIS CHAPTER

- Five Facts about Triangles
- Similar Figures
- Three Special Triangle Types
- Pythagorean Theorem
- Four Special Right Triangles
- Five Special Quadrilateral Types
- Polygon Angles
- Four Circle Formulas

HOW TO USE THIS CHAPTER

Maybe you already know all the plane geometry you need. You can find out by taking the Plane Geometry Diagnostic Test. The six plane geometry questions on the Diagnostic Test are typical of the Mathematics subject tests. Check your answers using the answer key following the test. No matter how you score, don't worry! The answer key also shows where to find a detailed explanation for each question. The "Find Your Study Plan" section that follows the test will suggest the next steps based on your performance on the Diagnostic.

Find Your Level

How you use this chapter really depends on how much time you have to prep. Find your level and pace below.

Standard Plan. No matter how well you do on the Plane Geometry Diagnostic Test, read the rest of this chapter and do all the practice problems.

Shortcut: Take the Plane Geometry Diagnostic Test and check your answers. If you can answer at least four of the six questions correctly, then you already know the material in this chapter well enough to move on.

Panic Plan. Take the Plane Geometry Diagnostic Test and check your answers. The material in this chapter is vital. Don't try to move on to solid geometry, coordinate geometry, or trigonometry until you feel comfortable with the material in this chapter.

PLANE GEOMETRY DIAGNOSTIC TEST

6 Questions (8 Minutes)

Directions: Solve the following problems. Fill in the oval corresponding to the best answer choice in the grid to the right of each question. (Answers are on page 68.)

DO YOUR FIGURING HERE

1. In Figure 1, the length of $\overline{WZ}$ is $3a + 15$, and the length of $\overline{WX}$ is $7a + 5$. If Y is the midpoint of $\overline{XZ}$, what is the length of $\overline{WY}$?

   (A) $5 + 4a$
   (B) $5 - 2a$
   (C) $25 + 4a$
   (D) $20 + 5a$
   (E) $10 + 5a$

   Ⓐ Ⓑ Ⓒ Ⓓ Ⓔ

2. In Figure 2, $\overline{QR} = \overline{RS}$. If the area of ΔRST is $\frac{c}{2}$, what is the area of ΔQSP?

   (A) $c\sqrt{2}$
   (B) $c\sqrt{3}$
   (C) c
   (D) $2c$
   (E) $3c$

   Ⓐ Ⓑ Ⓒ Ⓓ Ⓔ

3. In Figure 3, ΔJKL is equilateral, and ΔJKM is isosceles. If $\overline{KL} = 2$, what is the distance from L to M?

   (A) 0.572
   (B) 0.636
   (C) 0.667
   (D) 0.732
   (E) 0.767

   Ⓐ Ⓑ Ⓒ Ⓓ Ⓔ

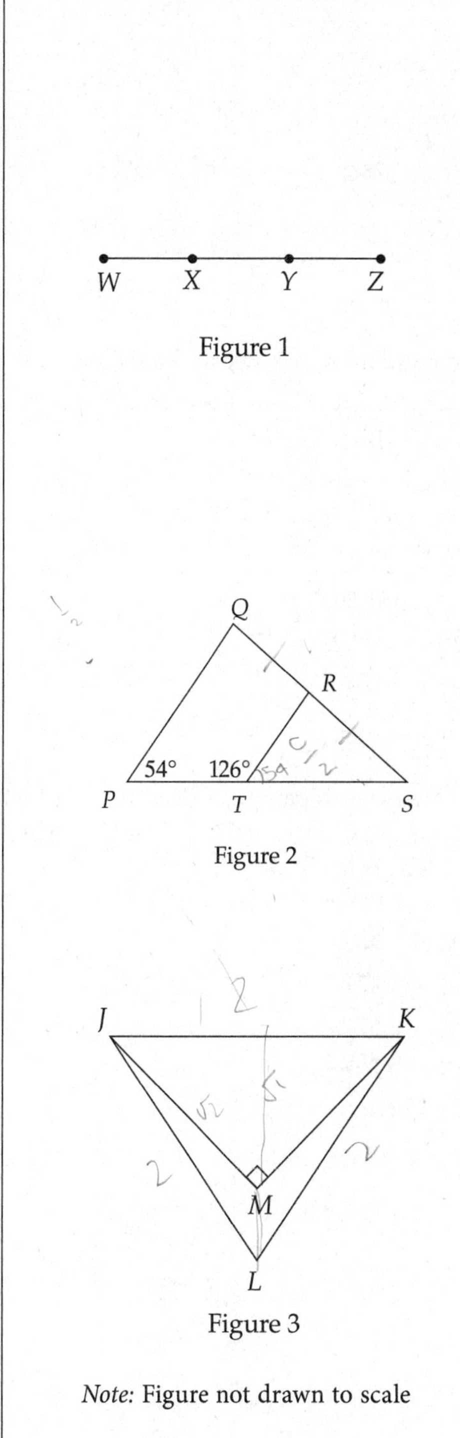

DO YOUR FIGURING HERE

4. In Figure 4, the perimeter of isosceles trapezoid $WXYZ$ is 62. If $\overline{ZY} = 7$ and $\overline{WX} = 25$, what is the length of diagonal WY?

(A) 15
(B) 17
(C) 20
(D) 25
(E) 26

Ⓐ Ⓑ Ⓒ Ⓓ Ⓔ

Figure 4

5. An equilateral triangle and a regular hexagon have the same perimeter. If the area of the hexagon is $72\sqrt{3}$, what is the area of the triangle?

(A) 62.354
(B) 83.138
(C) 101.823
(D) 103.923
(E) 124.708

Ⓐ Ⓑ Ⓒ Ⓓ Ⓔ

6. In Figure 5, rectangle $ABCD$ is inscribed in a circle. If the radius of the circle is 2 and $\overline{CD} = 2$, what is the area of the shaded region?

(A) 0.362
(B) 0.471
(C) 0.577
(D) 0.707
(E) 0.866

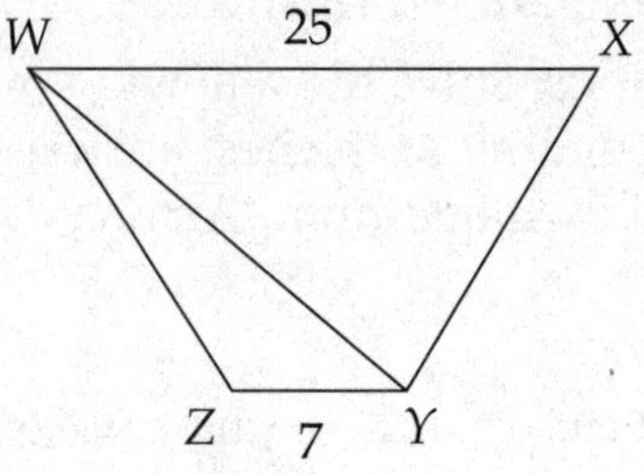

Ⓐ Ⓑ Ⓒ Ⓓ Ⓔ

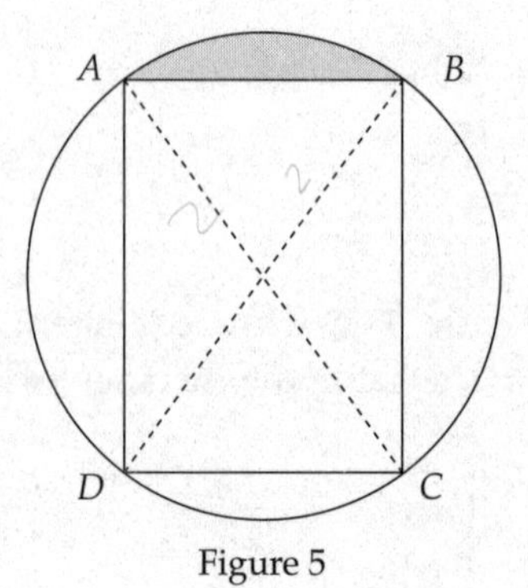

Figure 5

Find Your Study Plan

The answer key on the following page shows where in this chapter to find explanations for the questions you missed. Here's how you should proceed based on your Diagnostic Test score.

6: Superb! You really know your plane geometry. Unless you have lots of time and just love to read about plane geometry, you might consider skipping this chapter. You seem to know it all already. Just to make absolutely sure, you could look over the plane geometry facts, formulas, and strategies in the margins of this chapter. And if all you want is more plane geometry questions to try, go to the Follow-Up Test at the end of this chapter.

4–5: You're quite good at plane geometry. Some of these are especially difficult questions. If you're taking a "shortcut" or you're on the Panic Plan, you don't really have time to study this chapter, and you don't really need to. You might at least look over those pages that address the questions you didn't get right. You could also look over the plane geometry facts, formulas, and strategies in the margins of this chapter. If you just want to try more plane geometry questions, go to the Follow-Up Test at the end of the chapter.

0–3: You should read this chapter and do the Follow-Up Test at the end. You need to have a good command of the material in this chapter before moving on to later chapters on solid geometry, coordinate geometry, and trigonometry.

PLANE GEOMETRY TEST TOPICS

None of these questions would appear on the Math 2 test, but that's only because the Math 2 test makers assume you know plane geometry cold. To perform well on Math 2 solid geometry, coordinate geometry, and trigonometry questions, you must first be able to do the kind of plane geometry questions you'll see here. In this chapter, we'll use these questions to review the plane geometry you're expected to know for the Mathematics 2 subject test. We will also use these questions to demonstrate some effective problem-solving techniques, alternative methods, and test-taking strategies that apply to geometry questions.

Adding and Subtracting Segment Lengths

The simplest type of plane geometry you may have to do is like Example 1, which involves adding and subtracting segment lengths. It's typical of Math 2 that the lengths you have to add and subtract are algebraic expressions rather than numbers.

Example 1

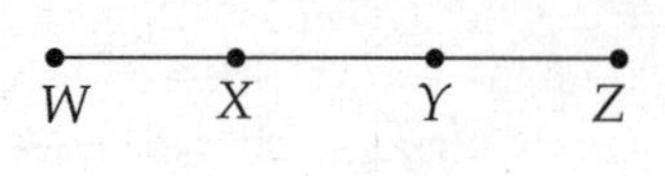

Figure 1

> **DIAGNOSTIC TEST ANSWER KEY**
>
> **1. E**
>
> See "Adding and Subtracting Segment Lengths."
>
> **2. D**
>
> See "Similar Triangles."
>
> **3. D**
>
> See "Three Special Triangle Types."
>
> **4. C**
>
> See "Hidden Special Triangles."
>
> **5. B**
>
> See "Polygons—Perimeter and Area."
>
> **6. A**
>
> See "Circles Combined with Other Figures."

In Figure 1, the length of $\overline{WZ}$ is $3a + 15$, and the length of $\overline{WX}$ is $7a + 5$. If Y is the midpoint of $\overline{XZ}$, what is the length of $\overline{WY}$?

(A) $5 + 4a$

(B) $5 - 2a$

(C) $25 + 4a$

(D) $20 + 5a$

(E) $10 + 5a$

Usually the best thing to do to start on a plane geometry question is to mark up the figure. Put as much of the information into the figure as you can. Doing so is a good way to organize your thoughts and avoid having to go back and forth between the figure and the question.

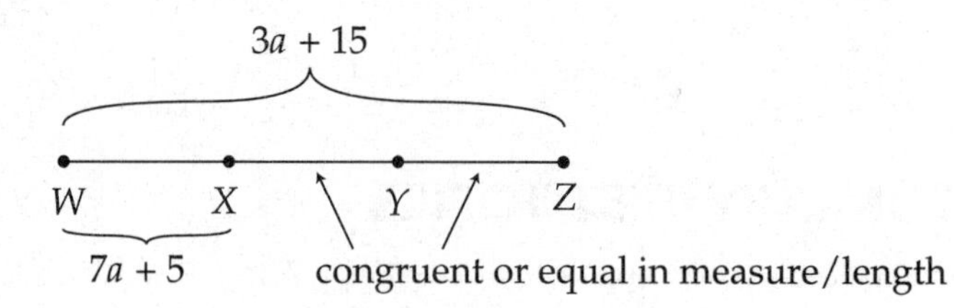

> **MARK UP THE FIGURE**
>
> Try to put all the given information into the figure so that you can see everything at a glance and don't have to keep going back and forth between the question stem and the figure.

Now you can plan your attack. First subtract $\overline{WX}$ from the whole length $\overline{WZ}$ to get $\overline{XZ}$:

$$\overline{XZ} = \overline{WZ} - \overline{WX} = (3a + 15) - (7a + 5) = 3a + 15 - 7a - 5 = -4a + 10$$

Then, because Y is the midpoint of $\overline{XZ}$, you can divide $\overline{XZ}$ by 2 to get $\overline{XY}$ and $\overline{YZ}$:

$$\overline{XY} = \overline{YZ} = \frac{\overline{XZ}}{2} = \frac{-4a+10}{2} = -2a+5$$

Watch out! That matches choice (B), but is not the final answer. What you're looking for is $\overline{WY}$, so you have to add:

$$\overline{WY} = \overline{WX} + \overline{XY} = (7a + 5) + (-2a + 5) = 5a + 10$$

The answer is (E).

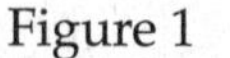

Basic Traits of Triangles

Most Math 2 plane geometry calculations are about closed figures: polygons and circles. And the test makers' favorite closed figure by far is the three-sided polygon; that is, the triangle. All three-sided polygons are interesting because they share so many characteristics, and certain special three-sided polygons—equilateral, isosceles, and right triangles—are interesting because of their special characteristics.

Let's look at the traits that all triangles share.

Sum of the interior angles: The three interior angles of any triangle add up to 180°.

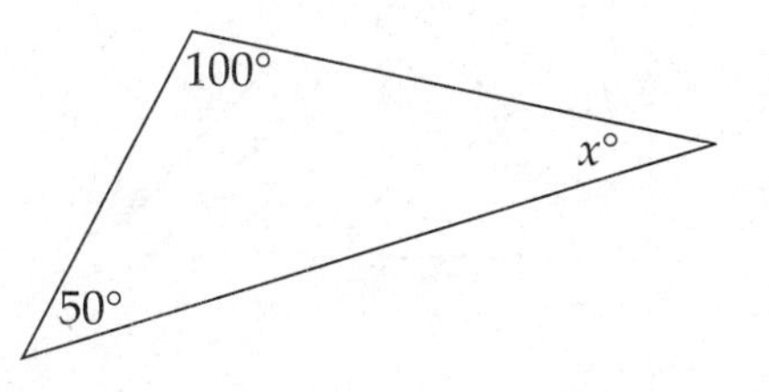

In the figure above, $x + 50 + 100 = 180$, so $x = 30$.

Measure of an exterior angle: The measure of an exterior angle of a triangle is equal to the sum of the measures of the remote interior angles.

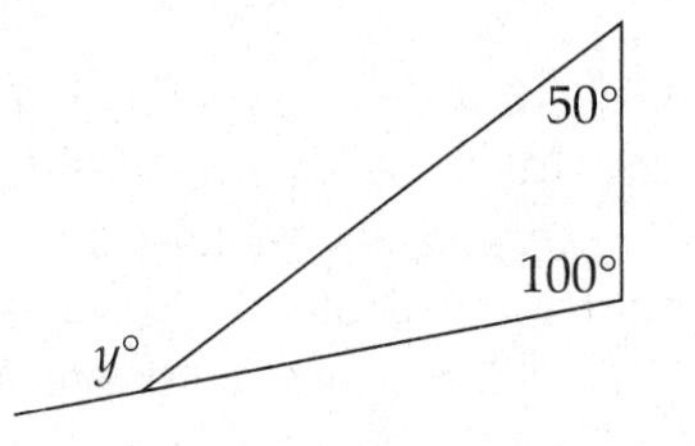

In the figure above, the measure of the exterior angle labeled $y°$ is equal to the sum of the measures of the remote interior angles: $y = 50 + 100 = 150$.

Sum of the exterior angles: The measures of the three exterior angles of any triangle add up to 360°.

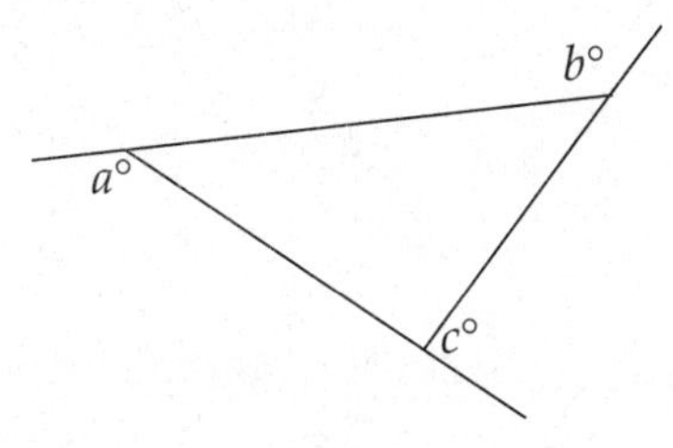

In the figure above, $a + b + c = 360$. (Note: In fact, the measures of the exterior angles of any polygon add up to 360°.)

FIVE FACTS ABOUT TRIANGLES

1. Interior angles add up to 180°.
2. Exterior angle equals sum of remote interior angles.
3. Exterior angles add up to 360°.
4. Area of a triangle = $\frac{1}{2}$ (base)(height).
5. Each side is greater than the difference and less than the sum of the other two sides.

Area formula: The general formula for the area of a triangle is always the same. The formula is

$$\textbf{Area of Triangle} = \frac{1}{2}\textbf{(base)(height)}$$

The height is the perpendicular distance between the side that's chosen as the base and the opposite vertex.

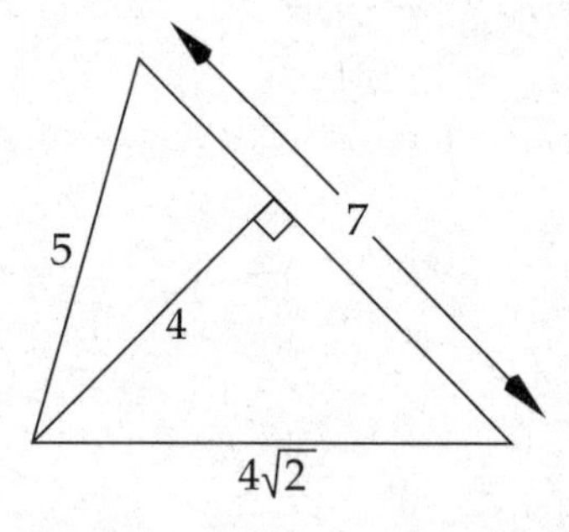

In the triangle above, 4 is the height when the side of length 7 is chosen as the base.

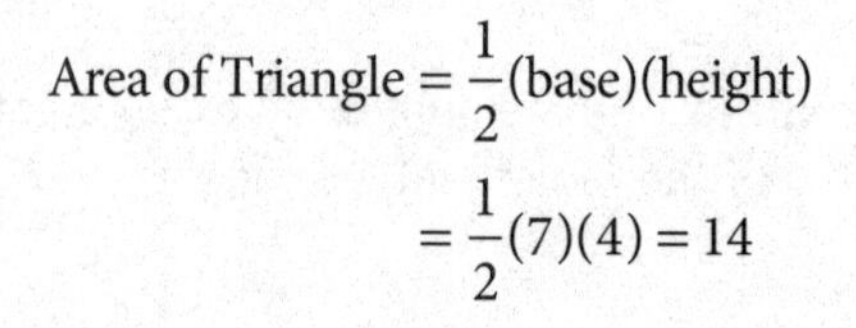

$$\text{Area of Triangle} = \frac{1}{2}\text{(base)(height)}$$
$$= \frac{1}{2}(7)(4) = 14$$

Triangle Inequality Theorem: The length of any one side of a triangle must be greater than the positive difference between, and less than the sum of, the lengths of the other two sides. For example, if it is given that the length of one side is 3 and the length of another side is 7, then the length of the third side must be greater than 7 – 3 = 4 and less than 7 + 3 = 10.

Similar Triangles

In Example 2, you're asked to express the area of one triangle in terms of the area of another.

SIMILAR FIGURES

The area ratio between similar figures is the square of the side ratio.

Example 2

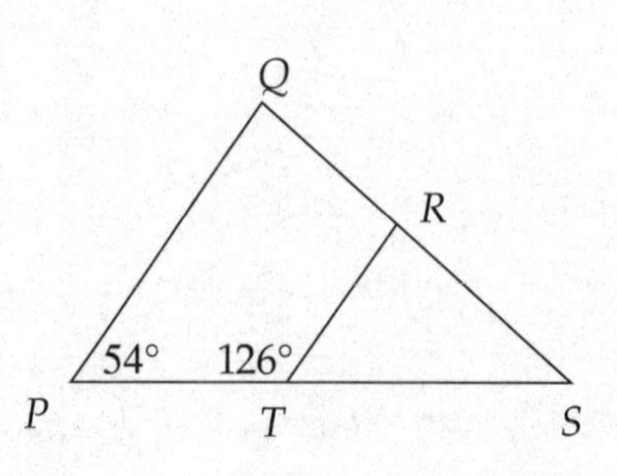

Figure 2

In Figure 2, $\overline{QR} = \overline{RS}$. If the area of ΔRST is $\frac{c}{2}$, what is the area of ΔQSP?

(A) $c\sqrt{2}$
(B) $c\sqrt{3}$
(C) c
(D) $2c$
(E) $3c$

You might wonder here how you're supposed to find the area of ΔQSP when you're given no lengths you can use for a base or an altitude. The only numbers you have are the angle measures. They must be there for some reason—the test makers never provide superfluous information. In fact, because the two angle measures provided add up to 180°, you know that $\overline{PQ}$ and $\overline{TR}$ are parallel. And that, in turn, tells you that ΔRST is similar to ΔQSP, because they have the same three angles.

Similar triangles are triangles that have the same shape: Corresponding angles are equal and corresponding sides are proportional. In this case, because it's given that $\overline{QR} = \overline{RS}$, you know that $\overline{QS}$ is twice $\overline{QR}$ and that corresponding sides are in a ratio of 2:1. Each side of the larger triangle is twice the length of the corresponding side of the smaller triangle. That doesn't mean, however, that the ratio of the areas is also 2:1. In fact, the area ratio is the square of the side ratio, and the larger triangle has four times the area of the smaller triangle.

$$\text{area of } \Delta QSP = 4 \times \text{area of } \Delta RST = 4 \times \frac{c}{2} = 2c$$

The answer is (D).

Alternative method: If you didn't see the similar triangles, or if you didn't know for sure how the area of the larger triangle is related to the area of the smaller triangle, you at least could have eliminated some answer choices based on appearances. Look at the figure and use your eyes to compare the areas. (We call this method eyeballing.) Doesn't it look as though the larger triangle has more than twice as much room inside it as the smaller triangle? That means that answer choices (A), (B), and (C) are all visibly too small. If you can narrow the choices down to two, it certainly pays to guess.

Eyeballing is never what you're supposed to do to answer a question, but if you don't see a better way, eyeballing's better than skipping.

EYEBALL THE FIGURE

You can assume that a figure is drawn to scale unless the problem says otherwise. So, when you're stuck on a geometry question and don't know what else to do, see if you can at least use your eyes to eliminate a few answer choices as visibly too small or too big.

Three Special Triangle Types

Three special triangle types deserve extra attention:

- Isosceles triangles
- Equilateral triangles
- Right triangles

Be sure you know not just the definitions of these triangle types, but more importantly their special characteristics: side relationships, angle relationships, and area formulas.

THREE SPECIAL TRIANGLE TYPES

1. **Isosceles:** Two equal sides and two equal angles.
2. **Equilateral:** Three equal sides and three 60° angles. If the length of one side is s, then:

   Area of Equilateral Triangle $= \frac{s^2\sqrt{3}}{4}$
3. **Right:** One right angle. You can use the legs to find the area:

   Area of Right Triangle $= \frac{1}{2}(\text{leg}_1)(\text{leg}_2)$

Isosceles triangle: An isosceles triangle is a triangle that has two equal sides. Not only are two sides equal, but the angles opposite the equal sides, called *base angles*, are also equal.

Equilateral triangle: An equilateral triangle is a triangle that has three equal sides. Since all the sides are equal, all the angles are also equal. All three angles in an equilateral triangle measure 60 degrees, regardless of the lengths of the sides. You can find the area of an equilateral triangle by dividing it into two 30-60-90 triangles, or you can use this formula in terms of the length of one side s:

$$\textbf{Area of Equilateral Triangle} = \frac{s^2\sqrt{3}}{4}$$

Right triangle: A right triangle is a triangle with a right angle. The two sides that form the right angle are called *legs*, and you can use them as the base and height to find the area of a right triangle.

$$\textbf{Area of Right Triangle} = \frac{1}{2}(\text{leg}_1)(\text{leg}_2)$$

Pythagorean theorem: If you know any two sides of a right triangle, you can find the third side by using the Pythagorean theorem:

$$(\text{leg}_1)^2 + (\text{leg}_2)^2 = (\text{hypotenuse})^2$$

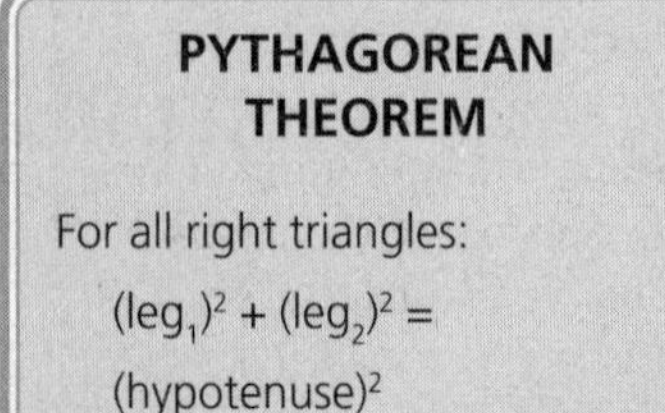
PYTHAGOREAN THEOREM

For all right triangles:

$(\text{leg}_1)^2 + (\text{leg}_2)^2 = (\text{hypotenuse})^2$

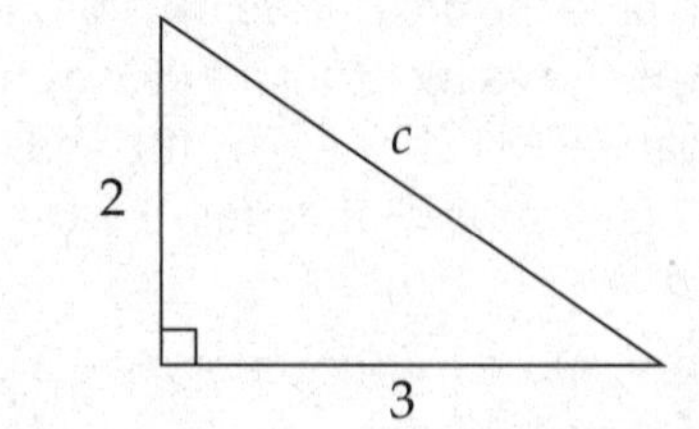

For example, if one leg is 2 and the other leg is 3, then

$$2^2 + 3^2 = c^2$$
$$c^2 = 4 + 9$$
$$c = \sqrt{13}$$

Pythagorean triplet: A Pythagorean triplet is a set of integers that fit the Pythagorean theorem. The simplest Pythagorean triplet is (3, 4, 5). In fact, any integers in a 3:4:5 ratio make up a Pythagorean triplet. And there are many other Pythagorean triplets: (5, 12, 13); (7, 24, 25); (8, 15, 17); (9, 40, 41); all their multiples; and infinitely many more.

3-4-5 triangles: If a right triangle's leg-to-leg ratio is 3:4, or if the leg-to-hypotenuse ratio is 3:5 or 4:5, then it's a 3-4-5 triangle, and you don't need to use the Pythagorean theorem to find the third side. Just figure out what multiple of 3-4-5 it is.

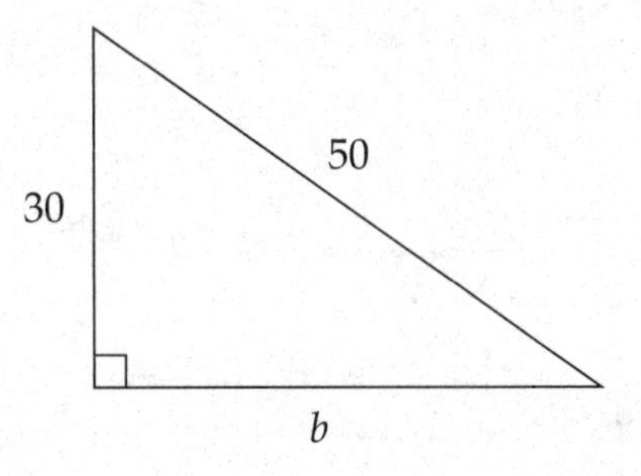

In the right triangle shown, one leg is 30 and the hypotenuse is 50. This is 10 times 3-4-5. The other leg is 40.

5-12-13 triangles: If a right triangle's leg-to-leg ratio is 5:12, or if the leg-to-hypotenuse ratio is 5:13 or 12:13, then it's a 5-12-13 triangle, and you don't need to use the Pythagorean theorem to find the third side. Just figure out what multiple of 5-12-13 it is.

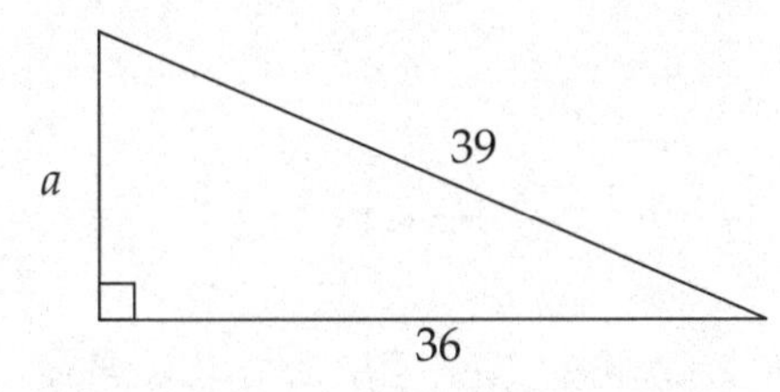

Here one leg is 36 and the hypotenuse is 39. This is 3 times 5-12-13. The other leg is 3×5 or 15.

45-45-90 triangles: The sides of a 45-45-90 triangle are in a ratio of $1:1:\sqrt{2}$.

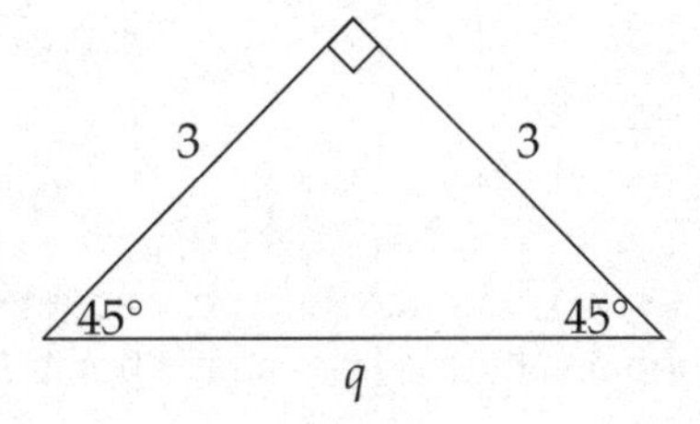

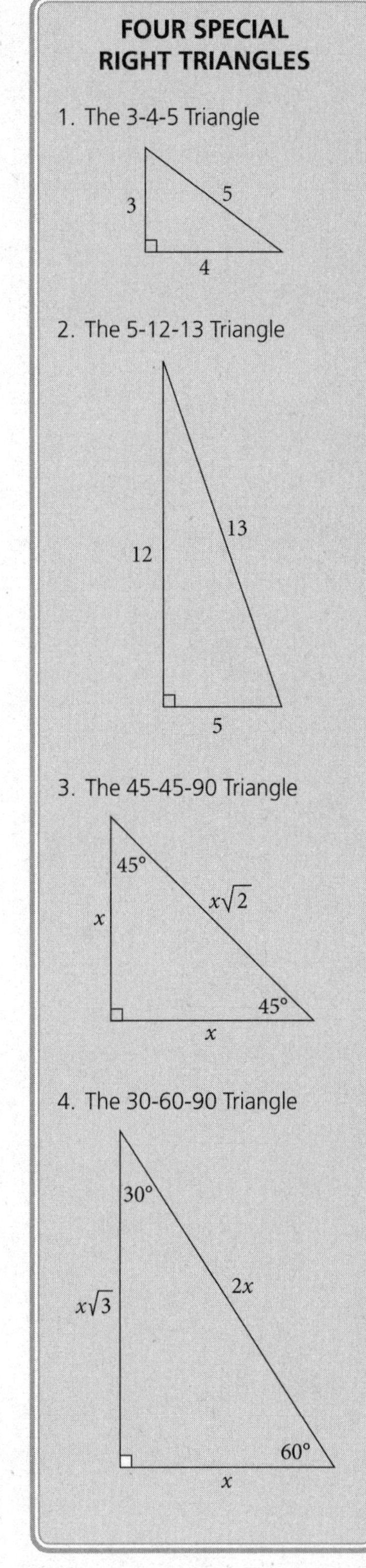

If one leg is 3, then the other leg is also 3, and the hypotenuse is equal to a leg times $\sqrt{2}$, or $3\sqrt{2}$.

30-60-90 triangles: The sides of a 30-60-90 triangle are in a ratio of $1:\sqrt{3}:2$. You don't need to use the Pythagorean theorem.

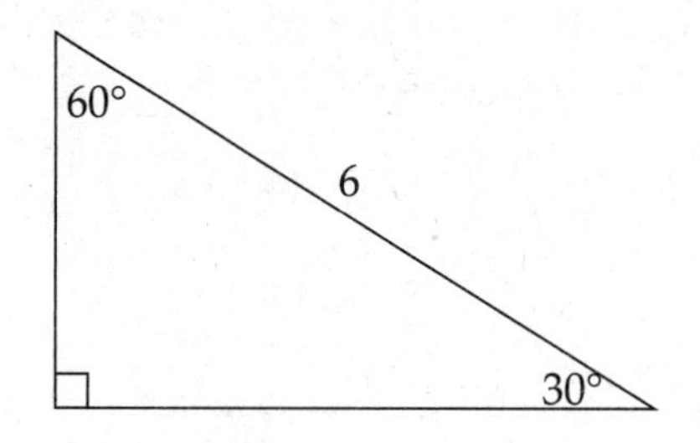

If the hypotenuse is 6, then the shorter leg is half that, or 3, and then the longer leg is equal to the short leg times $\sqrt{3}$, or $3\sqrt{3}$.

Example 3 includes one triangle that's equilateral and another that's both right and isosceles.

Example 3

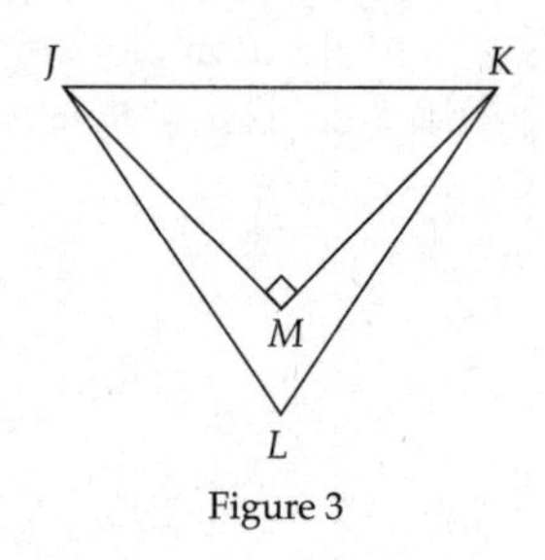

Figure 3

In Figure 3, ΔJKL is equilateral, and ΔJKM is isosceles. If $\overline{KL} = 2$, what is the distance from L to M?

(A) 0.572
(B) 0.636
(C) 0.667
(D) 0.732
(E) 0.767

To get the answer to this question, you need to know about equilateral, 45-45-90, and 30-60-90 triangles. If you drop an altitude from a new point, *N*, through *M* to *L*, you will

divide the equilateral triangle into two 30-60-90 triangles and divide the right isosceles (or 45-45-90) triangle into two smaller right isosceles triangles:

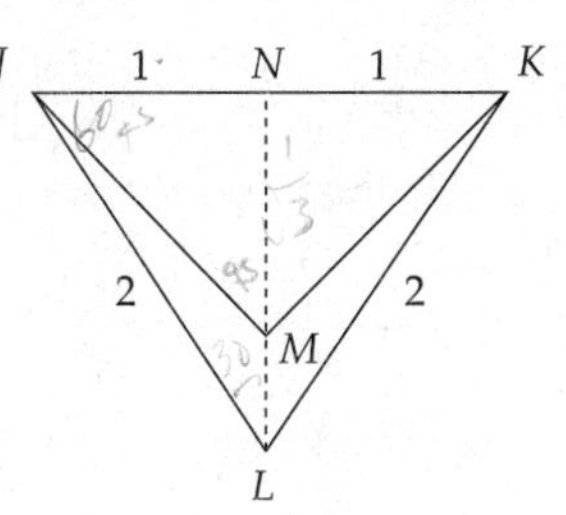

Using the side ratios for 30-60-90s and 45-45-90s, you know that $\overline{LN} = \sqrt{3}$ and that $\overline{MN} = 1$.

Therefore, $\overline{LM} = \overline{LN} - \overline{MN} = \sqrt{3} - 1 \approx 1.732 - 1 = 0.732$.

The answer is (D).

Hidden Special Triangles

It happens a lot that the key to solving a geometry problem is to add a line segment or two to the figure. Often what results is one or more special triangles. The ability to spot, even to create, special triangles comes in handy in a question like Example 4.

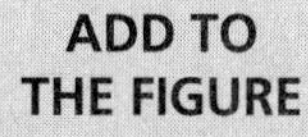

ADD TO THE FIGURE

Often the breakthrough on a plane geometry problem comes when you add a line segment or two to the figure. Perpendiculars can be especially useful. They can function as rectangle sides or triangle altitudes or right triangle legs.

Example 4

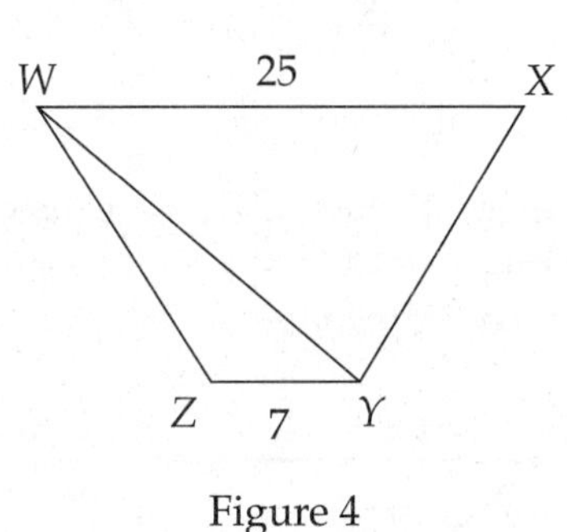

Figure 4

In Figure 4, the perimeter of isosceles trapezoid $WXYZ$ is 62. If $\overline{ZY} = 7$ and $\overline{WX} = 25$, what is the length of diagonal WY?

(A) 15
(B) 17
(C) 20
(D) 25
(E) 26

As you read the stem, you might wonder what an *isosceles trapezoid* is. If you've never heard the term before, you still might have been able to extrapolate its meaning from what you know of isosceles triangles. *Isosceles* means "having two equal sides." When applied to a trapezoid, the term *isosceles* tells you that the two nonparallel sides—the legs—are equal. In this case, that's $\overline{WZ}$ and $\overline{XY}$. If the total perimeter is 62, and the two marked sides add up to 25 + 7 = 32, then the two unmarked sides split the difference of 62 – 32 = 30. In other words, $\overline{WZ} = \overline{XY} = 15$.

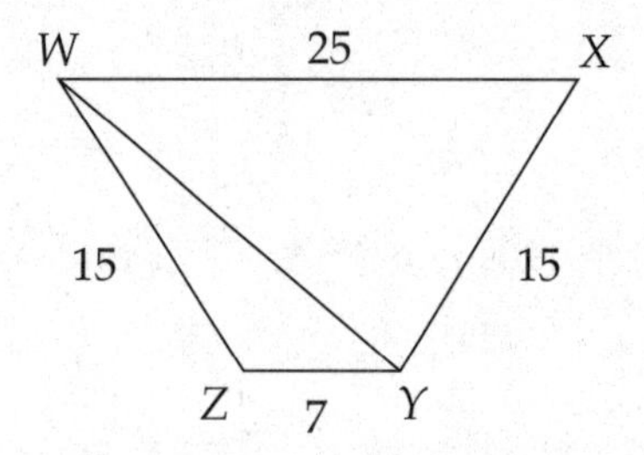

There aren't any special triangles yet. As so often happens, though, you can get some by constructing altitudes. Drop perpendiculars to points *Z* and *Y*, and you make two right triangles. The length 25 of side $\overline{WX}$ then gets split into 9, 7, and 9.

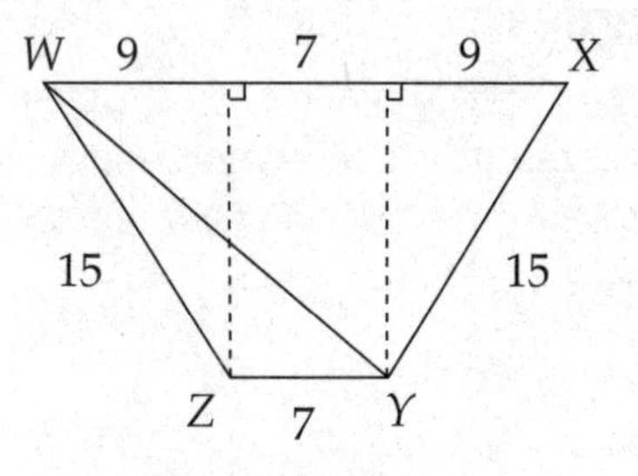

Now you can see that those right triangles are 3-4-5s (times 3) and that the height of the trapezoid is 12 because it is a leg of a 9 – x – 15 right triangle. Now look at the right triangle of which $\overline{WY}$ is the hypotenuse.

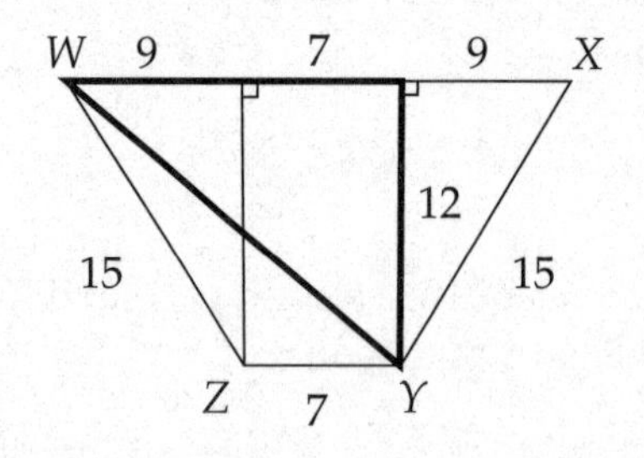

One leg is 7 + 9 = 16, and the other leg is 12. The resulting triangle of which $\overline{WY}$ is the hypotenuse is another 3-4-5 triangle (times 4). $\overline{WY} = 20$, and the answer is (C).

Special Quadrilaterals

The trapezoid is just one of five special quadrilaterals you need to be familiar with. As with triangles, there is some overlap among these categories, and some figures fit into none of these categories. Just as a 45-45-90 triangle is both right and isosceles, a quadrilateral with four equal sides and four right angles is not only a square but also a rhombus, a rectangle, and a parallelogram. It is wise to have a solid grasp of the definitions and special characteristics of these five quadrilateral types.

Trapezoids: A trapezoid is a four-sided figure with one pair of parallel sides and one pair of nonparallel sides.

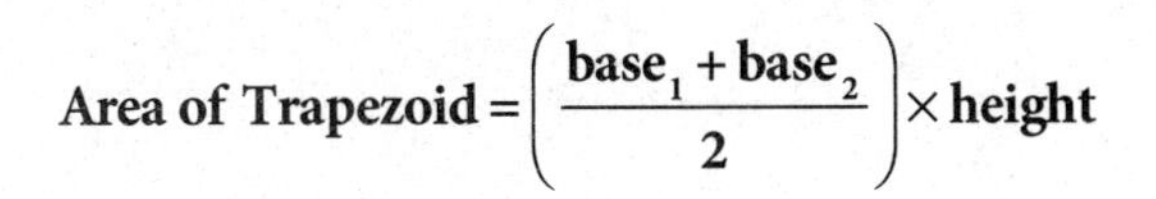

$$\textbf{Area of Trapezoid} = \left(\frac{\textbf{base}_1 + \textbf{base}_2}{2}\right) \times \textbf{height}$$

Think of this formula as the average of the bases (the two parallel sides) times the height (the length of the perpendicular altitude).

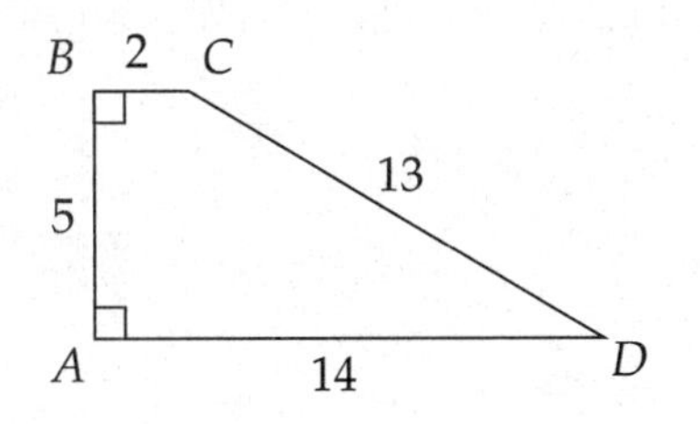

In the trapezoid *ABCD* above, you can use side *AB* for the height. The average of the bases is $\frac{2+14}{2} = 8$, so the area is 8 × 5, or 40.

Parallelograms: A parallelogram is a four-sided figure with two pairs of parallel sides. Opposite sides are equal. Opposite angles are equal. Consecutive angles add up to 180°.

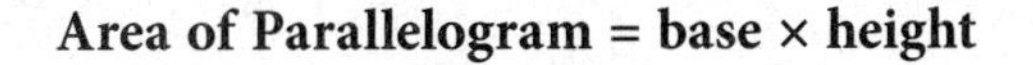

$$\textbf{Area of Parallelogram} = \textbf{base} \times \textbf{height}$$

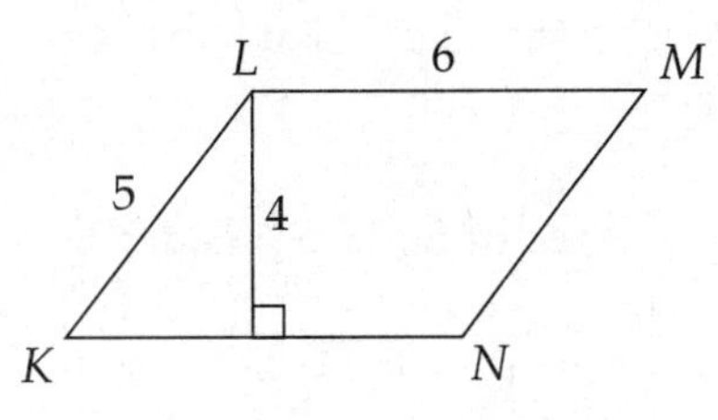

In parallelogram *KLMN* above, 4 is the height when *LM* or *KN* is used as the base. Base × height = 6 × 4 = 24.

Remember that to find the area of a parallelogram you need the height, which is the perpendicular distance from the base to the opposite side. You can use a side of a parallelogram for the height only when the side is perpendicular to the base, in which case you have a rectangle.

Rectangles: A rectangle is a four-sided figure with four right angles. Opposite sides are equal. Diagonals are equal. The perimeter of a rectangle is equal to the sum of the lengths of the four sides, which is equal to 2(length + width).

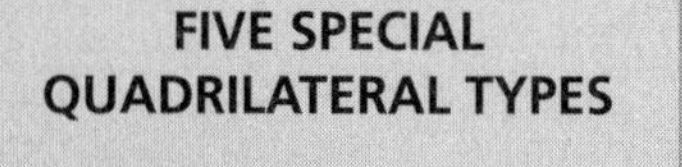

FIVE SPECIAL QUADRILATERAL TYPES

1. **Trapezoid:** One pair of parallel sides.
   Area of Trapezoid
   $= \frac{\text{base}_1 + \text{base}_2}{2} \times \text{height}$
2. **Parallelogram:** Two pairs of parallel sides.
   Area of Parallelogram
   = base × height
3. **Rectangle:** Four right angles.
   Perimeter of Rectangle
   = 2(length + width)
   Area of Rectangle
   = length × width
4. **Rhombus:** Four equal sides.
   Perimeter of Rhombus
   = 4 × side
   Area of Rhombus
   = base × height
5. **Square:** Four right angles and four equal sides.
   Perimeter of Square
   = 4 × side
   Area of Square = $(\text{side})^2$

Area of Rectangle = length × width

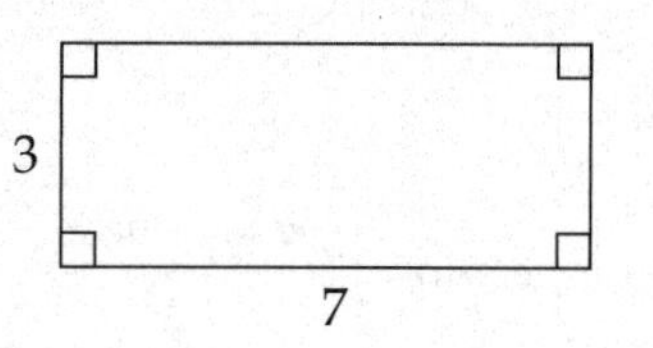

The area of a 7-by-3 rectangle is 7 × 3 = 21.

Rhombus: A rhombus is a four-sided figure with four equal sides.

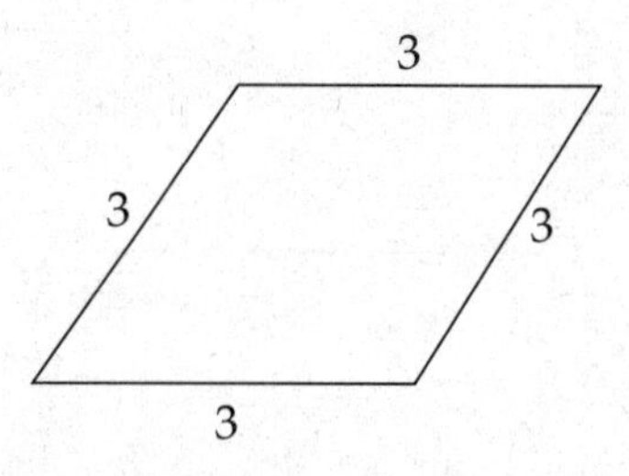

All four sides of the quadrilateral above have the same length, so it's a rhombus. A rhombus is also a parallelogram, so to find the area of a rhombus, you need its height. The more a rhombus "leans over," the smaller the height and therefore the smaller the area. The maximum area for a rhombus of a certain perimeter is that rhombus that has each pair of adjacent sides perpendicular, in which case you have a square.

Square: A square is a four-sided figure with four right angles and four equal sides. A square is also a rectangle, a parallelogram, and a rhombus. The perimeter of a square is equal to 4 times the length of one side.

Area of Square = $(\text{side})^2$

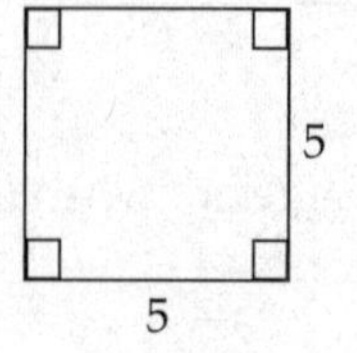

The square above, with sides of length 5, has an area of $5^2 = 25$.

Polygons—Perimeter and Area

The Math 2 test makers like to write problems that combine the concepts of perimeter and area. What you need to remember is that perimeter and area are not directly related. In Example 5, for instance, you have two figures with the same perimeter, but that doesn't mean they have the same area.

Example 5

5. A triangle and a regular hexagon have the same perimeter. If the area of the hexagon is $72\sqrt{3}$, what is the area of the triangle?

   (A) 62.354
   (B) 83.138
   (C) 101.823
   (D) 103.923
   (E) 124.708

SKETCH A FIGURE IF NONE IS PROVIDED

The best way to get a handle on a figureless geometry problem is usually to sketch a figure of your own. You don't have to be an artist. Just be neat and clear enough to get the picture.

Sketch what's described in the question.

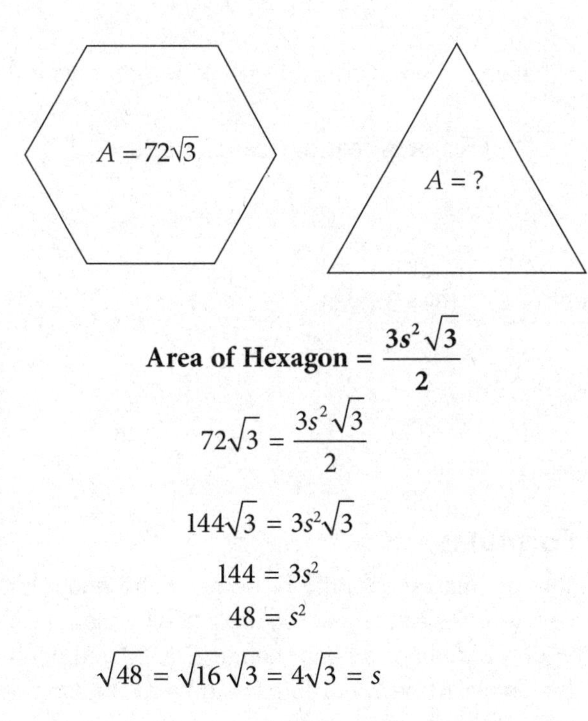

$$\textbf{Area of Hexagon} = \frac{3s^2\sqrt{3}}{2}$$

$$72\sqrt{3} = \frac{3s^2\sqrt{3}}{2}$$

$$144\sqrt{3} = 3s^2\sqrt{3}$$

$$144 = 3s^2$$

$$48 = s^2$$

$$\sqrt{48} = \sqrt{16}\sqrt{3} = 4\sqrt{3} = s$$

Each side of the regular hexagon has a length of $4\sqrt{3}$.

If you haven't memorized the formula for the area of a regular hexagon, just remember that it is comprised of six equilateral triangles. (Notice that the formula for the area of a regular hexagon is just 6 times the formula for the area of an equilateral triangle.) So even if you don't remember the formula, it should still be fairly easy to calculate the side lengths of the hexagon if you remember the side lengths of special right triangles.

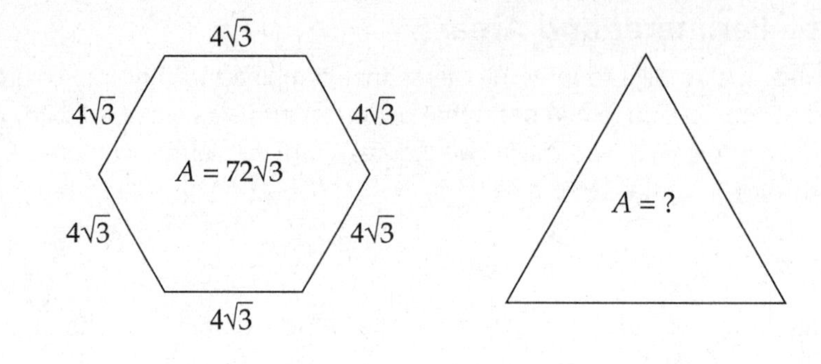

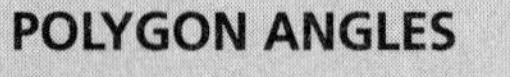

For any polygon of n sides:

Sum of Interior Angles $= (n - 2) \times 180°$

Sum of Exterior Angles $= 360°$

Use the fact that the perimeters of the hexagon and triangle are the same in order to figure out the length of each side of the triangle.

Perimeter of an equilateral hexagon =

$$6 \times s = 6 \times 4\sqrt{3} = 24\sqrt{3}$$

Side of an equilateral triangle =

$$\frac{\text{Perimeter}}{3} = \frac{24\sqrt{3}}{3} = 8\sqrt{3}$$

Use the formula for the area of an equilateral triangle to get to your answer.

Area of an equilateral triangle =

$$\frac{s^2\sqrt{3}}{4}$$

$$\frac{(8\sqrt{3})^2(\sqrt{3})}{4} = \frac{(64 \times 3)(\sqrt{3})}{4} = 16 \times 3 \times \sqrt{3} = 48\sqrt{3} \approx 83.138$$

The answer is (B).

Circles—Four Formulas

After the triangle, the test makers' favorite plane geometry figure is the circle. Circles don't come in as many varieties as triangles do. In fact, all circles are similar—they're all the same shape. The only difference among them is size. So you don't have to learn to recognize types or remember names. All you have to know about circles is how to find four things:

- Circumference
- Length of an arc
- Area
- Area of a sector

You could think of the task as one of memorizing four formulas, but you'll be better off in the end if you have some idea of where the arc and sector formulas come from and how they are related to the circumference and area formulas.

Circumference: Circumference is a measurement of length. You could think of it as the perimeter: It's the total distance around the circle. If the radius of the circle is r,

$$\textbf{Circumference} = 2\pi r$$

Since the diameter is twice the radius, you can easily express the formula in terms of the diameter d:

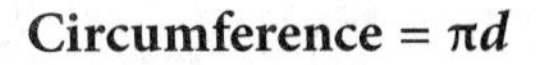

$$\textbf{Circumference} = \pi d$$

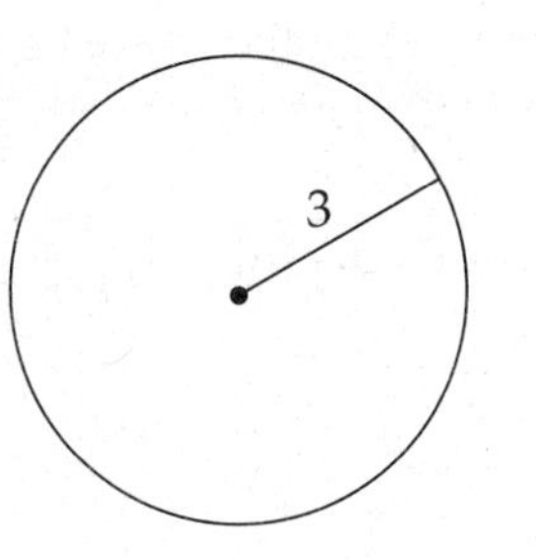

In the circle above, the radius is 3, so the circumference is $2\pi(3) = 6\pi$.

Length of an arc: An arc is a piece of the circumference. If n is the degree measure of the arc's central angle, then the formula is

$$\textbf{Length of an Arc} = \left(\frac{n}{360}\right)(2\pi r)$$

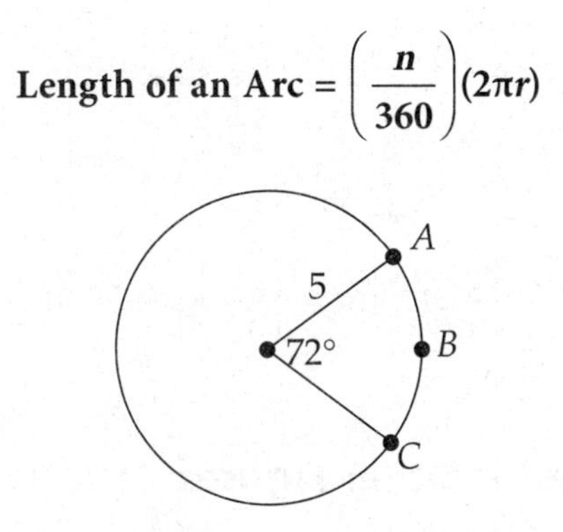

In the figure above, the radius is 5, and the measure of the central angle is 72°. The arc length is $\frac{72}{360}$ or $\frac{1}{5}$ of the circumference:

$$\left(\frac{72}{360}\right)(2\pi)(5) = \left(\frac{1}{5}\right)(10\pi) = 2\pi$$

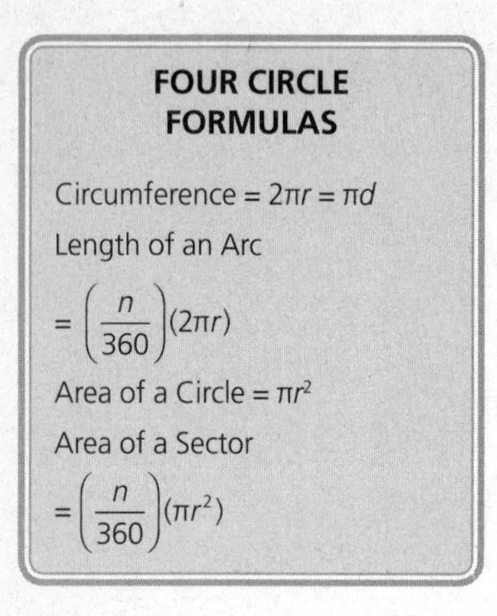

FOUR CIRCLE FORMULAS

Circumference = $2\pi r = \pi d$

Length of an Arc

$= \left(\frac{n}{360}\right)(2\pi r)$

Area of a Circle = πr^2

Area of a Sector

$= \left(\frac{n}{360}\right)(\pi r^2)$

Area: The area of a circle is usually found using this formula in terms of the radius r:

$$\textbf{Area of a Circle} = \pi r^2$$

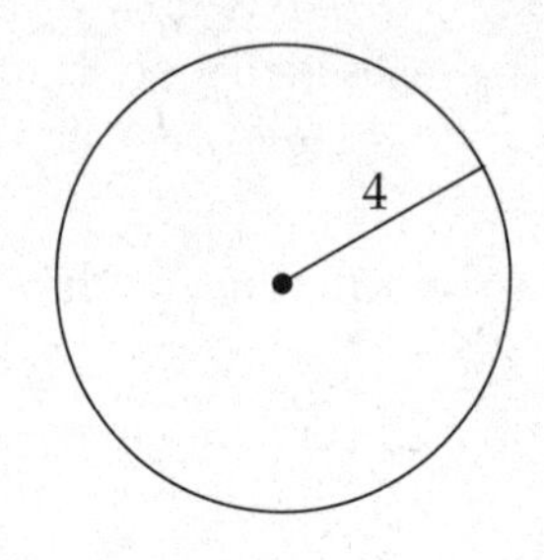

The area of the circle above is $\pi(4)^2 = 16\pi$.

Area of a sector: A sector is a piece of the area of a circle. If n is the degree measure of the sector's central angle, then the area formula is

$$\textbf{Area of a Sector} = \left(\frac{n}{360}\right)(\pi r^2)$$

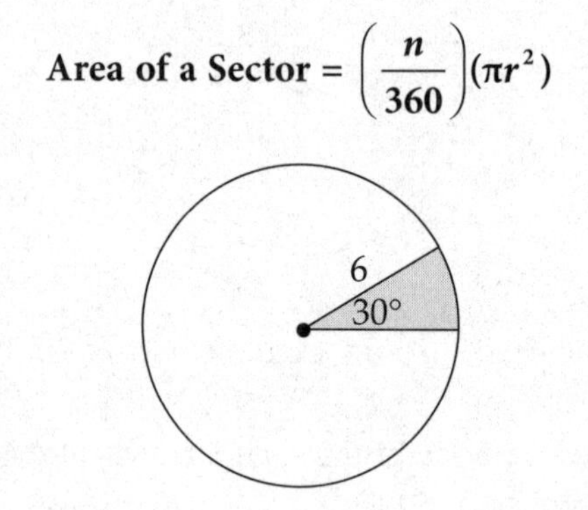

In the figure above, the radius is 6, and the measure of the sector's central angle is 30°. The sector has $\frac{30}{360}$ or $\frac{1}{12}$ of the area of the circle:

$$\left(\frac{30}{360}\right)(\pi)(6^2) = \left(\frac{1}{12}\right)(36\pi) = 3\pi$$

Circles Combined with Other Figures

Some of the most challenging plane geometry questions are those that combine circles with other figures.

Consider Example 6.

Example 6

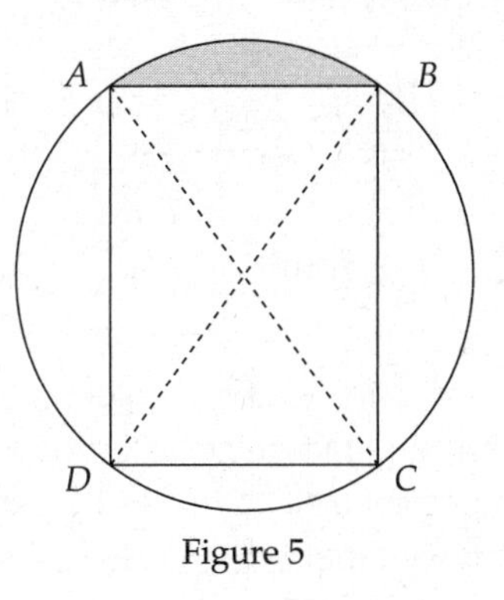

Figure 5

6. In Figure 5, rectangle *ABCD* is inscribed in a circle. If the radius of the circle is 2 and $\overline{CD} = 2$, what is the area of the shaded region?

 (A) 0.362
 (B) 0.471
 (C) 0.577
 (D) 0.707
 (E) 0.866

As usual, the key here is to add to the figure. And here, as is so often the case with circles, you should add radii. The equilateral triangles formed by $\overline{CD}$ and the radii, and by $\overline{AB}$ and the radii, have central angles of 60°.

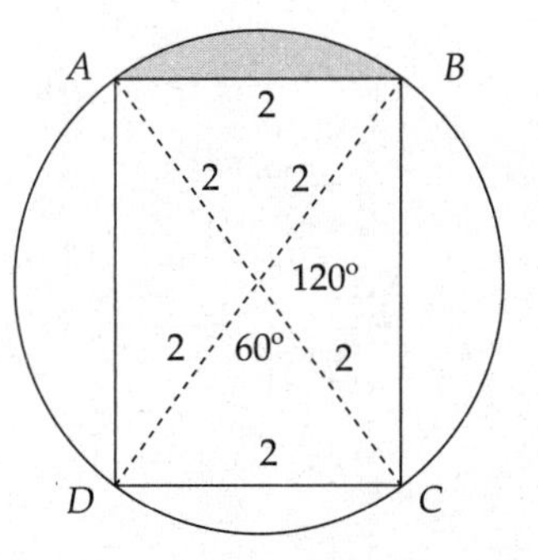

The shaded region is what's left of the 60° sector after you subtract the area of an equilateral triangle. To find the area of the shaded region, find the areas of the sector and triangle; then subtract the latter from the former. The sector is exactly one-sixth of the circle (because 60° is one-sixth of 360°). Thus,

$$\text{Area of the sector} = \frac{1}{6}\pi r^2 = \frac{1}{6}(\pi)(2^2) = \frac{4\pi}{6}$$

Use the formula for the area of an equilateral triangle to close in on the answer.

$$\textbf{Area of an equilateral triangle} = \frac{s^2\sqrt{3}}{4}$$

$$\frac{2^2\sqrt{3}}{4} = \frac{4\sqrt{3}}{4} = \sqrt{3}$$

The shaded area, then, is $\frac{4\pi}{6} - \sqrt{3} \approx 0.362$. The answer is (A).

Now that you've reviewed all the relevant plane geometry facts and formulas, seen some of the test makers' favorite plane geometry situations, and learned a few good Kaplan strategies, it's time to put yourself to the test again. If you were disappointed in your performance on the Plane Geometry Diagnostic Test, here's your chance to make up for it.

THINGS TO REMEMBER:

Five Facts about Triangles

1. Interior angles add up to 180°.
2. Exterior angle equals sum of remote interior angles.
3. Exterior angles add up to 360°.
4. Area of a triangle = $\frac{1}{2}$ (base)(height).
5. Each side is greater than the difference between and less than the sum of the other two sides.

Similar Figures

The area ratio between similar figures is the square of the side ratio.

Three Special Triangle Types

1. Isosceles: Two equal sides and two equal angles.
2. Equilateral: Three equal sides and three 60° angles. If the length of one side is s, then

   Area of Equilateral Triangle = $\frac{s^2\sqrt{3}}{4}$

3. Right: One right angle. You can use the legs to find the area:

   Area of Right Triangle = $\frac{1}{2}(\text{leg}_1)(\text{leg}_2)$

Pythagorean Theorem

For all right triangles,

$(\text{leg}_1)^2 + (\text{leg}_2)^2 = (\text{hypotenuse})^2$

Four Special Right Triangles

1. The 3-4-5 Triangle

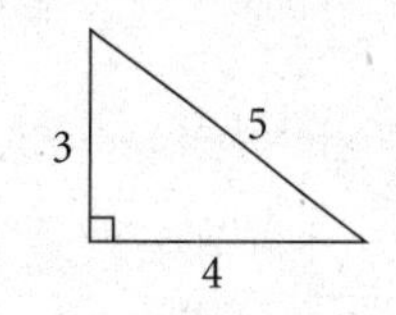

2. The 5-12-13 Triangle

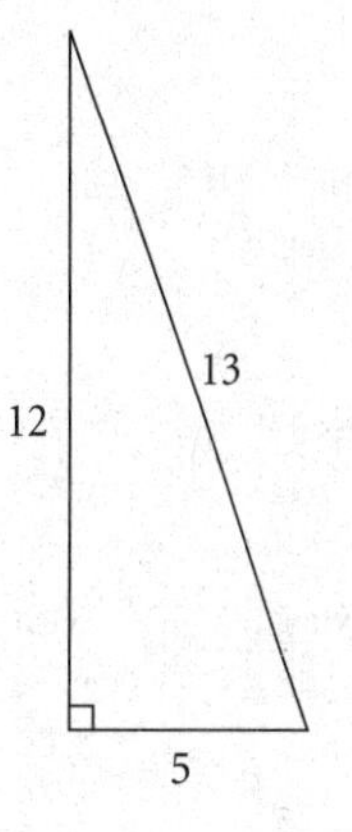

3. The 45-45-90 Triangle

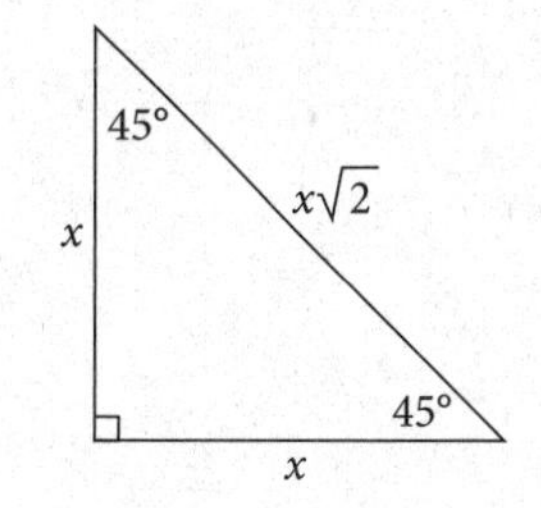

4. The 30-60-90 Triangle

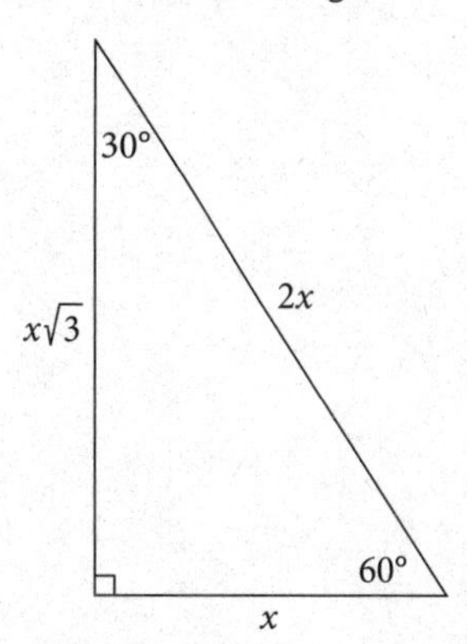

Five Special Quadrilateral Types

1. Trapezoid:
   Area of Trapezoid =
   $\frac{\text{base}_1 + \text{base}_2}{2} \times \text{height}$
2. Parallelogram:
   Area of Parallelogram = base × height
3. Rectangle:
   Perimeter of Rectangle = 2(length + width)

   Area of Rectangle =
   length × width
4. Rhombus: Four equal sides.
   Perimeter of Rhombus =
   4 × side

   Area of Rhombus =
   base × height
5. Square: Four right angles and four equal sides.
   Perimeter of Square =
   4 × side

   Area of Square = $(\text{side})^2$

Polygon Angles

For any polygon of n sides:
Sum of Interior Angles =
$(n - 2) \times 180°$

Sum of Exterior Angles = 360°

Four Circle Formulas

Circumference = $2\pi r = \pi d$

Length of an Arc =
$\left(\frac{n}{360}\right)(2\pi r)$

Area of a Circle = πr^2

Area of a Sector =
$\left(\frac{n}{360}\right)(\pi r^2)$

PLANE GEOMETRY FOLLOW-UP TEST

6 Questions (8 Minutes)

Directions: Solve the following problems. Fill in the oval corresponding to the best answer choice in the grid to the right of each question. (Answers and explanations begin on page 91.)

DO YOUR FIGURING HERE

1. In Figure 1, the ratio of $\overline{DE}$ to $\overline{EF}$ is 4 to 9. If $\overline{DF} = 2$, what is the distance from D to the midpoint of $\overline{EF}$?

(A) $\frac{9}{13}$

(B) $\frac{12}{13}$

(C) $\frac{14}{13}$

(D) $\frac{16}{13}$

(E) $\frac{17}{13}$

Ⓐ Ⓑ Ⓒ Ⓓ Ⓔ

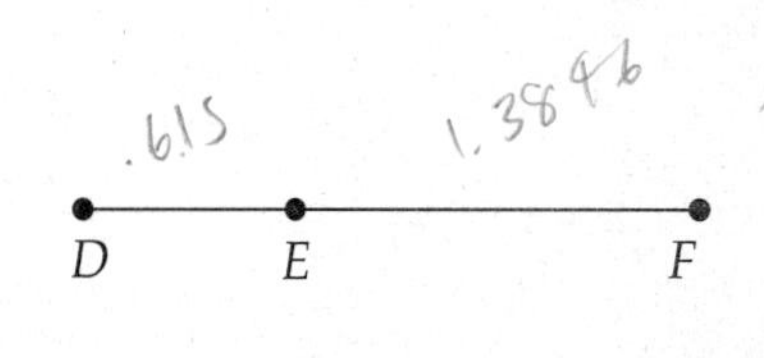

Figure 1

2. In ΔYVW in Figure 2, $\overline{VX}$ is the altitude to side $\overline{YW}$, and $\overline{ZW}$ is the altitude to side $\overline{YV}$. If $\overline{VX} = 3$, $\overline{YV} = 4$, and $\overline{ZW} = 5$, what is the length of side $\overline{YW}$?

(A) 4.157

(B) 5.303

(C) 6.667

(D) 6.928

(E) 8.660

Ⓐ Ⓑ Ⓒ Ⓓ Ⓔ

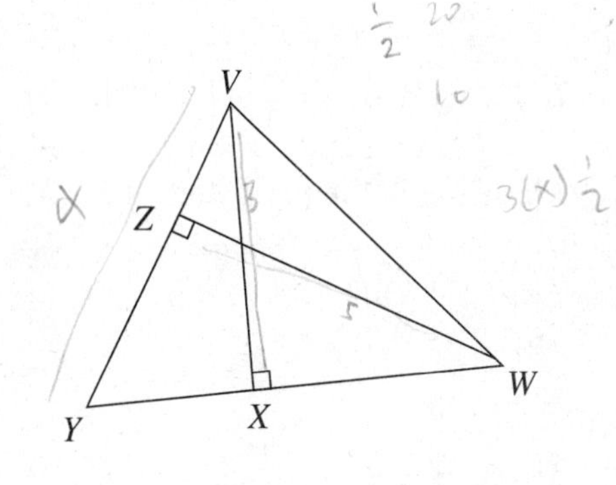

Figure 2

3. In Figure 3, $\overline{EG} \parallel \overline{DH}$, and the lengths of segments $\overline{DE}$ and $\overline{EF}$ are as marked. If the area of ΔEFG is a, what is the area of ΔDFH in terms of a?

(A) $\frac{4a}{5}$

(B) $\frac{16a}{25}$

(C) $\frac{16a}{20}$

(D) $\frac{5a}{4}$

(E) $\frac{25a}{16}$

Ⓐ Ⓑ Ⓒ Ⓓ Ⓔ

4. In Figure 4, if $\overline{XY} = 3$, what is the area of ΔXYZ?

(A) 1.949
(B) 3.074
(C) 5.324
(D) 7.529
(E) 12.728

Ⓐ Ⓑ Ⓒ Ⓓ Ⓔ

5. In Figure 5, the area of parallelogram $JKLM$ is 28 and $\overline{JK} = 7$. If $\overline{KN}$ is perpendicular to $\overline{ML}$ and if N is the midpoint of $\overline{ML}$, what is the perimeter of $JKLM$?

(A) 24.6302
(B) 23.25
(C) 28
(D) 31.596
(E) 33.941

Ⓐ Ⓑ Ⓒ Ⓓ Ⓔ

DO YOUR FIGURING HERE

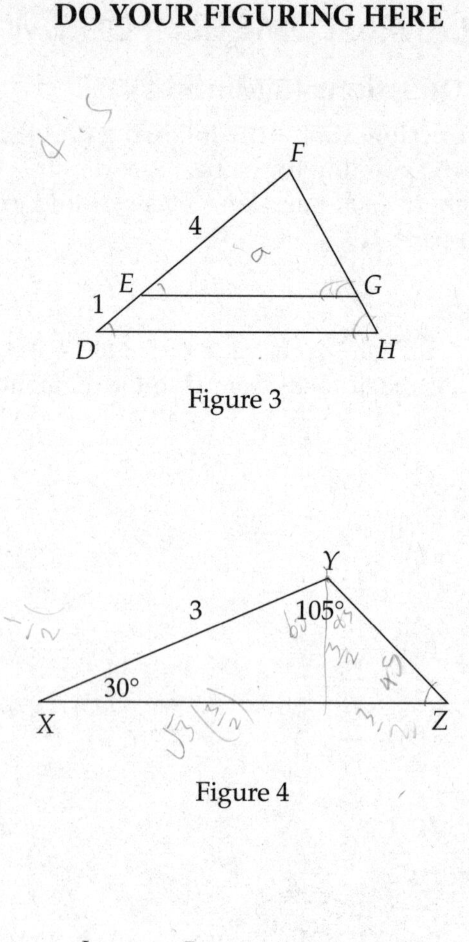

Figure 3

Figure 4

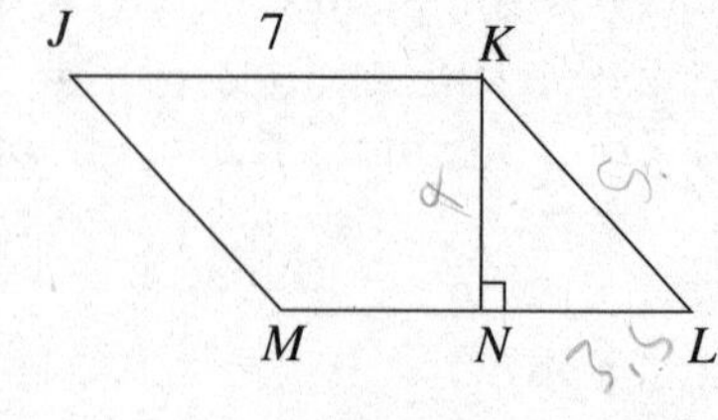

Figure 5

6. In Figure 6, points P, Q, and R lie on the circumference of the circle centered at O. If $\angle OPQ$ measures $x°$ and $\angle QRO$ measures $y°$, what is the measure of $\angle POR$ in terms of x and y?

(A) $(360 + x + y)°$
(B) $(360 - x - y)°$
(C) $(360 - 2x - 2y)°$
(D) $(180 + x + y)°$
(E) $(180 - 2x - 2y)°$

Ⓐ Ⓑ Ⓒ Ⓓ Ⓔ

DO YOUR FIGURING HERE

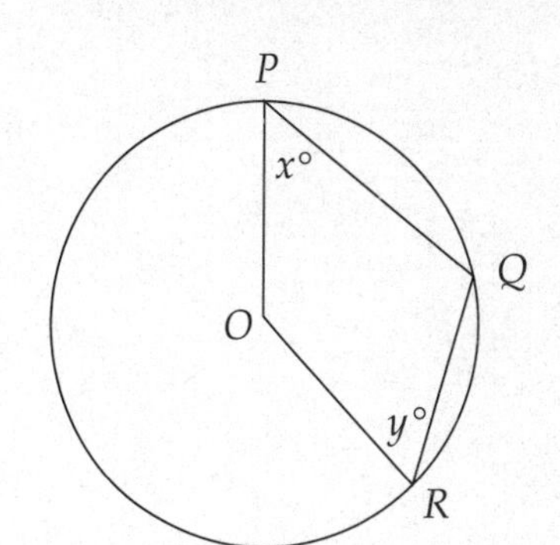

Figure 6

FOLLOW-UP TEST—ANSWERS AND EXPLANATIONS

1. E

Mark up the figure:

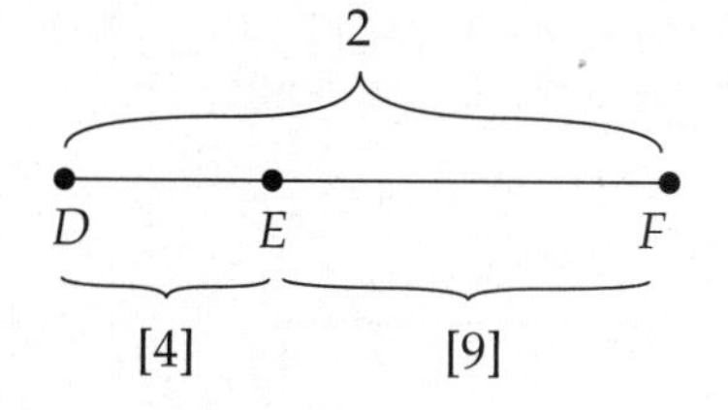

The brackets around the 4 and the 9 are meant to show that they're proportions, not actual lengths. Since the ratio of $\overline{DE}$ to $\overline{EF}$ is 4 to 9, you can say that 4 parts out of 13 are in $\overline{DE}$ and 9 parts out of 13 are in $\overline{EF}$.

Because the whole length $\overline{DF} = 2$, $\overline{DE} = \frac{4}{13} \times 2 = \frac{8}{13}$ and $\overline{EF} = \frac{9}{13} \times 2 = \frac{18}{13}$. The midpoint of $\overline{EF}$ divides it in half, so each half length is half of $\frac{18}{13}$, which is $\frac{9}{13}$. The length you're looking for is $\frac{8}{13} + \frac{9}{13} = \frac{17}{13}$.

2. C

Put the given lengths into the figure:

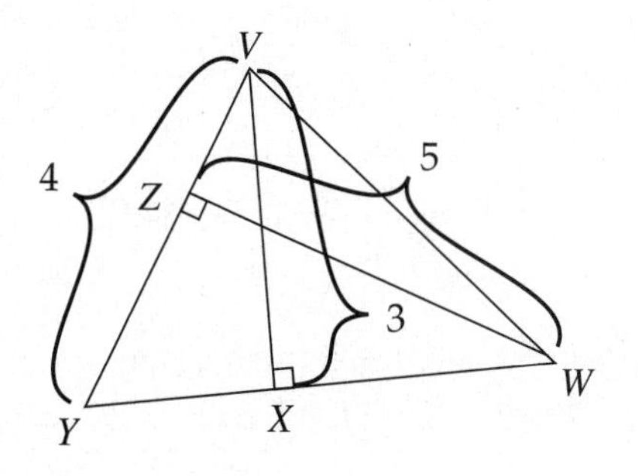

One-half the product of $\overline{YW}$ and $\overline{VX}$ is the same as one-half the product of $\overline{YV}$ and $\overline{ZW}$. Or, ignoring the one-halfs, you can simply say that the products are equal.

$$\frac{1}{2}(\overline{YW})(\overline{VX}) = \frac{1}{2}(\overline{YV})(\overline{ZW})$$

$$(\overline{YW})(\overline{VX}) = (\overline{YV})(\overline{ZW})$$

$$(\overline{YW})(3) = (4)(5)$$

$$(\overline{YW}) = \frac{20}{3} \approx 6.667$$

3. E

The only information that's in the question and not in the figure is that $\overline{EG}$ and $\overline{DH}$ are parallel and that the area of ΔEFG is a. That $\overline{EG}$ and $\overline{DH}$ are parallel tells you that ΔEFG and ΔDFH are similar—they have the same angles, and their sides are in proportion. Because the ratio of $\overline{DF}$ to $\overline{EF}$ is $\frac{5}{4}$, the ratio of any pair of corresponding sides will also be $\frac{5}{4}$. But that's not the ratio of the areas. Remember that the ratio of the areas of similar triangles is the square of the ratio of the sides. Here the side ratio is $\frac{5}{4}$, so the area ratio is $\frac{5^2}{4^2} = \frac{25}{16}$. If the area of the small triangle is a, then the area of the large one is $\frac{25a}{16}$.

4. B

Drop an altitude, and you'll reveal two hidden special right triangles:

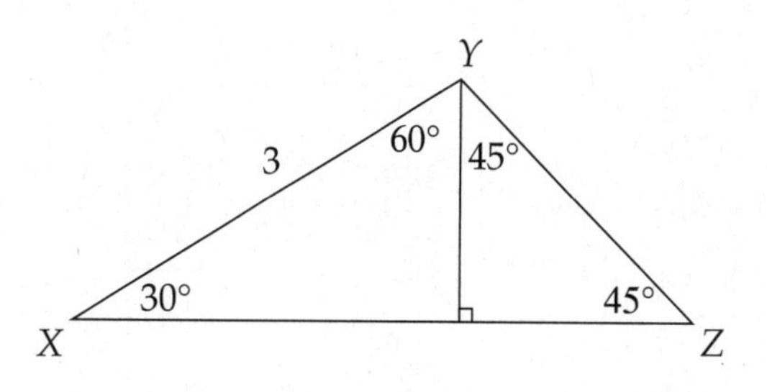

Now you can use the known side ratios of 45-45-90 and 30-60-90 triangles to get all the lengths you need. If the hypotenuse of a 30-60-90 is 3, then the shorter leg is $\frac{3}{2}$ and the longer leg is $\frac{3\sqrt{3}}{2}$:

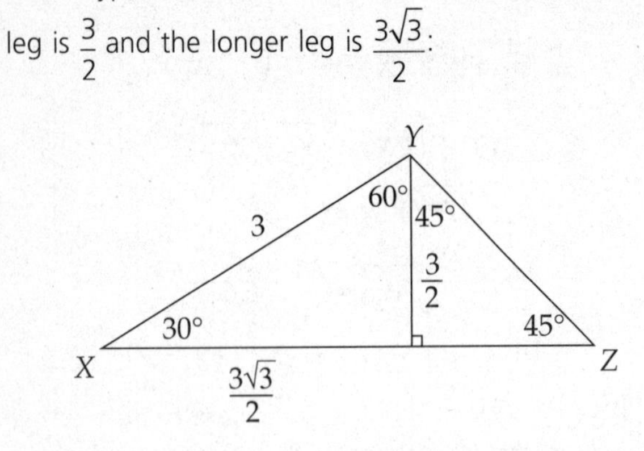

Now that you know one of the legs of the 45-45-90, you know the length of the other leg.

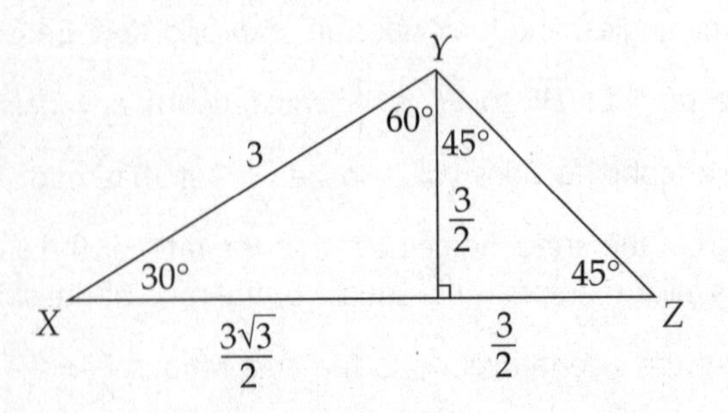

Now you have the base and the height. The base is $\frac{3\sqrt{3}+3}{2}$, and the height is $\frac{3}{2}$. Thus,

$$\text{Area of triangle} = \frac{1}{2}(\text{base})(\text{height})$$

$$= \left(\frac{1}{2}\right)\left(\frac{3\sqrt{3}+3}{2}\right)\left(\frac{3}{2}\right)$$

$$= \frac{3(3\sqrt{3}+3)}{8} = \frac{9\sqrt{3}+9}{8}$$

$$\approx 3.074$$

5. A

Because *JKLM* is a parallelogram, opposite sides are equal and $\overline{ML} = 7$. Midpoint *N* divides that 7 into two equal pieces of $\frac{7}{2}$ each. The area of the parallelogram, which is equal to the base times the height, is given as 28, and because the base is 7, altitude $\overline{KN} = 4$. Put all this into the figure.

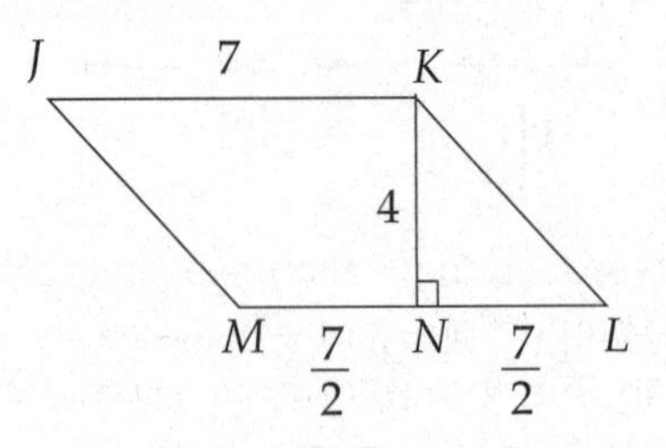

Use the Pythagorean theorem to determine the length of $\overline{KL}$.

$$\overline{LN}^2 + \overline{KN}^2 = \overline{KL}^2$$

$$\left(\frac{7}{2}\right)^2 + 4^2 = \overline{KL}^2$$

$$\frac{49}{4} + 16 = \overline{KL}^2$$

$$\frac{49}{4} + \frac{64}{4} = \overline{KL}^2$$

$$\frac{113}{4} = \overline{KL}^2$$

$$\sqrt{\frac{113}{4}} = \overline{KL} \approx 5.3151$$

Now add all the sides of the parallelogram to find the perimeter: 7 + 7 + 5.3151 + 5.3151 = 24.6302.

6. C

As is so often true when circles are combined with other figures, the key to solving this question is to draw in a radius:

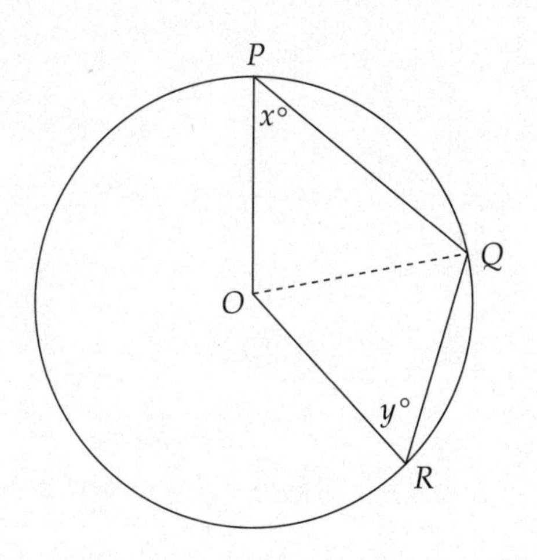

All the radii of a circle are equal, so within each of the two triangles you just created, the angles opposite the radii are equal:

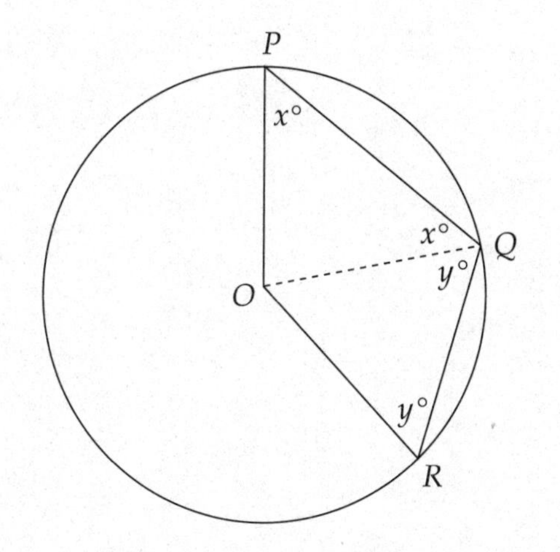

Now that you know two angles in each triangle, you can figure out the third:

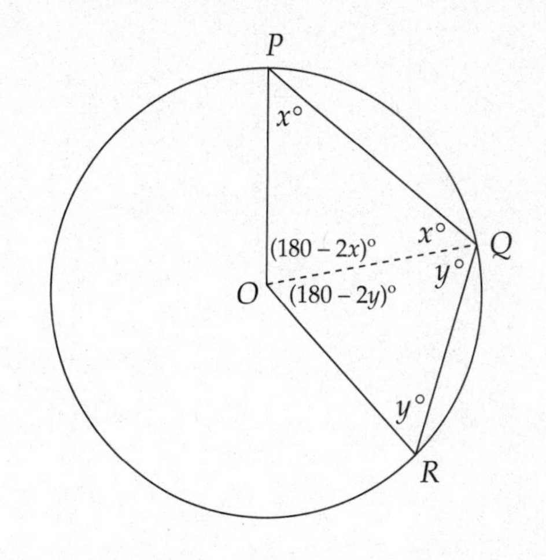

The angle you're looking for is $(180 - 2x)° + (180 - 2y)° = (360 - 2x - 2y)°$.

Chapter 6: **Solid Geometry**

- Volume and surface
- Test makers' favorites
- Rotating polygons

The material in this chapter is relevant to both Mathematics subject tests. Level 1 and Level 2 both include a few solid geometry questions, but this is not a big and vital category on either test.

SOLID GEOMETRY FACTS AND FORMULAS IN THIS CHAPTER

- Five Formulas You Don't Need to Memorize
- Surface Area of a Rectangular Solid
- Distance between Opposite Vertices of a Rectangular Solid
- Volume Formulas for Uniform Solids

HOW TO USE THIS CHAPTER

You may already know all the solid geometry you need. Find out by taking the Solid Geometry Diagnostic Test. The five questions on the Diagnostic Test are typical of those on the Mathematics subject tests. Check your answers using the answer key following the test. No matter how you score, don't worry! The answer key also shows where to find a detailed explanation for each question. The "Find Your Study Plan" section that follows the test will suggest next steps based on your performance on the Diagnostic.

Find Your Level

How you use this chapter really depends on how much time you have to prep. Find your level and pace below.

Standard Plan. Take the Solid Geometry Diagnostic Test to find out how much you already know about solid geometry. Then read the rest of the chapter and do the Follow-Up Test at the end.

Shortcut: Take the Solid Geometry Diagnostic Test and check your answers. If you can answer most of the questions correctly, you should skip this chapter.

Panic Plan. Skip this chapter.

SOLID GEOMETRY DIAGNOSTIC TEST

DO YOUR FIGURING HERE

5 Questions (6 Minutes)

Directions: Solve the following problems. Fill in the oval corresponding to the best answer choice in the grid to the right of each question. (Answers are on page 99.)

Reference Information: Use the following formulas as needed.

Right circular cone: If r = radius and h = height, then Volume = $\frac{1}{3}\pi r^2 h$; and if c = circumference of the base and ℓ = slant height, then Lateral Area = $\frac{1}{2}c\ell$.

Sphere: If r = radius, then Volume = $\frac{4}{3}\pi r^3$ and Surface Area = $4\pi r^2$.

Pyramid: If B = area of the base and h = height, then Volume = $\frac{1}{3}Bh$.

1. A right circular cone and a sphere have equal volumes. If the cone has radius x and height $2x$, what is the radius of the sphere?

(A) x (B) $\frac{x}{\sqrt[3]{2}}$ (C) $\sqrt[3]{2}$ (D) $\frac{1}{\sqrt[3]{2}}$ (E) $\sqrt[3]{2}x$

Ⓐ Ⓑ Ⓒ Ⓓ Ⓔ

2. In the rectangular solid in Figure 1, what is the distance from vertex R to vertex T?

(A) $2\sqrt{19}$
(B) $2\sqrt{17}$
(C) $8\sqrt{2}$
(D) $10\sqrt{2}$
(E) 30

Ⓐ Ⓑ Ⓒ Ⓓ Ⓔ

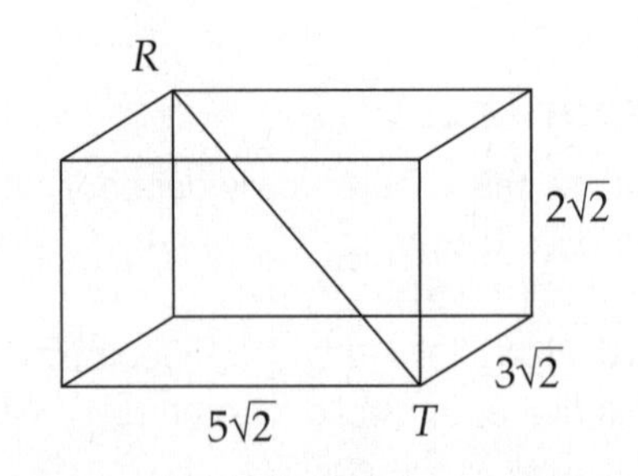

Figure 1

3. In Figure 2, the bases of the right uniform solid are triangles with sides of lengths 5, 12, and 13. If the surface area of the solid is 360, what is the volume?

(A) 30
(B) 180
(C) 300
(D) 600
(E) 780

Ⓐ Ⓑ Ⓒ Ⓓ Ⓔ

4. In Figure 3, the measure of angle *QPR* is 90 degrees, and the area of triangle *PQR* is 6. If triangle *PQR* is rotated 360° about side *PR*, what is the total surface area of the resulting solid?

(A) 24.00
(B) 50.27
(C) 75.40
(D) 97.39
(E) 147.65

Ⓐ Ⓑ Ⓒ Ⓓ Ⓔ

5. Figure 4 shows a rectangular box and a cylindrical marker, which has diameter 0.5" and height 9". If the box is filled with as many markers as possible, what percentage of the space will be unused?

(A) 21.5%
(B) 24.6%
(C) 29.2%
(D) 31.8%
(E) 70.7%

Ⓐ Ⓑ Ⓒ Ⓓ Ⓔ

DO YOUR FIGURING HERE

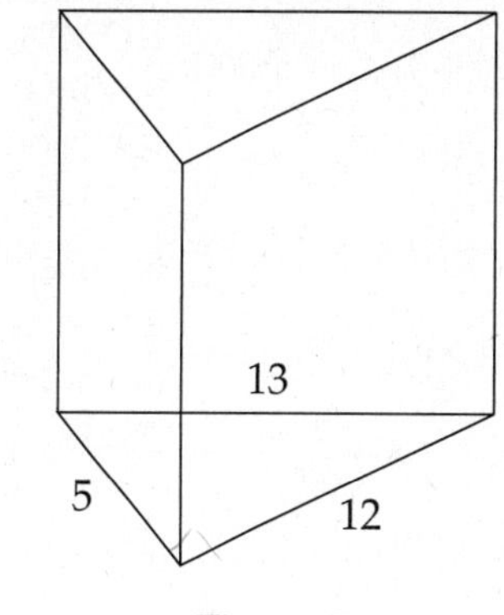

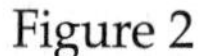
Figure 2

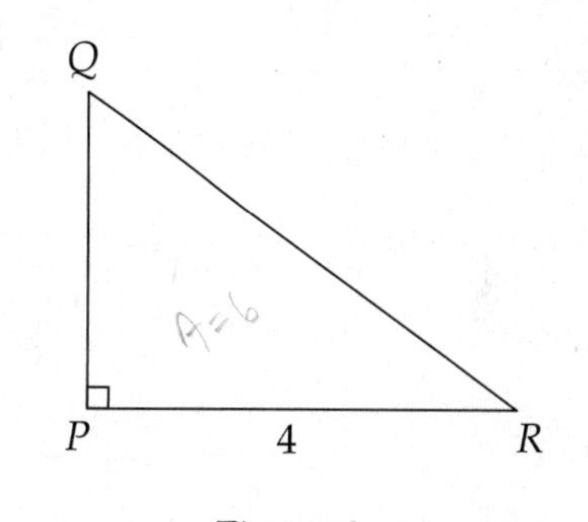

Figure 3

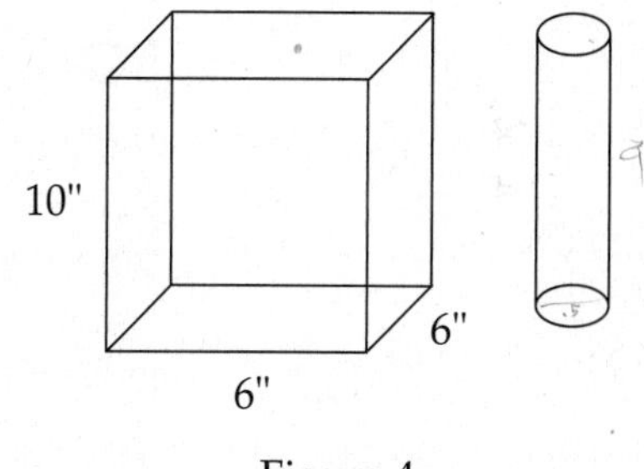

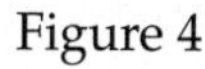
Figure 4

Note: Figure not drawn to scale

Find Your Study Plan

The answer key on the following page shows where in this chapter to find explanations for the questions you missed. Here's how you should proceed based on your Diagnostic Test score.

5: Superb! You have a solid grasp of the essentials of solid geometry, and if you're pressed for time, you might consider skipping this chapter. If you want to try your hand at a few more solid geometry questions, go straight to the Follow-Up Test at the end of this chapter.

3–4: Good! If you can get three or four of these relatively difficult questions right, then you have a decent understanding of solid geometry. If you're pressed for time, you might consider skipping this chapter. Alternatively, you might want to look over the parts of this chapter that deal with the one or two questions you didn't get right. If you just want to try your hand at a few more solid geometry questions, go straight to the Follow-Up Test at the end of this chapter.

0–2: Solid geometry is not your forte right now, but since it accounts for only three or four questions on the test, it may not be worth worrying about. If you're pressed for time and are confident with the other test topics, you might consider skipping this chapter. But if you have the time, you should read the rest of this chapter, study the examples, and then try the Follow-Up Test at the end of the chapter.

SOLID GEOMETRY TEST TOPICS

The questions in the Solid Geometry Diagnostic Test are typical of those on the Math 2 test. In this chapter, we'll use these questions to review the solid geometry you're expected to know. We will also use these questions to demonstrate some effective problem-solving techniques, alternative methods, and test-taking strategies that apply to solid geometry questions.

Five Formulas You Don't Need to Memorize

Some people find solid geometry intimidating because of all those formulas. On Math 2, however, some of the scariest formulas are given to you in the directions. It's good to know what formulas are included in the directions so you don't waste time memorizing them—and so that you won't forget to memorize all the relevant formulas that are not included in the directions. These are the formulas that are printed in the directions.

Lateral area of a cone: Given base circumference c and slant height ℓ,

$$\textbf{Lateral Area of Cone} = \frac{1}{2}c\ell$$

The lateral area of a cone is the area of the part that extends from the vertex to the circular base. It does not include the circular base.

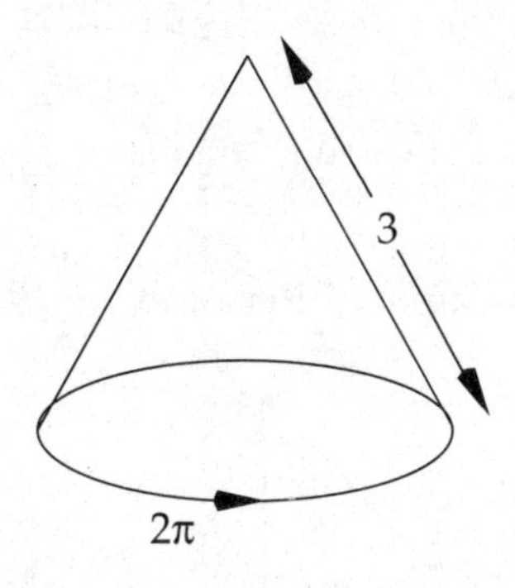

For example, in the figure above, $c = 2\pi$ and $\ell = 3$, so

$$\text{Lateral Area} = \frac{1}{2}(2\pi)(3) = 3\pi$$

Volume of a cone: Given base radius r and height h,

$$\textbf{Volume of Cone} = \frac{1}{3}\pi r^2 h$$

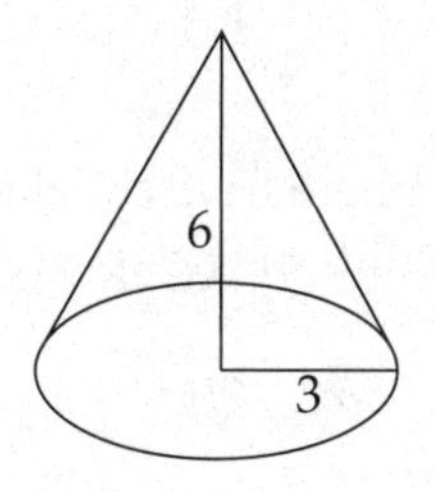

For example, in the figure above, $r = 3$, and $h = 6$, so

$$\text{Volume} = \frac{1}{3}\pi(3^2)(6) = 18\pi$$

Surface area of a sphere: Given radius r,

$$\textbf{Surface Area of Sphere} = 4\pi r^2$$

For example, if the radius of a sphere is 2, then

$$\text{Surface Area} = 4\pi(2^2) = 16\pi$$

Volume of a sphere: Given radius r,

$$\textbf{Volume of Sphere} = \frac{4}{3}\pi r^3$$

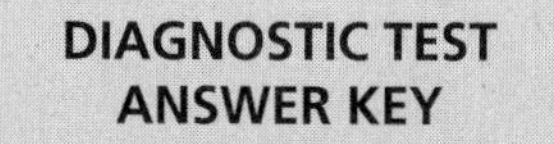

DIAGNOSTIC TEST ANSWER KEY

1. B
See "Five Formulas You Don't Need to Memorize."

2. A
See "The Test Makers' Favorite Solid."

3. C
See "Uniform Solids."

4. C
See "Picturing Solids—Rotating Polygons."

5. C
See "Picturing Solids—Maximums and Counting."

For example, if the radius of a sphere is 2, then

$$\text{Volume} = \frac{4}{3}\pi(2)^3 = \frac{32\pi}{3}$$

Volume of a pyramid: Given base area B and height h,

$$\textbf{Volume of Pyramid} = \frac{1}{3}Bh$$

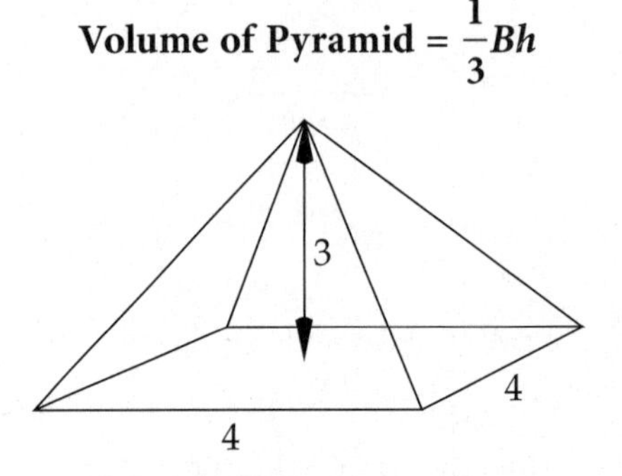

FIVE FORMULAS YOU DON'T NEED TO MEMORIZE

These solids formulas are included in the Math 2 test directions.

Lateral Area of a Cone = $\frac{1}{2}c\ell$

Volume of a Cone = $\frac{1}{3}\pi r^2 h$

Surface Area of a Sphere = $4\pi r^2$

Volume of a Sphere = $\frac{4}{3}\pi r^3$

Volume of a Pyramid = $\frac{1}{3}Bh$

For example, in the figure above, $h = 3$ and, because the base is a square, $B = 16$. Thus,

$$\text{Volume} = \frac{1}{3}(16)(3) = 16$$

Here's a question that uses some of these formulas.

Example 1

A right circular cone and a sphere have equal volumes. If the cone has radius x and height $2x$, what is the radius of the sphere?

(A) x (B) $\frac{x}{\sqrt[3]{2}}$ (C) $\sqrt[3]{2}$ (D) $\frac{1}{\sqrt[3]{2}}$ (E) $\sqrt[3]{2}x$

As you can see, this question is hard enough even with the formulas provided. This is no mere matter of plugging values into a formula and cranking out the answer. This question is more algebraic than that and takes a little thought. It's really a word problem. It describes in words a mathematical (in this case, geometric) situation that can be translated into algebra.

The pivot in this situation is that the cone and sphere have equal volumes. You're looking for the radius of the sphere in terms of x, the radius of the cone, and fortunately you can express both volumes in terms of those two variables. Be careful. Both formulas include an r, but they're not the same r's—the cone and the sphere have different radii.

$$\text{Volume of cone} = \frac{1}{3}\pi r^2 h = \frac{1}{3}\pi(x)^2(2x) = \frac{2}{3}\pi x^3$$

$$\text{Volume of sphere} = \frac{4}{3}\pi r^3$$

Now write an equation that says that the expressions for the two volumes are equal to each other, and solve for *r*:

$$\frac{4}{3}\pi r^3 = \frac{2}{3}\pi x^3$$

$$4r^3 = 2x^3$$

$$2r^3 = x^3$$

$$r^3 = \frac{x^3}{2}$$

$$r = \sqrt[3]{\frac{x^3}{2}} = \frac{x}{\sqrt[3]{2}}$$

The answer is (B).

So the direct way to do this question is the algebraic way. But as you can see, the algebra is quite convoluted, and the matter of the different *r*'s can be confusing. There's another way to do this question: Pick Numbers. All the given measures are in terms of *x*, and all the answer choices are in terms of *x*, so to make things simpler, you could just pick a number for *x*, plug it into the question, and see what you get. Pick a number that's easy to work with.

PICK NUMBERS

When the answer choices are algebraic expressions, it is often quickest to pick numbers for the unknowns, plug those numbers into the stem and see what you get, and then plug those same numbers into the answer choices to find matches.

Warning: When you pick numbers, you have to check all the answer choices. Sometimes more than one works with the number(s) you pick, in which case you have to pick numbers again.

If $x = 2$, then the cone has radius 2 and height 4.

$$\text{Volume of cone} = \frac{1}{3}\pi r^2 h = \frac{1}{3}\pi(2)^2(4) = \frac{16}{3}\pi$$

$$\text{Volume of sphere} = \frac{4}{3}\pi r^3$$

So what does *r* have to be for the volumes to be equal?

$$\frac{4}{3}\pi r^3 = \frac{16}{3}\pi$$

$$r^3 = 4$$

$$r = \sqrt[3]{4}$$

BRUSH UP ON YOUR ALGEBRA

You can't escape algebra! The Mathematics subject tests are both full of algebra. Make sure your algebra skills are in peak condition by Test Day.

When $x = 2$, *r* turns out to be $\sqrt[3]{4}$. Now plug $x = 2$ into the answer choices. Without too much effort, you can tell that (A), (C) and (E) are wrong. What about (B) and (D)?

(B) $\frac{x}{\sqrt[3]{2}} = \frac{2}{\sqrt[3]{2}} \times \frac{\sqrt[3]{2}}{\sqrt[3]{2}} \times \frac{\sqrt[3]{2}}{\sqrt[3]{2}}$ (rationalize the denominator)

$$= \frac{2\sqrt[3]{4}}{2} = \sqrt[3]{4}$$

Rationalize the denominator of (D) in the same way, and of course it doesn't turn out to be $\sqrt[3]{4}$ (since it differs from (B) by a factor of x, or 2, in this case). This alternative method is still somewhat algebraic, but this way the algebra's a lot less convoluted.

The Test Makers' Favorite Solid

The test makers' favorite solid is the rectangular solid. That's the official geometric term for a box, which has 6 rectangular faces and 12 edges that meet at right angles at 8 vertices.

SURFACE AREA OF A RECTANGULAR SOLID

If the length is ℓ, the width is w, and the height is h,

the formula is:

Surface Area = $2\ell w + 2wh + 2\ell h$

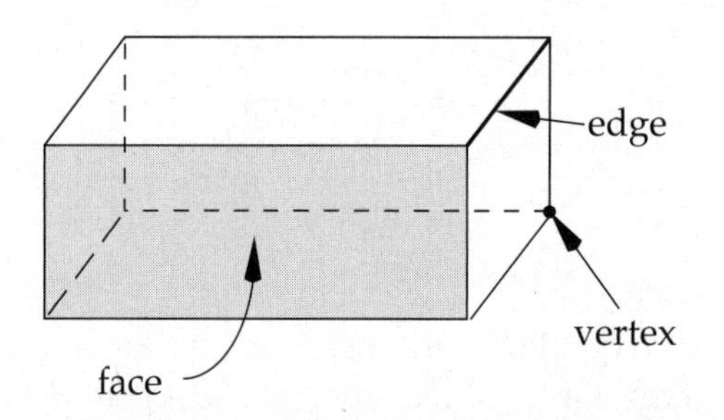

The surface area of a rectangular solid is simply the sum of the areas of the faces. That's what the formula "Surface Area = $2\ell w + 2wh + 2\ell h$" says. If the length is ℓ, the width is w, and the height is h, then two rectangular faces have area ℓw, two have area wh, and two have area ℓh. The total surface area is the sum of those three pairs of areas.

Instead of the surface area, you may be asked to find the distance between opposite vertices of a rectangular solid. Look at Example 2.

Example 2

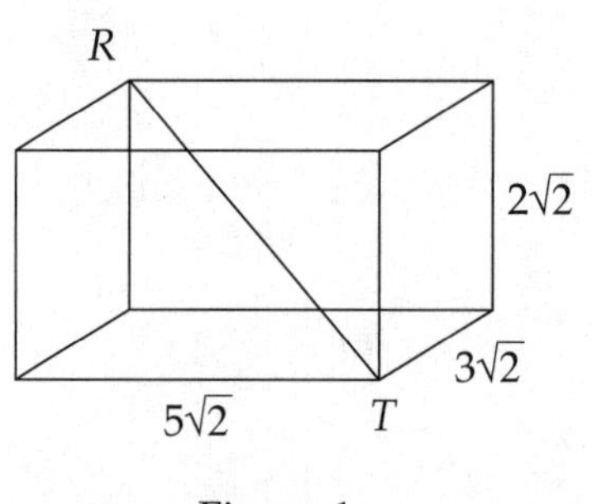

Figure 1

In the rectangular solid in Figure 1, what is the distance from vertex R to vertex T?

(A) $2\sqrt{19}$
(B) $2\sqrt{17}$
(C) $8\sqrt{2}$
(D) $10\sqrt{2}$
(E) 30

One way to find this distance is to apply the Pythagorean theorem twice. First plug the dimensions of the base into the Pythagorean theorem to find the diagonal of the base:

$$\text{Diagonal of base} = \sqrt{(3\sqrt{2})^2 + (5\sqrt{2})^2} = \sqrt{18+50} = \sqrt{68} = 2\sqrt{17}$$

Notice that the base diagonal combines with an edge and with the segment $\overline{RT}$—the segment you're looking for—to form a right triangle:

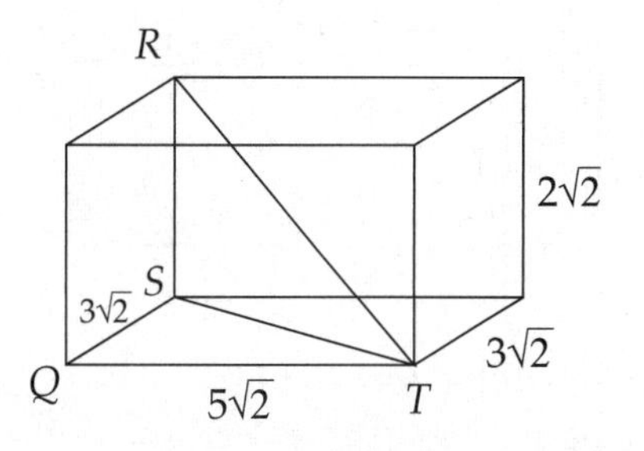

So you can plug the base diagonal and the height into the Pythagorean theorem to find $\overline{RT}$:

$$\overline{RT} = \sqrt{(2\sqrt{2})^2 + (2\sqrt{17})^2} = \sqrt{8+68} = \sqrt{76} = 2\sqrt{19}$$

The answer is (A).

The following two triangles show how the Pythagorean theorem applies.

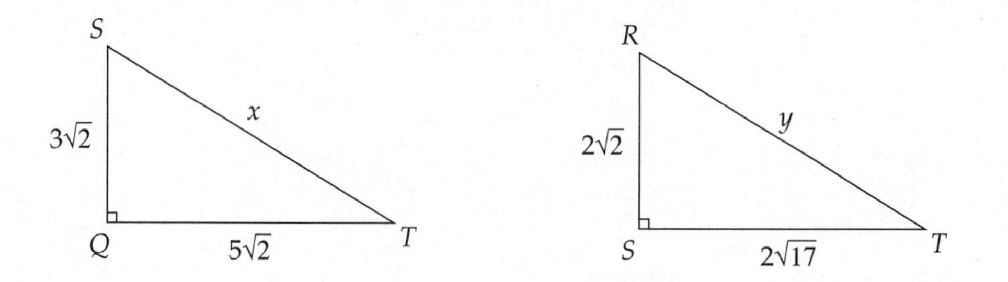

DISTANCE BETWEEN OPPOSITE VERTICES OF A RECTANGULAR SOLID

Distance = $\sqrt{l^2 + w^2 + h^2}$

Another way to find this distance is to use a formula, which you could say is just the Pythagorean theorem taken to another dimension. If the length is ℓ, the width is w, and the height is h, the formula is

$$\textbf{Distance} = \sqrt{\ell^2 + w^2 + h^2}$$

Uniform Solids

A rectangular solid is one type of *uniform solid*. A uniform solid is what you get when you take a plane and move it, without tilting it, through space. Here are some uniform solids.

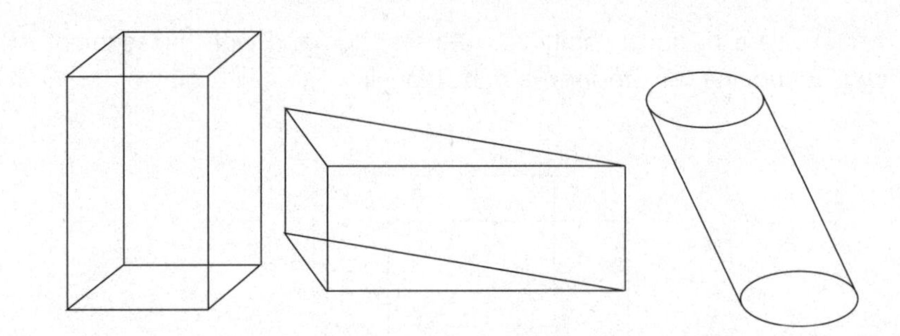

The way these solids are drawn, the top and bottom faces are parallel and congruent. These faces are called the *bases*. You can think of each of these solids as the result of sliding the base through space. The perpendicular distance through which the base slides is called the *height*.

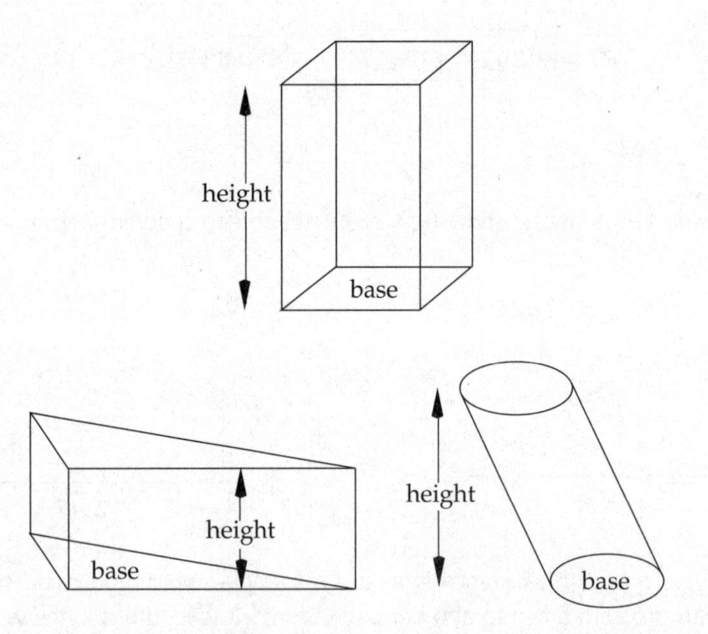

In every one of the above cases—indeed, in the case of *any* uniform solid—the volume is equal to the area of the base times the height. So, you can say that for any uniform solid, given the area of the base B and the height h,

Volume of a Uniform Solid = Bh

Volume of a rectangular solid: A rectangular solid is a uniform solid whose base is a rectangle and whose height is perpendicular to its base.

Given the length ℓ, width *w*, and height *h*, the area of the base is ℓ*w*, and so the volume formula is

VOLUME FORMULAS FOR UNIFORM SOLIDS

Volume of a Uniform Solid = Bh

Volume of a Rectangular Solid = lwh

Volume of a Cube = e^3

Volume of a Cylinder = $\pi r^2 h$

Volume of a Rectangular Solid = ℓ*wh*

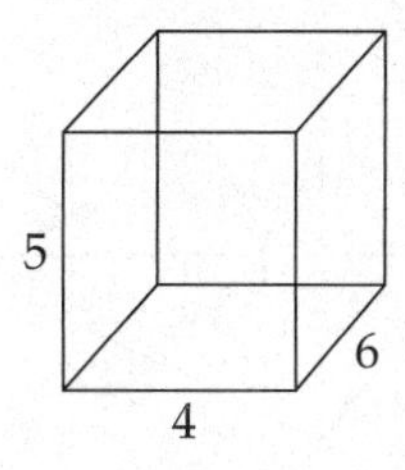

The volume of a 4-by-5-by-6 box is

$$4 \times 5 \times 6 = 120$$

Volume of a cube: A cube is a rectangular solid with length, width, and height all equal. If *e* is the length of an edge of a cube, the volume formula is

Volume of a Cube = e^3

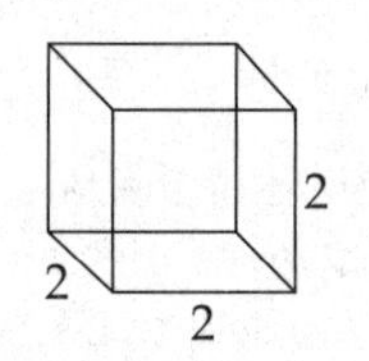

The volume of this cube is $2^3 = 8$.

Volume of a cylinder: A *cylinder* is a uniform solid whose base is a circle. Given base radius *r* and height *h*, the area of the base is πr^2, so the volume formula is

Volume of a Cylinder = $\pi r^2 h$

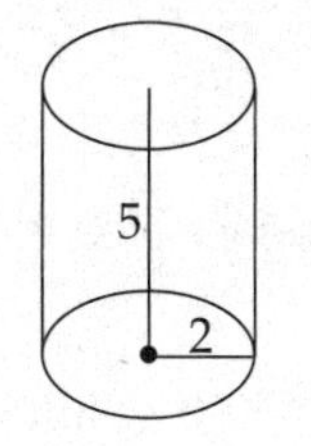

In the cylinder above, $r = 2$ and $h = 5$. Therefore,

$$\text{Volume} = \pi(2^2)(5) = 20\pi$$

Example 3 gives you the volume of a uniform solid and asks for the surface area.

Example 3

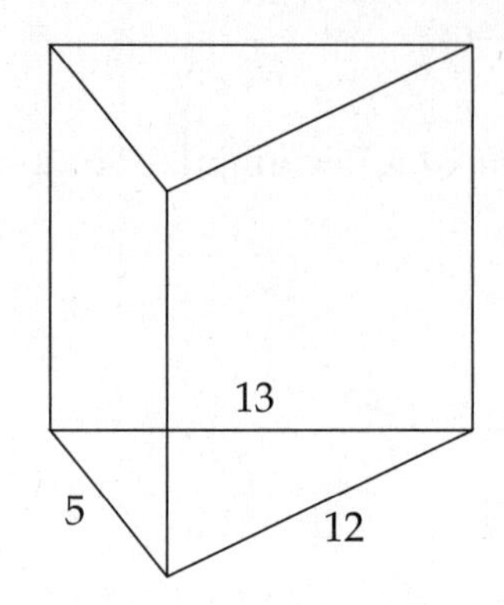

In the figure above, the bases of the right uniform solid are triangles with sides of lengths 5, 12, and 13. If the surface area of the solid is 360, what is the volume?

(A) 30
(B) 180
(C) 300
(D) 600
(E) 780

The surface area is the sum of the areas of the faces. To find the areas of the faces, you need to figure out what kinds of polygons they are so that you'll know what formulas to use. Start with the bases, which are said to be "triangles with sides of lengths 5, 12, and 13." If these side lengths don't ring a bell in your head, go back and study your Pythogorean triplets. This is a 5-12-13 triangle, which means it's a right triangle, which means that you can use the legs as the base and height to find the area:

$$\text{Area of Right Triangle} = \frac{1}{2}(\text{leg}_1)(\text{leg}_2) = \frac{1}{2}(5)(12) = 30$$

That's the area of each of the bases. Now call the height of the solid h. The other three faces are rectangles with areas $5 \times h$, $12 \times h$, and $13 \times h$. The total surface area, then, is $30 + 30 + 5h + 12h + 13h = 60 + 30h$.

Since the problem tells us that the surface area is 360,

$$60 + 30h = 360$$

$$h = 10$$

Volume = base × height = 30 × 10 = 300. The answer is (C).

Picturing Solids

You might have thought that solid geometry questions are difficult because of the formulas. As you have seen, however, using formulas is by definition routine. It's the nonroutine problems that can be the most challenging. Those are the solid geometry problems that require you to visualize, like Example 4.

Example 4

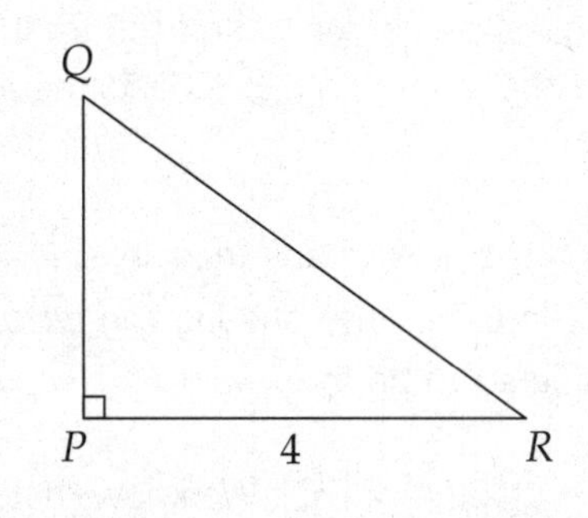

In the figure above, the measure of angle QPR is 90 degrees, and the area of triangle PQR is 6. If triangle PQR is rotated 360° about side PR, what is the total surface area of the resulting solid?

(A) 24.00
(B) 50.27
(C) 75.40
(D) 97.39
(E) 147.65

Think about what shape the top and bottom will create when each is rotated about its midpoint.

Can you visualize what the resulting solid looks like? It can be a little difficult to sketch 3-D geometry. You're probably better off if you can just "see" it in your head. It'll look something like this:

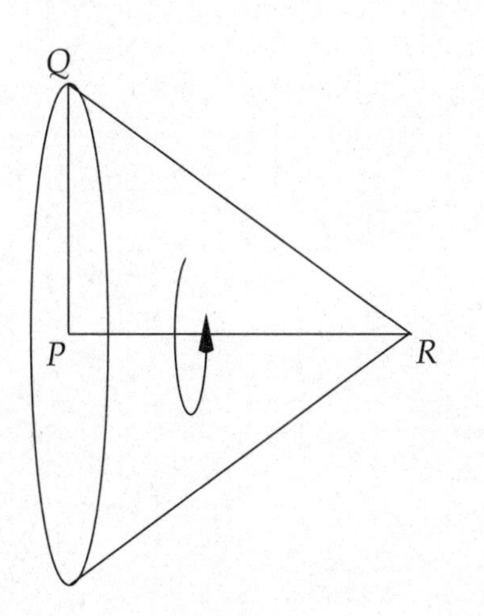

The resulting solid will be a right circular cone. The total surface area of a right circular cone is the sum of the area of the circular base plus the lateral area. If we call the radius of the circular base r, then the area of the circular base is πr^2. If we call the circumference of the circular base c and the slant height ℓ, then the lateral area is $\frac{1}{2}c\ell$.

When right triangle *PQR* is rotated about *PR*, the resulting right circular cone will have a circular base with a radius that is leg *PQ* of right triangle *PQR* and a slant height that is hypotenuse *QR* of this triangle. So in order to find the total surface area of the right circular cone, we need to find the length of leg *PQ* and the length of hypotenuse *QR* of right triangle *PQR*.

Let's first find the length of *PQ*. The area of any triangle is $\frac{1}{2}$ × base × height. The area of a right triangle is $\frac{1}{2} \times \text{leg}_1 \times \text{leg}_2$, because one leg can be considered to be the base and the other leg can be considered to be the height.

The area of right triangle *PQR* is $\frac{1}{2} \times (PR) \times (PQ)$. We know that $PR = 4$ and that the area of right triangle *PQR* is 6. So $\frac{1}{2} \times 4 \times (PQ) = 6$. Let's solve this equation for *PQ*. We have $2 \times (PQ) = 6$, so $PQ = 3$. The radius of the base of the right circular cone is 3. The area of the circular base is $\pi(3^2) = 9\pi$.

Now let's find the hypotenuse *QR* of right triangle *PQR*. This triangle has legs of length 3 and 4, so this is a 3-4-5 right triangle. So the length of *QR* is 5. Thus, the slant height ℓ of the right circular cone is 5. The circumference c of the circular base is $2\pi r = 2\pi(3) = 6\pi$. Thus, $c = 6\pi$ and $\ell = 5$. The lateral area is $\frac{1}{2}c\ell = \frac{1}{2}(6\pi)(5) = (3\pi)(5) = 15\pi$. The area of the circular base is 9π and the lateral area is 15π. The total surface area is $9\pi + 15\pi = 24\pi$. Now $24\pi \approx 75.3982236862$. To the nearest hundredth, $24\pi \approx 75.40$. Choice (C) is correct.

A typical SAT Subject Test: Mathematics Level 2 will include one or two questions that entail visualizing. Here's another one in Example 5.

Example 5

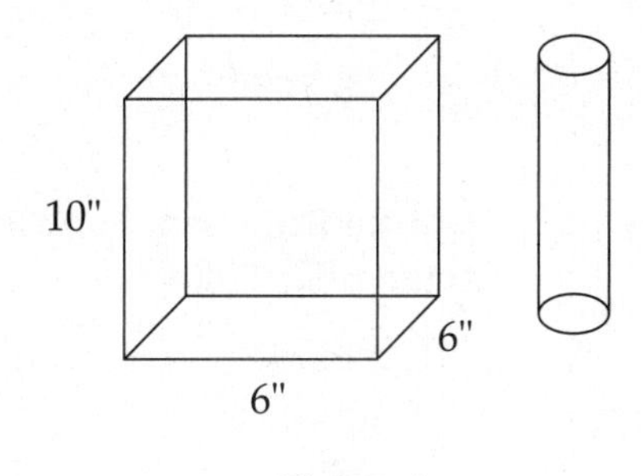

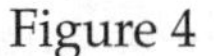

Figure 4

Note: Figure not drawn to scale

Figure 4 shows a rectangular box and a cylindrical marker, which has diameter 0.5" and height 9". If the box is filled with as many markers as possible, what percentage of the space will be unused?

(A) 21.5%
(B) 24.6%
(C) 29.2%
(D) 31.8%
(E) 70.7%

Since the markers are too long to be turned horizontal, they must be gathered together vertically. Put them in rows as long and wide as possible.

Since each marker has diameter 0.5", you can fit 12 across and 12 back, for a total of 144 markers. Now all that remains is to find the volume of the markers and the volume of the box.

The volume of the box = $6 \times 6 \times 10 = 360$.

The volume of one marker = $\pi r^2 h = \pi(0.25)^2 \times 9 \approx 1.77$.

The volume of all 144 markers $\approx 144 \times 1.77 \approx 254.88$.

The volume unused is $360 - 254.88 \approx 105.12$.

The percent unused is $\frac{105.12}{360} \approx 29.2\%$, so the answer is (C).

You've covered the solid geometry topics that you're likely to encounter on Math 2. You've reviewed the facts and formulas and learned some useful strategies. Now it's time for you to try another set of typical Math 2 solid geometry questions.

THINGS TO REMEMBER:

Five Formulas You Don't Need to Memorize

(These solids formulas are included in the Math 2 directions.)

Lateral Area of a Cone = $\frac{1}{2}c\ell$

Volume of a Cone = $\frac{1}{3}\pi r^2 h$

Surface Area of a Sphere = $4\pi r^2$

Volume of a Sphere = $\frac{4}{3}\pi r^3$

Volume of a Pyramid = $\frac{1}{3}Bh$

Surface Area of a Rectangular Solid

If the length is ℓ, the width is w, and the height is h, the formula is:

Surface Area = $2\ell w + 2wh + 2\ell h$

Distance Between Opposite Vertices of a Rectangular Solid

Distance = $\sqrt{\ell^2 + w^2 + h^2}$

Volume Formulas for Uniform Solids

Volume of a Uniform Solid = Bh

Volume of a Rectangular Solid = ℓwh

Volume of a Cube = e^3

Volume of a Cylinder = $\pi r^2 h$

SOLID GEOMETRY FOLLOW-UP TEST

DO YOUR FIGURING HERE

5 Questions (6 Minutes)

Directions: Solve the following problems. Fill in the oval corresponding to the best answer choice in the grid to the right of each question. (Answers and explanations begin on page 113.)

Reference Information: Use the following formulas as needed.

Right circular cone: If r = radius and h = height, then Volume $= \frac{1}{3}\pi r^2 h$; and if c = circumference of the base and ℓ = slant height, then Lateral Area $= \frac{1}{2}c\ell$.

Sphere: If r = radius, then Volume $= \frac{4}{3}\pi r^3$ and Surface Area $= 4\pi r^2$.

Pyramid: If B = area of the base and h = height, then Volume $= \frac{1}{3}Bh$.

1. In Figure 1, the lateral area of the right circular cone is 60π. If the radius of the base is 6, what is the volume of the cone?

   (A) 96π
   (B) 108π
   (C) 120π
   (D) 184π
   (E) 288π

   Ⓐ Ⓑ Ⓒ Ⓓ Ⓔ

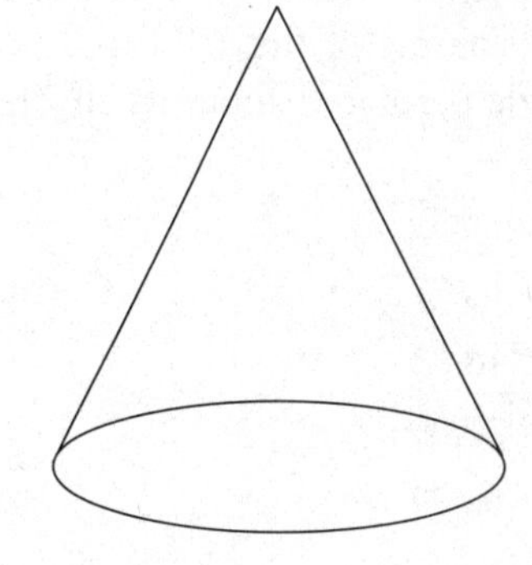

Figure 1

2. In Figure 2, d is the distance from vertex F to vertex G. The base is square, and the height is twice the width. What is the volume of the solid in terms of d?

   (A) $12d^3\sqrt{6}$
   (B) $10d^3\sqrt{5}$
   (C) $6d^3\sqrt{2}$
   (D) $\frac{2d^3\sqrt{5}}{25}$
   (E) $\frac{d^3\sqrt{6}}{18}$

   Ⓐ Ⓑ Ⓒ Ⓓ Ⓔ

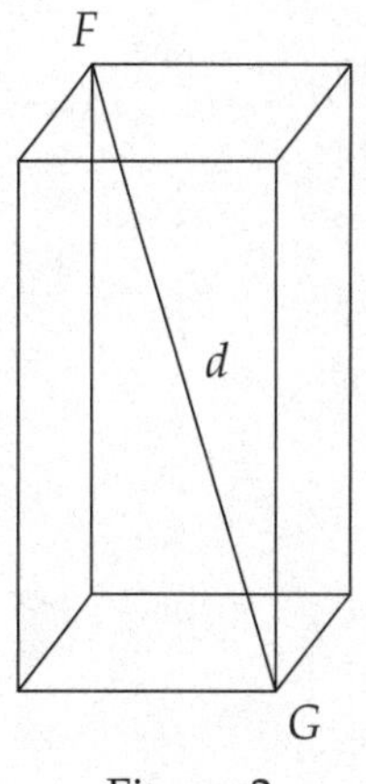

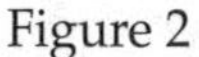
Figure 2

DO YOUR FIGURING HERE

3. A cube with edges of length b is divided into 8 equal smaller cubes. What is the difference between the combined surface area of the 8 smaller cubes and the surface area of the original cube?

(A) $\frac{3}{2}b^2$

(B) $\frac{3}{4}b^2$

(C) $\frac{9}{2}b^2$

(D) $6b^2$

(E) $12b^2$

Ⓐ Ⓑ Ⓒ Ⓓ Ⓔ

4. When a right triangle of area 4 is rotated 360° about its longer leg, the solid that results has a volume of 16. What is the volume of the solid that results when the same right triangle is rotated about its shorter leg?

(A) 4.39
(B) 8.77
(C) 16.93
(D) 35.09
(E) 50.27

Ⓐ Ⓑ Ⓒ Ⓓ Ⓔ

5. The pyramid in Figure 3 is composed of a square base of area 16 and four isoceles triangles, in which each base angle measures 60°. What is the volume of the pyramid?

(A) 7.39
(B) 9.24
(C) 15.08
(D) 21.33
(E) 24.05

Ⓐ Ⓑ Ⓒ Ⓓ Ⓔ

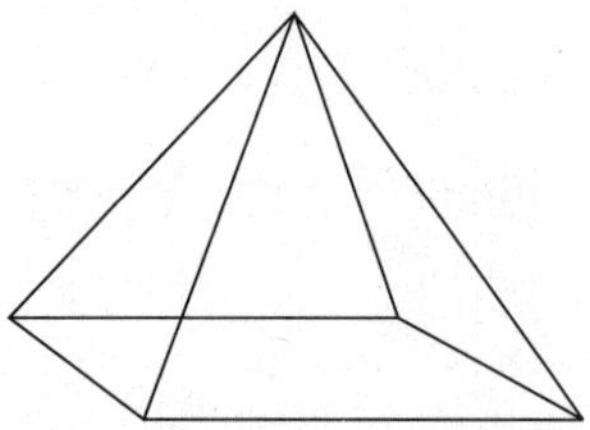

Figure 3

FOLLOW-UP TEST—ANSWERS AND EXPLANATIONS

1. A

The formula for the lateral surface area of a cone is in terms of c = circumference and ℓ = slant height. You can use the given base radius to solve for the slant height:

$$\text{Lateral area} = \frac{1}{2}c\ell = \frac{1}{2}(2\pi r)\ell =$$

$$\frac{1}{2}(2\pi \times 6)\ell = 6\pi\ell = 60\pi$$

$$6\ell = 60$$

$$\ell = 10$$

You can think of the slant height as the hypotenuse of a right triangle whose legs are the base radius and height of the cone.

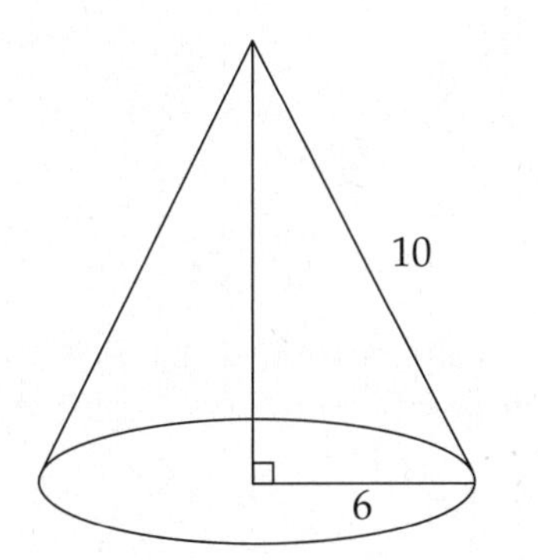

Now you can see that the triangle is a 3-4-5 and that $h = 8$. Plug $r = 6$ and $h = 8$ into the volume formula:

$$\text{Volume} = \frac{1}{3}\pi r^2 h = \frac{1}{3}\pi(6)^2(8) = 96\pi$$

2. E

Let the length and width be x and the height be $2x$. Solve for x in terms of d.

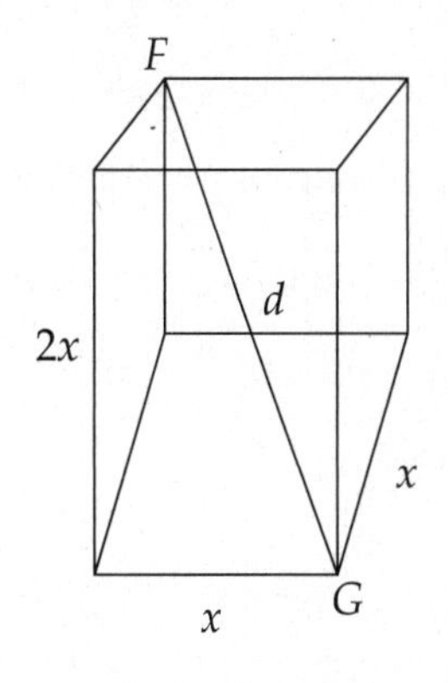

Use the formula for the distance between opposite vertices:

$$d = \sqrt{\ell^2 + w^2 + h^2}$$

$$d = \sqrt{x^2 + x^2 + (2x)^2} = \sqrt{6x^2} = \sqrt{6}x$$

$$x = \frac{d}{\sqrt{6}}$$

So for the dimensions of the solid we have $\ell = w = \frac{d}{\sqrt{6}}$, $h = \frac{2d}{\sqrt{6}}$.

$$\text{Volume} = \ell \times w \times h = \left(\frac{d}{\sqrt{6}}\right)\left(\frac{d}{\sqrt{6}}\right)\left(\frac{2d}{\sqrt{6}}\right)$$

$$= \frac{2d^3}{6\sqrt{6}}$$

$$= \frac{d^3}{3\sqrt{6}}$$

$$= \frac{d^3}{3\sqrt{6}} \times \frac{\sqrt{6}}{\sqrt{6}}$$

$$= \frac{d^3\sqrt{6}}{18}$$

You also could Pick Numbers on this problem. Choose a value for x and then find the volume and the value of d. Plug d into the answer choices and see which one matches.

3. D

When a cube of edge length b is divided into 8 identical smaller cubes, the edge of each of the smaller cubes is $\frac{b}{2}$.

The surface area of the original cube is $6 \times b \times b = 6b^2$.

The surface area of one of the smaller cubes is $6 \times \frac{b}{2} \times \frac{b}{2} = \frac{6b^2}{4}$ or $\frac{3b^2}{2}$. There are eight small cubes, so their combined surface area is $8\left(\frac{3b^2}{2}\right) = 12b^2$. The difference is $12b^2 - 6b^2 = 6b^2$.

4. D

When you write the equations for the volumes, try to plug in the area of the triangle, since that is known.

When you rotate a right triangle about a leg, you get a cone:

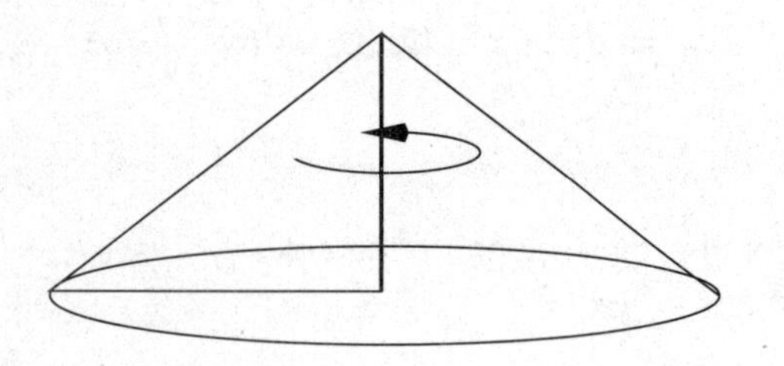

One leg becomes the base radius, and the other becomes the height of the cone. Call the short leg a and the long leg b and plug them into the cone volume formula:

$$\text{Volume of Cone} = \frac{1}{3}\pi r^2 h = \frac{1}{3}\pi a^2 b$$

It's also given that the area of the right triangle is 4. Therefore,

$$\text{Area of Right Triangle} = \frac{1}{2}(ab) = 4$$

Plug $ab = 8$ into the expression for the volume of the cone and you can solve for a:

$$\frac{1}{3}\pi a^2 b = 16$$

$$\frac{1}{3}\pi a(ab) = 16$$

$$\frac{1}{3}\pi a(8) = 16$$

$$\frac{8\pi}{3}a = 16$$

$$a = 16\left(\frac{3}{8\pi}\right) = \frac{6}{\pi}$$

Then you can plug $a = \frac{6}{\pi}$ into the equation $ab = 8$ to solve for b:

$$ab = 8$$

$$\left(\frac{6}{\pi}\right)b = 8$$

$$b = 8\left(\frac{\pi}{6}\right) = \frac{4\pi}{3}$$

When the triangle is rotated about its shorter leg, a becomes the height and b becomes the base radius. Thus,

$$\text{Volume} = \frac{1}{3}\pi r^2 h = \frac{1}{3}\pi b^2 a = \frac{\pi}{3}b(ab)$$

$$= \left(\frac{\pi}{3}\right)\left(\frac{4\pi}{3}\right)(8) = \frac{32\pi^2}{9} \approx 35.09$$

5. C

The sides of the pyramid are actually equilateral (the base angles are each 60°, which means the last angle is also 60°). To find the volume of a pyramid, you need the area of the base, which is given here as 16, and you need the height, which you have to figure out.

Imagine a triangle that includes the height and one of the lateral edges:

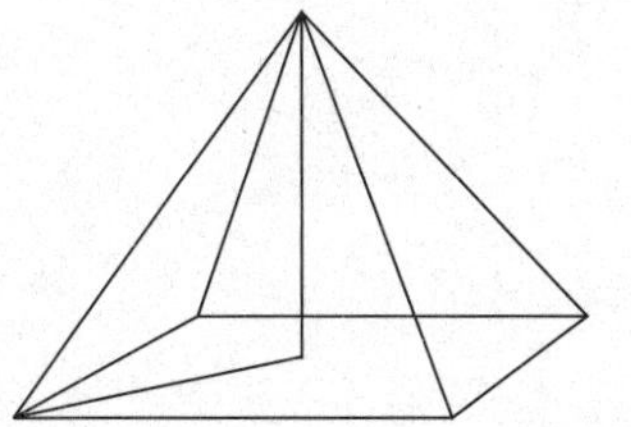

This triangle is a right triangle. The hypotenuse is the same as a side of one of the equilateral triangles, which is the same as a side of the square, which is the square root of 16, or 4:

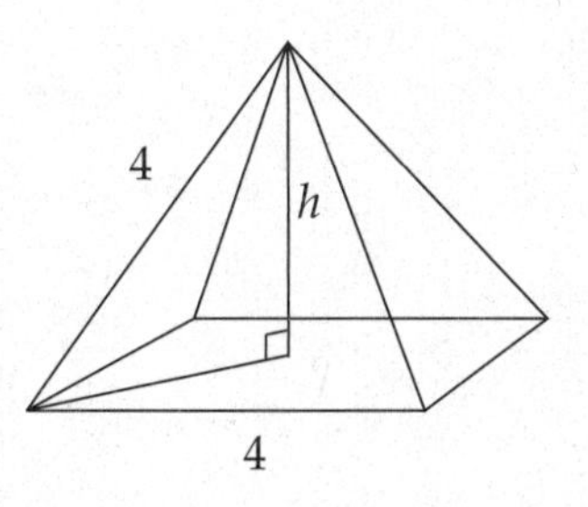

One of the legs of this right triangle is half of a diagonal of the square base—that is, half of $4\sqrt{2}$, or $2\sqrt{2}$.

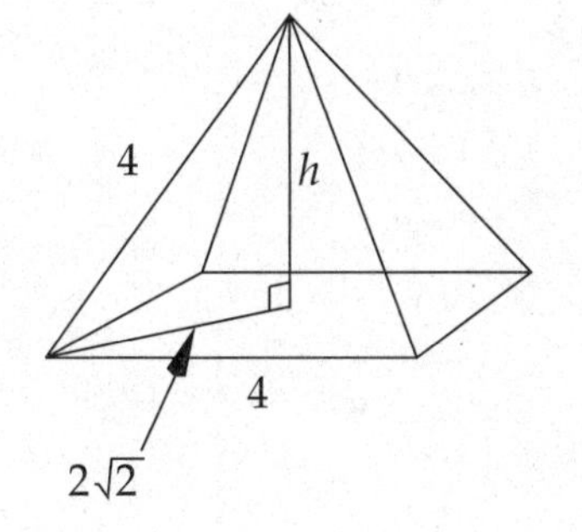

Now you can use the Pythagorean theorem to find the height:

$$(2\sqrt{2})^2 + h^2 = 16$$

$$8 + h^2 = 16$$

$$h^2 = 8$$

$$h = \sqrt{8} = 2\sqrt{2}$$

Finally, you have what you need to use the volume formula:

$$\text{Volume} = \frac{1}{3}Bh = \frac{1}{3}(16)(2\sqrt{2}) = \frac{32\sqrt{2}}{3} \approx 15.08$$

Chapter 7: **Coordinate Geometry**

- Midpoints, distances, and slopes
- Absolute values
- Inequalities
- Circles, parabolas, ellipses, and hyperbolas

A typical Math 2 test has about six coordinate geometry questions. But that's counting just the questions that are primarily about coordinate geometry. A lot of trigonometry and functions questions assume an understanding of coordinate geometry. In fact, the material in this chapter is relevant to approximately *30 percent of Math 2 questions*.

COORDINATE GEOMETRY FACTS AND FORMULAS IN THIS CHAPTER

- Midpoint
- Distance Formula
- Slope-Intercept Equation Form
- Slope Formula
- Positive and Negative Slopes
- Slopes of Parallel and Perpendicular Lines
- Circle
- Parabola
- Ellipse
- Hyperbola

HOW TO USE THIS CHAPTER

Maybe you already know all the coordinate geometry you need. You can find out by taking the Coordinate Geometry Diagnostic Test. Check your answers using the answer key following the test. No matter how you score, don't worry! The answer key also shows where to find a detailed explanation for each question. The "Find Your Study Plan" section that follows the test will suggest next steps based on your performance on the Diagnostic.

Find Your Level

How you use this chapter really depends on how much time you have to prep. Find your level and pace below.

Standard Plan. Do everything in this chapter. It's all relevant to Math 2.

> *Shortcut: Take the Coordinate Geometry Diagnostic Test and check your answers. If you can answer most of the questions correctly, you should skip this chapter.*

Panic Plan. Look through the chapter quickly and make sure you're comfortable with the material. If you're not comfortable with coordinate geometry, you should probably spend at least a little time with this chapter before moving on.

COORDINATE GEOMETRY DIAGNOSTIC TEST

6 Questions (8 Minutes)

Directions: Solve problems 1–6. Fill in the oval corresponding to the best answer choice in the grid to the right of each question. (Answers are on page 122.)

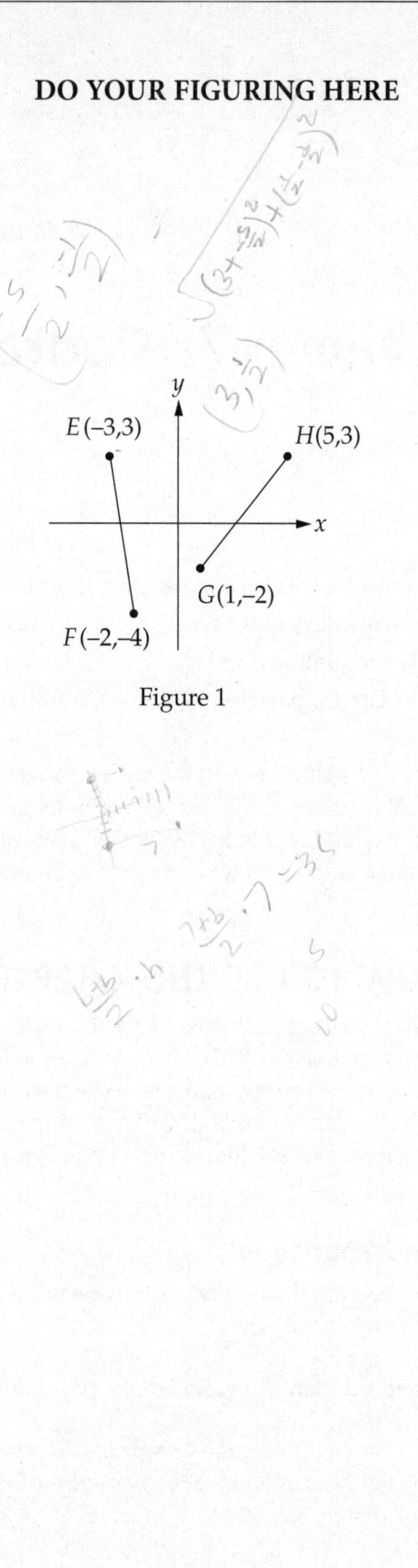

Figure 1

1. In Figure 1, what is the distance from the midpoint of $\overline{EF}$ to the midpoint of $\overline{GH}$?

   (A) 5.408
   (B) 5.454
   (C) 5.568
   (D) 5.590
   (E) 5.612

   A B C D E

2. If points (0,4), (0,–3), (7,–3), and (j,4) are consecutive vertices of a trapezoid of area 35, what is the value of j?

   (A) 3 (B) 4 (C) 7 (D) 10 (E) 11

   A B C D E

3. Which of the following lines has no point of intersection with the line $y = -\frac{1}{3}x + \sqrt{2}$?

   (A) $y = \frac{1}{3}x - \sqrt{2}$
   (B) $y = -3x + \sqrt{2}$
   (C) $y = -\frac{1}{3}x - \sqrt{2}$
   (D) $y = 3x - \sqrt{2}$
   (E) $y = 3x + \sqrt{2}$

   A B C D E

4. Which of the following shaded regions shows the graph of the inequality $y \leq |x - 2|$?

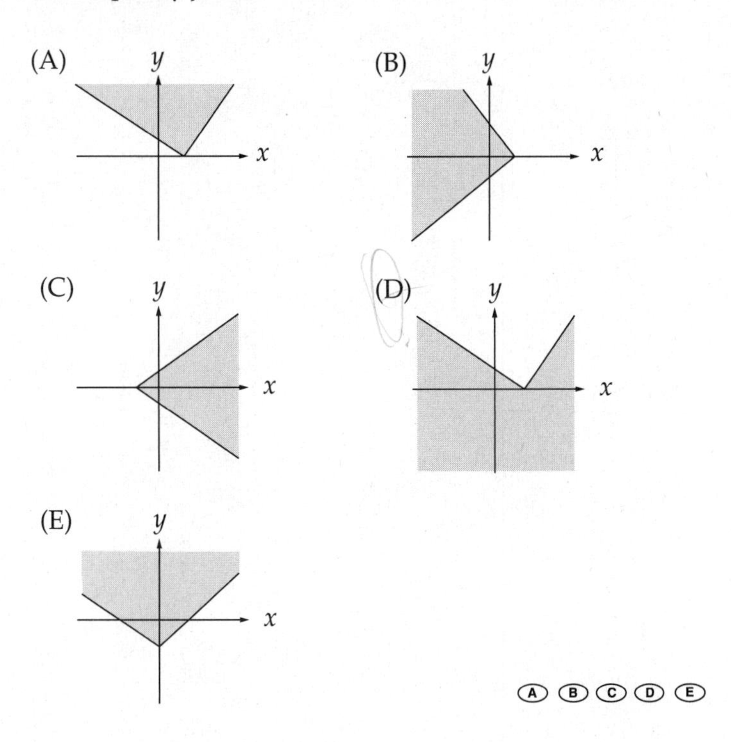

DO YOUR FIGURING HERE

5. Which of the following equations describes the set of all points (x,y) in the coordinate plane that are a distance of $\sqrt{3}$ from the point $(2,-5)$?

(A) $(x - 2)^2 + (y - 5)^2 = 3$
(B) $(x - 2)^2 + (y + 5)^2 = 3$
(C) $(x + 2)^2 + (y + 5)^2 = 3$
(D) $(x + 2)^2 - (y - 5)^2 = 3$
(E) $(x - 2)^2 - (y + 5)^2 = 3$

Ⓐ Ⓑ Ⓒ Ⓓ Ⓔ

6. Which of the following is an equation of an ellipse centered at the origin and with axial intersections at (0,±3) and (±2,0)?

   (A) $\frac{x}{2}+\frac{y}{3}=1$

   (B) $\frac{x}{2}+\frac{y}{3}=2$

   (C) $\frac{x}{3}+\frac{y}{2}=2$

   (D) $\frac{x^2}{4}+\frac{y^2}{9}=1$

   (E) $\frac{x^2}{4}+\frac{y^2}{9}=2$

Ⓐ Ⓑ Ⓒ Ⓓ Ⓔ

DO YOUR FIGURING HERE

Find Your Study Plan

The answer key on the next page shows where in this chapter to find explanations for the questions you missed. Here's how you should proceed based on your Diagnostic Test score.

5–6: Superb! If there's not much time before Test Day, then you might consider skipping this chapter—you seem to know it all already! If you just want to try your hand at some more coordinate geometry questions, go to the Follow-Up Test at the end of this chapter.

3–4: Good. You have a decent grasp of coordinate geometry. But this topic is fundamental enough that you might want to read at least those parts of this chapter that relate to the questions you were unable to answer correctly.

0–2: You need to work on coordinate geometry. This topic is fundamental. Read the rest of this chapter, study the examples, and see if you can do better on the Follow-Up Test at the end of the chapter.

COORDINATE GEOMETRY TEST TOPICS

The questions in the Coordinate Geometry Diagnostic Test are typical of those on the Math 2 test. In this chapter, we'll use these questions to review the coordinate geometry you're expected to know. We will also use these questions to demonstrate some effective problem-solving techniques, alternative methods, and test-taking strategies that apply to coordinate geometry questions.

Midpoints and Distances

Some of the more basic Math 2 coordinate geometry questions are ones that concern themselves with the layout of the grid, the location of points, distances between them, midpoints, and so on.

DIAGNOSTIC TEST ANSWERS KEY

1. D
See "Midpoints and Distances"

2. A
See "Geometry on the Grid"

3. C
See "Slope-Intercept Equation Form"

4. D
See "Absolute Value and Inequalities"

5. B
See "Circles and Parabolas"

6. D
See "Ellipses and Hyperbolas"

MIDPOINT

To find the midpoint between (x_1, y_1) and (x_2, y_2), average the x-coordinates and average the y-coordinates:

$$\text{Midpoint} = \left(\frac{x_1 + x_2}{2}, \frac{y_1 + y_2}{2}\right)$$

DISTANCE FORMULA

To find the distance between (x_1, y_1) and (x_2, y_2):

$$\text{Distance} = \sqrt{(x_2 - x_1)^2 + (y_2 - y_1)^2}$$

Example 1 involves both distances and midpoints.

Example 1

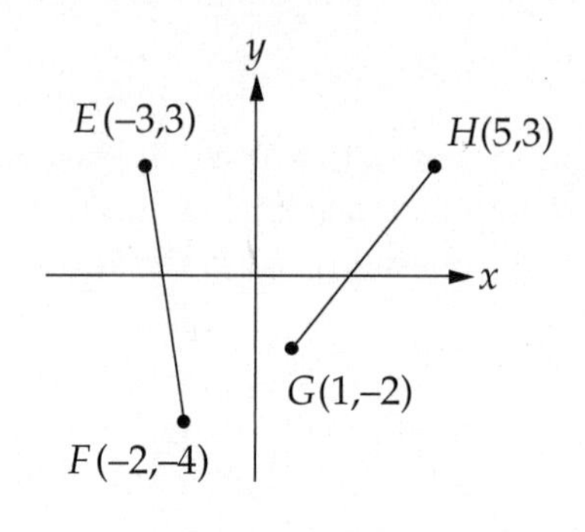

Figure 1

In Figure 1, what is the distance from the midpoint of $\overline{EF}$ to the midpoint of $\overline{GH}$?

(A) 5.408 (B) 5.454 (C) 5.568 (D) 5.590 (E) 5.612

To find the midpoint of a segment, average the x-coordinates and average the y-coordinates of the endpoints:

$$\text{midpoint of } \overline{EF} = \left(\frac{-3+(-2)}{2}, \frac{3+(-4)}{2}\right) = \left(-\frac{5}{2}, -\frac{1}{2}\right)$$

$$\text{midpoint of } \overline{GH} = \left(\frac{1+5}{2}, \frac{3+(-2)}{2}\right) = \left(3, \frac{1}{2}\right)$$

To find the distance between two points, use the distance formula:

$$\begin{aligned}\text{Distance} &= \sqrt{(x_2 - x_1)^2 + (y_2 - y_1)^2} \\ &= \sqrt{\left(3 - \left(-\frac{5}{2}\right)\right)^2 + \left(\frac{1}{2} - \left(-\frac{1}{2}\right)\right)^2} \\ &= \sqrt{\left(\frac{11}{2}\right)^2 + (1)^2} \\ &= \sqrt{30.25 + 1} \\ &= \sqrt{31.25} \approx 5.590\end{aligned}$$

The answer is (D).

Geometry on the Grid

Finding midpoints and distances is basically a geometry thing. Here's another example of a coordinate geometry question that's essentially plane geometry transferred to the coordinate plane.

Example 2

If points (0,4), (0,–3), (7,–3), and (j,4) are consecutive vertices of a trapezoid of area 35, what is the value of j?

(A) 3 (B) 4 (C) 7 (D) 10 (E) 11

Sketch a diagram to help you comprehend the situation. Plot the three given points and connect them to make two of the sides of the trapezoid:

PICTURE IT

Whenever you come to a geometry question with no figure, a sketch can make an abstract-looking problem concrete and manageable.

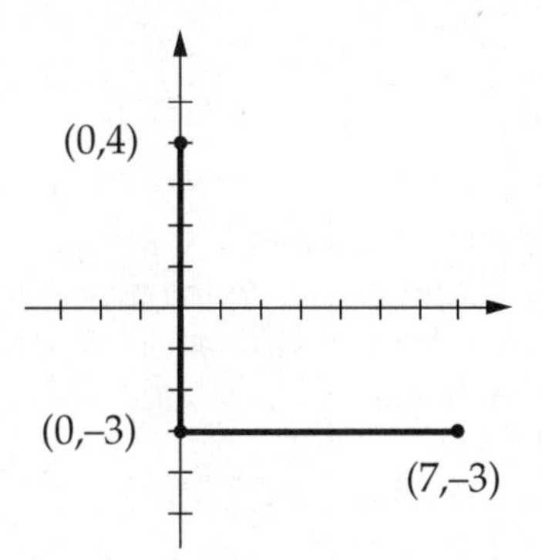

The fourth vertex has a y-coordinate of 4, so it must be somewhere along the line $y = 4$:

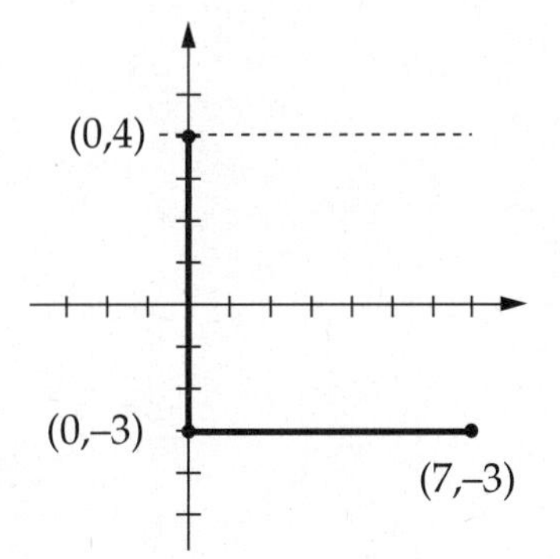

That will make the top and bottom sides of the trapezoid the parallel bases. The formula for the area of a trapezoid is

$$\textbf{Area of Trapezoid} = \left(\frac{b_1 + b_2}{2}\right) \times h$$

Here it's given that the area is 35, and you can see from the figure that one base is 7 and the height is also 7. That's enough to solve for the other base:

$$35=\left(\frac{7+b_2}{2}\right)\times 7$$
$$5=\frac{7+b_2}{2}$$
$$10=7+b_2$$
$$3=b_2$$

The top base is 3, so the coordinates of the fourth point are (3,4).

The answer is (A).

Slope-Intercept Form

SLOPE-INTERCEPT FORM

For any equation in the form $y = mx + b$:

m = slope

b = y-intercept

With the topic of slope-intercept form, we move from coordinate geometry that is primarily geometric into coordinate geometry that is primarily algebraic. Slopes and intercepts are descriptions of lines and points on the grid, but the processes of finding and/or using slopes and/or intercepts are generally algebraic processes. There's no need, for instance, to sketch a diagram for a question like Example 3.

Example 3

Which of the following lines has no point of intersection with the line $y = -\frac{1}{3}x + \sqrt{2}$?

(A) $y = \frac{1}{3}x - \sqrt{2}$

(B) $y = -3x + \sqrt{2}$

(C) $y = -\frac{1}{3}x - \sqrt{2}$

(D) $y = 3x - \sqrt{2}$

(E) $y = 3x + \sqrt{2}$

SLOPE FORMULA

Slope is defined as:

$$\frac{\text{change in } y}{\text{change in } x}$$

To find the slope of a line containing the points (x_1, y_1) and (x_2, y_2):

$$\text{Slope} = \frac{y_2 - y_1}{x_2 - x_1}$$

What does *has no intersection with* mean? It means that the lines are parallel, which in turn means that the lines have the same slope. If you know the slope-intercept form, you're able to spot the correct answer instantly.

When an equation is in the form $y = mx + b$, the letter m represents the slope, and the letter b represents the y-intercept. The equation in the stem is $y = -\frac{1}{3}x + \sqrt{2}$. That's in slope-intercept form, so the coefficient of x is the slope.

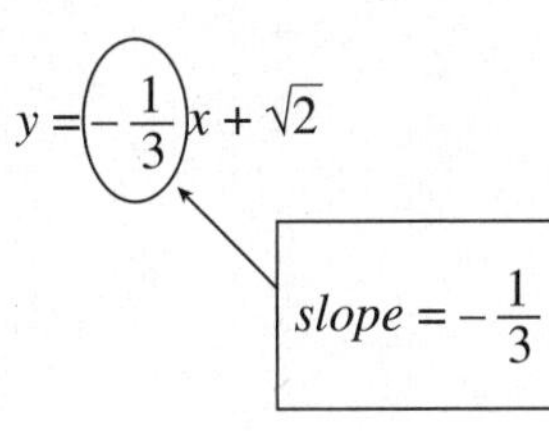

DON'T JUST MEMORIZE—INTERNALIZE

The best scores go to the test takers who don't just memorize methods and formulas but who really understand the underlying math.

Now look for the answer choice with the same slope. Conveniently, all the answer choices are presented in slope-intercept form, so spotting the one with $m = -\frac{1}{3}$ is a snap. The answer is (C).

People who are good at memorizing methods and formulas are not necessarily the ones who get the best scores on Math 2. It's people who have a deeper understanding of mathematics. If you really want to ace coordinate geometry questions, it's not enough to memorize the midpoint formula, the distance formula, the slope definition, the slope-intercept form, and so on. What you want is to have a real grasp of what slope is and what perpendicular, parallel, positive, negative, zero, and undefined slopes tell you.

Slope is a description of the "steepness" of a line. Lines that go uphill (from left to right) have positive slopes:

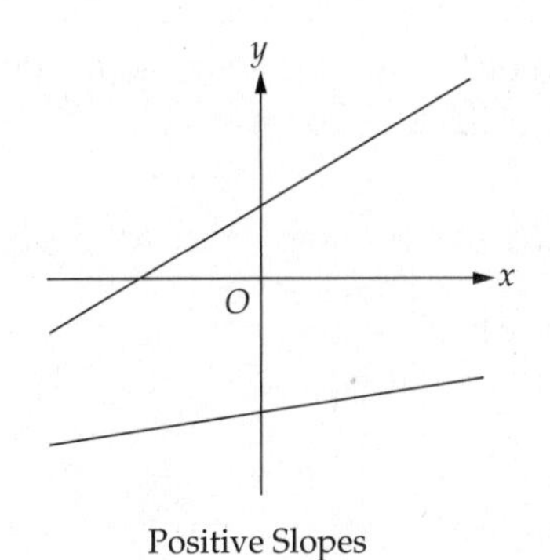

Positive Slopes

POSITIVE AND NEGATIVE SLOPES

Lines that go uphill from left to right have positive slopes. The steeper the uphill grade, the greater the slope.

Lines that go downhill from left to right have negative slopes. The steeper the downhill grade, the less the slope.

Lines that go downhill have negative slopes:

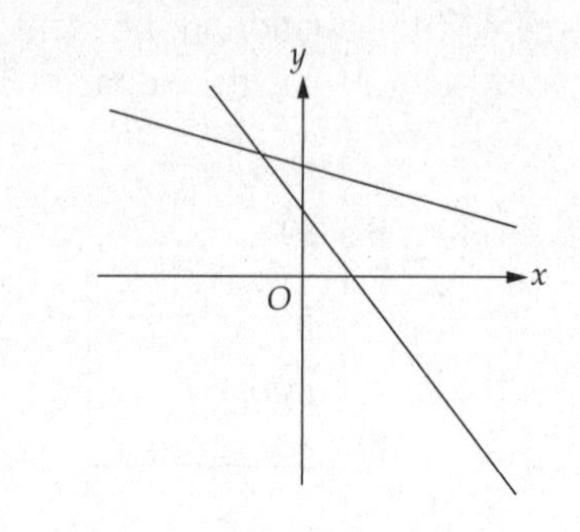

Negative Slopes

Lines parallel to the x-axis have slope = 0, and lines parallel to the y-axis have undefined slope.

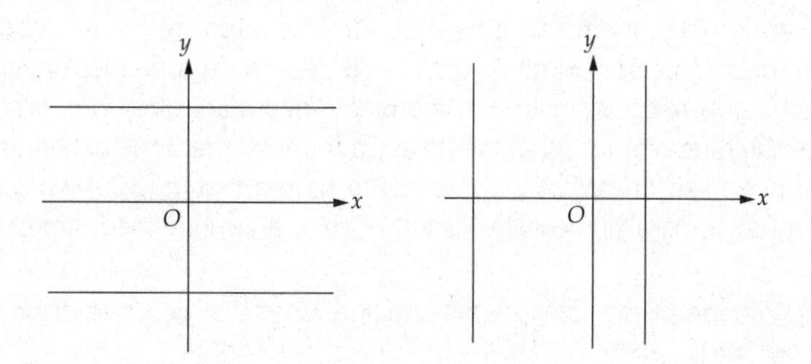

Lines that are parallel to each other have the same slope, and lines that are perpendicular to each other have negative-reciprocal slopes.

SLOPES OF PARALLEL AND PERPENDICULAR LINES

Parallel lines have the same slope.

Perpendicular lines have negative-reciprocal slopes.

In the figure below, the two parallel lines both have slope = 2, and the line that's perpendicular to them has slope = $-\frac{1}{2}$:

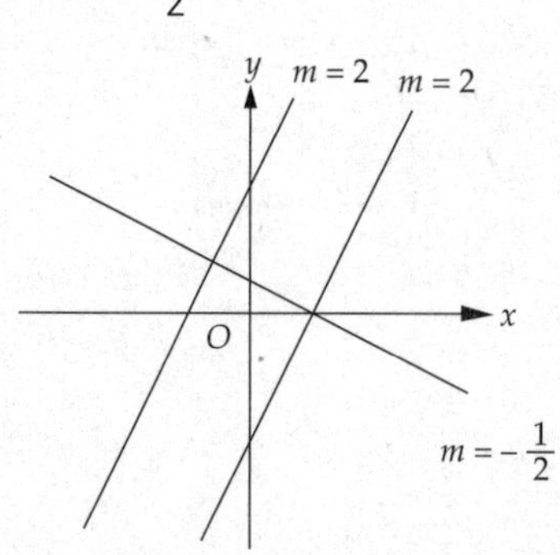

Parallel and Perpendicular Lines

Absolute Value and Inequalities

Among the more mind-bending coordinate geometry questions you'll find on the Math 2 test are ones like Example 4 that entail graphing absolute value and inequalities.

Example 4

Which of the following shaded regions shows the graph of the inequality $y \leq |x - 2|$?

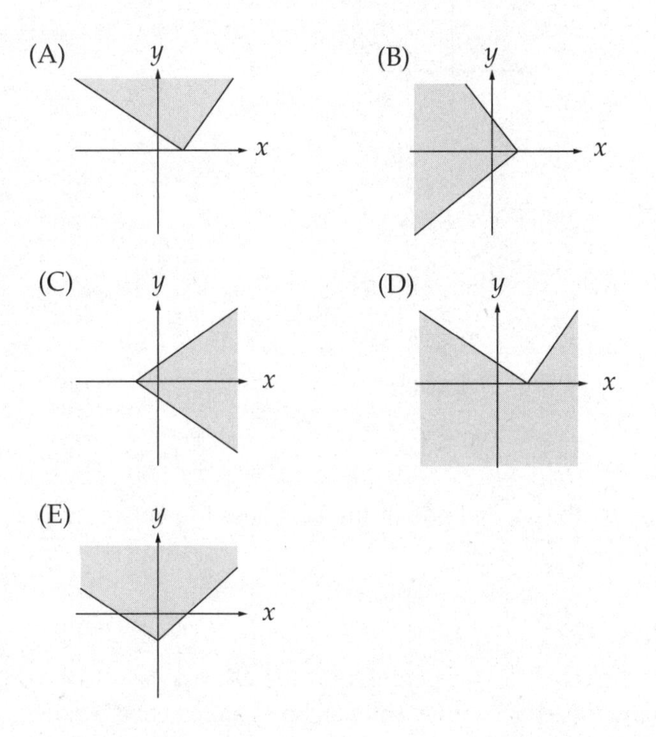

The way to handle an inequality is to think of it as an equation first, plot the line, and then figure out which side of the line to shade. This inequality is extra complicated because of the absolute value. When you graph an equation with absolute value, you generally get a line with a sharp bend, as in all of the answer choices above. To find the graph of an absolute value equation, figure out where that bend is. In this case, $|x - 2|$ has a turning-point value of 0, which happens when $x = 2$. So the bend is at the point (2,0). That narrows the choices down to (A), (B), and (D).

Next, figure out which side gets shaded. Pick a convenient point on either side and see whether that point's coordinates fit the given inequality. The point (3,0) is an easy one to work with.

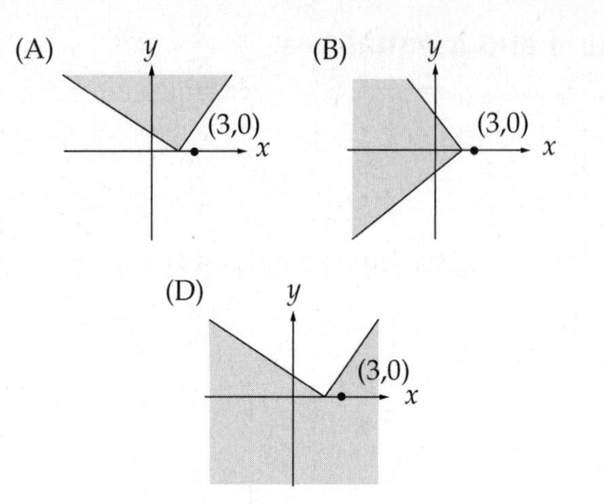

PICK A POINT

When you're trying to find an equation that fits a graph, pick a point or two from the graph and try them in the equations.

Do those coordinates satisfy the inequality?

$$y \leq |x - 2|$$
$$0 \overset{?}{\leq} |3 - 2|$$
$$0 \overset{?}{\leq} |1|$$
$$0 \leq 1$$

Yes. The point (3,0) must be on the shaded side of the bent line. The answer is (D).

Circles and Parabolas

You'll find circles, parabolas, ellipses, and hyperbolas on the Math 2 test. Like slope-intercept questions, questions concerning circles and parabolas are essentially algebraic. Answering circles questions like Example 5 is often just a matter of recalling and applying the appropriate equation.

CIRCLE

The equation of a circle centered at (h,k) and with radius r is:

$(x - h)^2 + (y - k)^2 = r^2$

Example 5

Which of the following equations describes the set of all points (x,y) in the coordinate plane that are a distance of $\sqrt{3}$ from the point (2,–5)?

(A) $(x - 2)^2 + (y - 5)^2 = 3$
(B) $(x - 2)^2 + (y + 5)^2 = 3$
(C) $(x + 2)^2 + (y + 5)^2 = 3$
(D) $(x + 2)^2 - (y - 5)^2 = 3$
(E) $(x - 2)^2 - (y + 5)^2 = 3$

To use the formula for the equation of a circle, you need the coordinates (h,k) of the center. Here they're given: $(2,-5)$. And you need the radius r. Here $r = \sqrt{3}$. So the equation is

$$(x - h)^2 + (y - k)^2 = r^2$$
$$(x - 2)^2 + (y + 5)^2 = 3$$

The answer is (B).

PARABOLA

The graph of the general quadratic equation $y = ax^2 + bx + c$ is a parabola with an axis of symmetry parallel to the y-axis.

There are all kinds of parabolas, and there's no simple, general parabola formula for you to memorize. You should know, however, that the graph of the general quadratic equation $y = ax^2 + bx + c$ is a parabola. It's one that opens either on top or on bottom, with an axis of symmetry parallel to the y-axis. Here, for example are parabolas representing the equations $y = x^2 - 2x + 1$ (dotted) and $y = -x^2 + 4$ (straight).

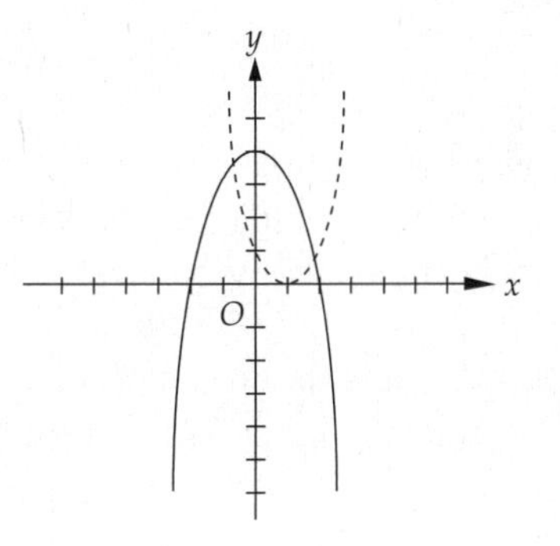

Ellipses and Hyperbolas

You might well encounter an ellipse on the Math 2 test. Example 6 tests your memory of the relevant formula.

Example 6

Which of the following is an equation of an ellipse centered at the origin and with axial intersections at $(0,\pm3)$ and $(\pm2,0)$?

(A) $\frac{x}{2} + \frac{y}{3} = 1$

(B) $\frac{x}{2} + \frac{y}{3} = 2$

(C) $\frac{x}{3} + \frac{y}{2} = 2$

(D) $\frac{x^2}{4} + \frac{y^2}{9} = 1$

(E) $\frac{x^2}{4} + \frac{y^2}{9} = 2$

When an ellipse is centered at the origin and has major and minor axes along the *x*- and *y*-axes, the equation of the ellipse is

$$\frac{x^2}{a^2}+\frac{y^2}{b^2}=1$$

where the axial intersections are (±*a*,0) and (0,±*b*). In Example 6, *a* = 2 and *b* = 3, so the equation is

$$\frac{x^2}{2^2}+\frac{y^2}{3^2}=1$$
$$\frac{x^2}{4}+\frac{y^2}{9}=1$$

The answer is (D).

ELLIPSE

The equation of an ellipse centered at the origin and with axial intersections at (±*a*,0) and (0,±*b*) is:

$$\frac{x^2}{a^2}+\frac{y^2}{b^2}=1$$

Note that if you don't remember the formula, you can still find the answer. It just takes a little longer. You have four points whose coordinates will satisfy the correct equation: (0,3), (0,–3), (2,0), and (–2,0). The only choice that works with all of these points is (D).

Every once in a while, a hyperbola turns up on the Math 2 test. Here's an equation of a hyperbola. It looks a lot like the equation for an ellipse, except that there's a minus sign.

$$\frac{x^2}{a^2}-\frac{y^2}{b^2}=1$$

HYPERBOLA

The equation of a hyperbola centered at the origin and with foci on the *x*-axis is:

$$\frac{x^2}{a^2}-\frac{y^2}{b^2}=1$$

The hyperbola

$$\frac{(x-h)^2}{a^2}-\frac{(y-k)^2}{b^2}=1$$

is centered at the point (*h*,*k*).

You've covered the coordinate geometry topics that you're likely to encounter on the Math 2 test. You've reviewed the facts and formulas and learned some useful strategies. You may also see a question on one of the following.

Coordinates in Three Dimensions

It's possible that you'll see a question involving a third dimension in coordinate geometry. Doing coordinate geometry in three dimensions is very similar to working in two dimensions.

	2 dimensions	3 dimensions
Midpoint	$\left(\frac{x_1+x_2}{2},\frac{y_1+y_2}{2}\right)$	$\left(\frac{x_1+x_2}{2},\frac{y_1+y_2}{2},\frac{z_1+z_2}{2}\right)$
Distance	$\sqrt{(x_2-x_1)^2+(y_2-y_1)^2}$	$\sqrt{(x_2-x_1)^2+(y_2-y_1)^2+(z_2-z_1)^2}$

Now that you have seen how to solve these typical (and a few not-so-typical) coordinate geometry questions, it's time to try some more on your own. Take the following Coordinate Geometry Follow-Up Test.

THINGS TO REMEMBER:

Midpoint = $\left(\frac{x_1 + x_2}{2}, \frac{y_1 + y_2}{2}\right)$

Distance = $\sqrt{(x_2 - x_1)^2 + (y_2 - y_1)^2}$

Slope-Intercept Form

For any equation in the form $y = mx + b$:

m = slope

b = y-intercept

Slope Formula

$$\text{slope} = \frac{y_2 - y_1}{x_2 - x_1}$$

Positive and Negative Slopes

Lines that go uphill from left to right have positive slopes. The steeper the uphill grade, the greater the slope.

Lines that go downhill from left to right have negative slopes. The steeper the downhill grade, the less the slope.

Slopes of Parallel and Perpendicular Lines

Parallel lines have the same slope.

Perpendicular lines have negative-reciprocal slopes.

Circle

The equation of a circle centered at (h, k) and with radius r is:

$$(x - h)^2 + (y - k)^2 = r^2$$

Parabola

The graph of the general quadratic equation $y = ax^2 + bx + c$ is a parabola with an axis of symmetry parallel to the y-axis.

Ellipse

The equation of an ellipse centered at the origin and with axial intersections at $(\pm a, 0)$ and $(0, \pm b)$ is:

$$\frac{x^2}{a^2} + \frac{y^2}{b^2} = 1$$

Hyperbola

The equation of a hyperbola centered at the origin and with foci on the x-axis is:

$$\frac{x^2}{a^2} - \frac{y^2}{b^2} = 1$$

COORDINATE GEOMETRY FOLLOW-UP TEST

DO YOUR FIGURING HERE

6 Questions (8 Minutes)

Directions: Solve problems 1–6. Fill in the oval corresponding to the best answer choice in the grid to the right of each question. (Answers and explanations begin on page 135.)

1. The graph of the equation $x^2 + y^2 = 169$ includes how many points (x, y) in the coordinate plane where x and y are both negative integers?

   (A) None
   (B) One
   (C) Two
   (D) Three
   (E) Infinitely many

   Ⓐ Ⓑ Ⓒ Ⓓ Ⓔ

2. In Figure 1, if the midpoints of segments $\overline{GH}$, $\overline{JK}$, and $\overline{LM}$ are connected, what is the area of the resulting triangle?

   (A) 20 (B) 23 (C) 26 (D) 33 (E) 37.5

   Ⓐ Ⓑ Ⓒ Ⓓ Ⓔ

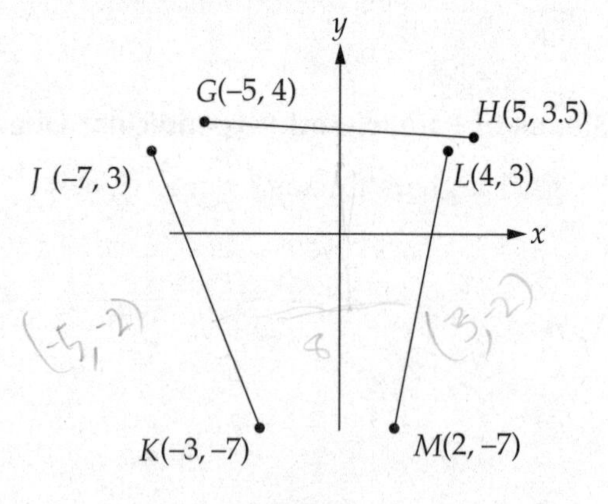

Figure 1

3. If the line $y = 3x - 15$ intersects the line $y = mx + 8$ in the third quadrant, which of the following must be true?

   (A) m is positive
   (B) m is negative
   (C) $m = 0$
   (D) $0 < m < 1$
   (E) m is undefined

   Ⓐ Ⓑ Ⓒ Ⓓ Ⓔ

DO YOUR FIGURING HERE

4. Which of the following lines is perpendicular to $y = -3x + 2$ and has the same y-intercept as $y = 3x - 2$?

(A) $y = -\frac{1}{3}x$

(B) $y = -\frac{1}{3}x + 2$

(C) $y = -\frac{1}{3}x - 2$

(D) $y = \frac{1}{3}x - 2$

(E) $y = \frac{1}{3}x + 2$

Ⓐ Ⓑ Ⓒ Ⓓ Ⓔ

5. The shaded portion of Figure 2 shows the graph of which of the following?

(A) $x\left(y - \frac{2}{3}x\right) \geq 0$

(B) $x\left(y - \frac{3}{2}x\right) \geq 0$

(C) $x\left(y + \frac{3}{2}x\right) \geq 0$

(D) $x\left(y + \frac{2}{3}x\right) \geq 0$

(E) $x\left(y + \frac{3}{2}x\right) \leq 0$

Ⓐ Ⓑ Ⓒ Ⓓ Ⓔ

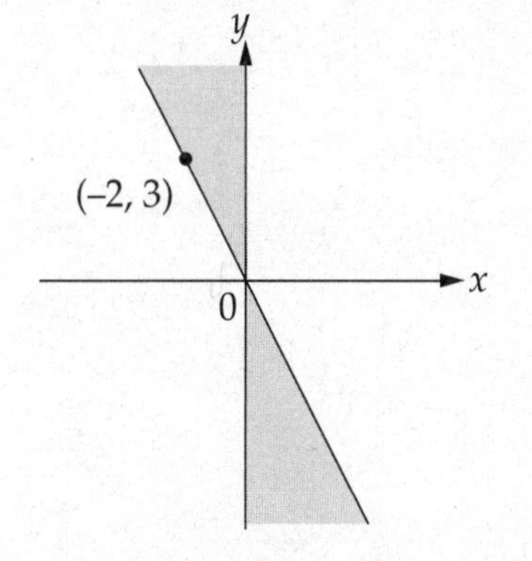

Figure 2

6. Which of the following is a point at which the ellipse $\frac{x^2}{9} + \frac{y^2}{16} = 1$ intersects the y-axis?

(A) (0,–3)

(B) (0,–4)

(C) (0,–8)

(D) (0,–9)

(E) (0,–16)

Ⓐ Ⓑ Ⓒ Ⓓ Ⓔ

FOLLOW-UP TEST—ANSWERS AND EXPLANATIONS

1. C

The graph of $x^2 + y^2 = 169$ is a circle centered at the origin and with a radius of 13. The square of that radius must equal the sum of the squares of the coordinates of any point on the circle. Remember, keep your eyes peeled for 5-12-13 triangles. The points shown below are the only 2 points on the circle where both *x* and *y* are negative integers.

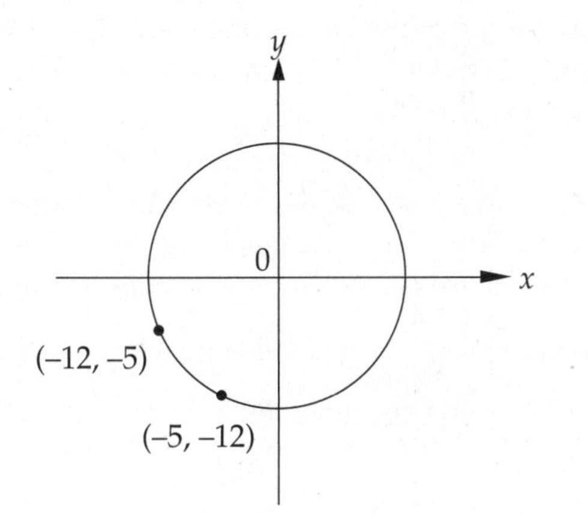

2. B

First find the three midpoints.

$$\text{midpoint of } \overline{GH} = \left(\frac{-5+5}{2}, \frac{4+3.5}{2}\right) = (0, 3.75)$$

$$\text{midpoint of } \overline{JK} = \left(\frac{-7+(-3)}{2}, \frac{3+(-7)}{2}\right) = (-5, -2)$$

$$\text{midpoint of } \overline{LM} = \left(\frac{4+2}{2}, \frac{3+(-7)}{2}\right) = (3, -2)$$

So the triangle looks like this:

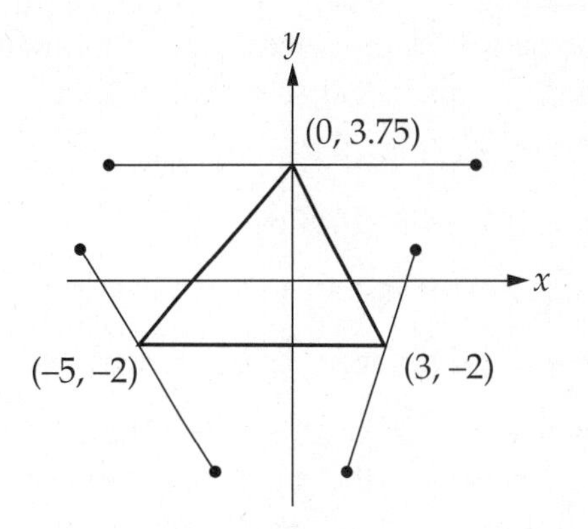

The base is 8 and the height is 5.75, so the area is 23.

3. A

A key to solving this problem quickly is to draw a diagram. Notice that the *y*-intercept of the first equation is –15 and the *y*-intercept of the second equation is 8.

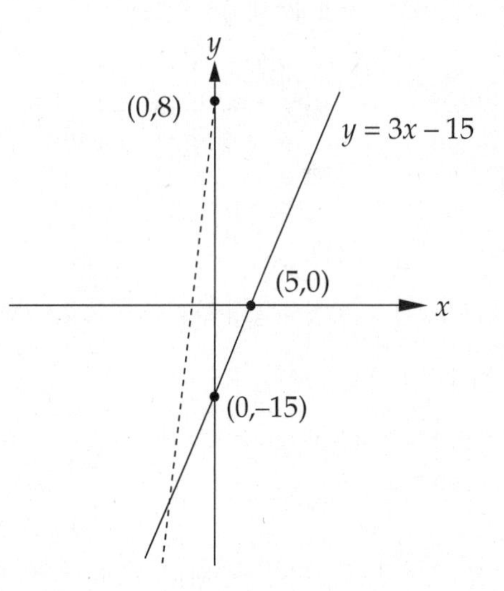

To get an idea of the value of *m*, imagine that the second line has the same slope as the first so that the two lines are parallel. Decreasing the slope of the second line would make the line more horizontal, causing the two lines to intersect in the first quadrant, which does not match the information given in the problem. However, increasing the slope of the second line would make the line more vertical, causing the two lines to intersect in the third quadrant as stated in the problem. Therefore, m has to be greater than 3, which means m has to be positive.

Once you have a graph drawn, you can also just draw in a line that does intersect the first in the third quadrant, then compare its slope to the answer choices. It is very important to understand some basics about slope on the Math 2 test.

4. D

A line that's perpendicular to $y = -3x + 2$ has a slope that's the negative reciprocal of -3, which is $\frac{1}{3}$. That narrows the choices to (D) and (E). The y-intercept of $y = 3x - 2$ is -2, and that's the y-intercept in (D).

5. E

First find the equation of the line to help you eliminate answer choices. The line passes through point $(-2,3)$ and the origin, so its equation is $y = -\frac{3}{2}x$.

According to the graph, you're looking for the answer choice that has the following properties:

$x \le 0$ and $y \ge -\frac{3}{2}x$ (the shaded portion of the second quadrant)

IN ADDITION

$x \ge 0$ and $y \le -\frac{3}{2}x$ (the shaded portion of the fourth quadrant).

To fulfill the first set of requirements, it is necessary that $x \le 0$ and $y + \frac{3}{2}x \ge 0$.

The product of these two expressions has to be negative, because the product of a negative and a positive is always negative. Therefore, $(x)\left(y+\frac{3}{2}x\right) \le 0$.

To fulfill the second set of requirements, it is necessary that $x \ge 0$ and $y + \frac{3}{2}x \le 0$.

Once again, the product of these two expressions has to be negative, resulting in the same inequality.

Therefore, $x\left(y+\frac{3}{2}x\right) \le 0$ fits the shaded portions of the graph.

6. B

You might think at first glance that what you're supposed to do is graph the ellipse and somehow figure out where the ellipse crosses the y-axis. But hold on a second! What does a point of intersection represent in coordinate geometry?

It represents a pair of numbers that satisfies both equations. The y-axis is the graph of the equation $x = 0$, so what you're looking for is a solution for both the equation $\frac{x^2}{9} + \frac{y^2}{16} = 1$ and the equation $x = 0$. Plug $x = 0$ into the ellipse equation, and you get

$$\frac{0^2}{9} + \frac{y^2}{16} = 1$$

$$\frac{y^2}{16} = 1$$

$$y^2 = 16$$

$$y = \pm 4$$

When x is 0, y is either 4 or -4, so the y-intercepts are $(0,4)$ and $(0,-4)$, and the answer is (B). So, what looked like a coordinate geometry question was really a simultaneous equations question. Don't be fooled by disguises!

Chapter 8: **Trigonometry**

- Right Angles and SOHCAHTOA
- Identities: Level 1 and Level 2
- Cotangent, secant, cosecant
- Identities: Functions of sums, etc.
- Radians
- Amplitude and period
- Law of sines and law of cosines

Trigonometry is a huge topic on Math 2. A typical test includes about ten trigonometry questions—that's one-fifth of the test. Bottom line: You have to be able to do at least some of the trigonometry questions to get a good score on Math 2.

HOW TO USE THIS CHAPTER

Maybe you already know all the trigonometry you need. You can find out by taking the Trigonometry Diagnostic Test. Check your answers using the answer key following the test. No matter how you score, don't worry! The answer key also shows where to find a detailed explanation for each question. The "Find Your Study Plan" section that follows the test will suggest next steps based on your performance on the Diagnostic.

Find Your Level

How you use this chapter really depends on how much time you have to prep. Find your level and pace below.

Standard Plan. Do everything in this chapter. It's all relevant to the Math 2 test.

Shortcut: Take the Trigonometry Diagnostic Test and check your answers. The "Find Your Study Plan" section that follows the test will suggest next steps based on your Diagnostic Test score.

Panic Plan. Make sure you can do at least some of the types of trigonometry questions that you see in this chapter.

TRIGONOMETRY DIAGNOSTIC TEST

6 Questions (8 Minutes)

Directions: Solve problems 1–6. Fill in the oval corresponding to the best answer choice in the grid to the right of each question. (Answers are on page 142.)

DO YOUR FIGURING HERE

1. In the right triangle in Figure 1, if $\theta = 42°$, what is the value of x?

   (A) 4.9 (B) 5.8 (C) 6.7 (D) 8.1 (E) 9.0

   Ⓐ Ⓑ Ⓒ Ⓓ Ⓔ

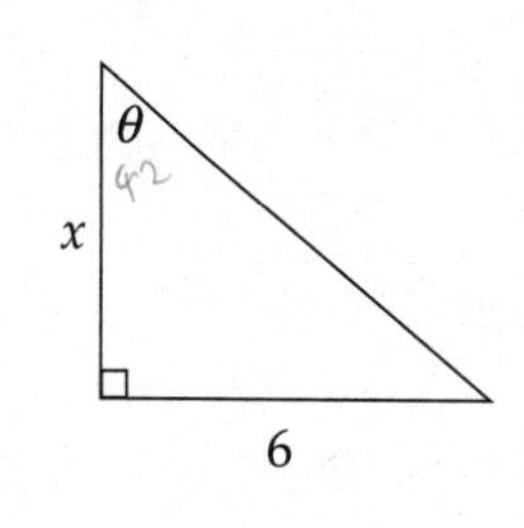

Figure 1

2. $12\sqrt{3}-(8\cos x)\left(\dfrac{3\sqrt{3}}{2}\cos x\right)=$

   (A) $\sin^2 x$

   (B) $12\sqrt{3}\sin^2 x$

   (C) $12\sqrt{3} - 12\sqrt{3}\cos x$

   (D) $12\sqrt{3}\cos^2 x$

   (E) $12\sqrt{3} - \dfrac{19\sqrt{3}}{2}\cos x$

   Ⓐ Ⓑ Ⓒ Ⓓ Ⓔ

3. If $\cos 2A = \dfrac{7}{19}$, what is the value of $\dfrac{1}{\cos^2 A - \sin^2 A}$?

   (A) 0.18 (B) 0.37 (C) 0.74 (D) 1.36 (E) 2.71

   Ⓐ Ⓑ Ⓒ Ⓓ Ⓔ

DO YOUR FIGURING HERE

4. Which of the following is an equation of the graph shown in Figure 2?

   (A) $y = \frac{1}{2}\sin 2x$

   (B) $y = \frac{1}{2}\cos 2x$

   (C) $y = 2\sin\left(\frac{x}{2}\right)$

   (D) $y = 2\cos\left(\frac{x}{2}\right)$

   (E) $y = 2\cos 2x$

   A B C D E

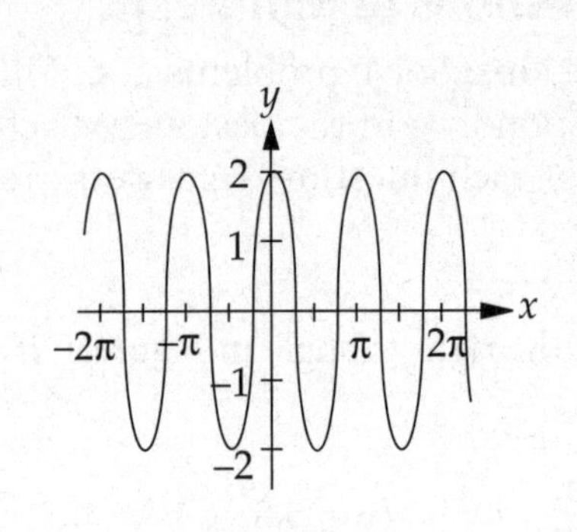

Figure 2

5. If $6\sin^2\theta - \sin\theta = 1$ and $0 \le \theta \le \pi$, what is the value of $\sin\theta$?

   (A) $\frac{1}{6}$ (B) $\frac{1}{3}$ (C) $\frac{1}{2}$ (D) 19 (E) 30

   A B C D E

6. What is the length of side BC in Figure 3?

   (A) 7.3 (B) 7.7 (C) 8.1 (D) 8.5 (E) 8.9

   A B C D E

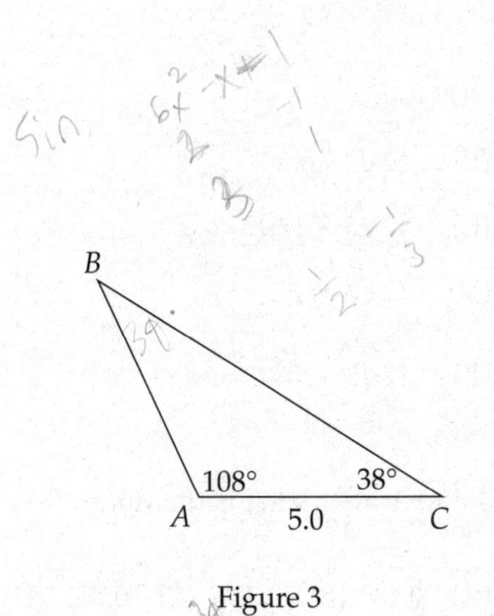

Figure 3

Find Your Study Plan

The answer key on the next page shows where in this chapter to find explanations for the questions you missed. Here's how you should proceed based on your Diagnostic Test score.

6: Superb! You're already good enough with trigonometry for the Math 2 test. If you're taking a "shortcut," you might consider skipping this chapter. Or you could just go straight to the Follow-Up Test at the end of the chapter.

4–5: Good. You're probably already good enough with trigonometry for the Math 2 test. If you're taking a "shortcut," you might consider skipping this chapter. But you should at least look at the parts of this chapter that discuss any questions you did not get right. Then, if you have time, you could move on to the Follow-Up Test at the end of the chapter.

0–3: Trigonometry may be a problem area for you, so you'd better spend some time with this chapter.

TRIGONOMETRY TEST TOPICS

As the Trigonometry Diagnostic Test showed you, you have a lot of definitions and identities to remember: the sine, cosine, and tangent of acute angles measured in degrees; cotangent, secant, and cosecant; angles greater than 90° and angles measured in radians; graphing trigonometric functions; arcsin, arccos, and arctan; half-angle and double-angle identities; and the laws of sines and cosines. We'll use the questions on the diagnostic to review these trigonometry topics.

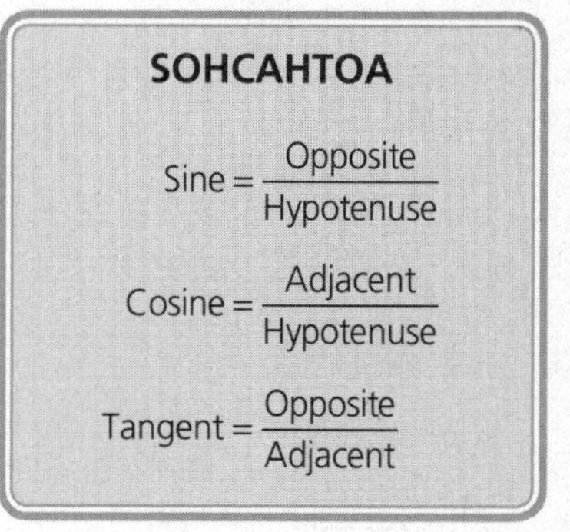

Right Triangles and SOHCAHTOA

To remember the definitions of sine, cosine, and tangent as they apply to right triangles, use the mnemonic SOHCAHTOA. That and a calculator are all you need to answer Example 1.

Example 1

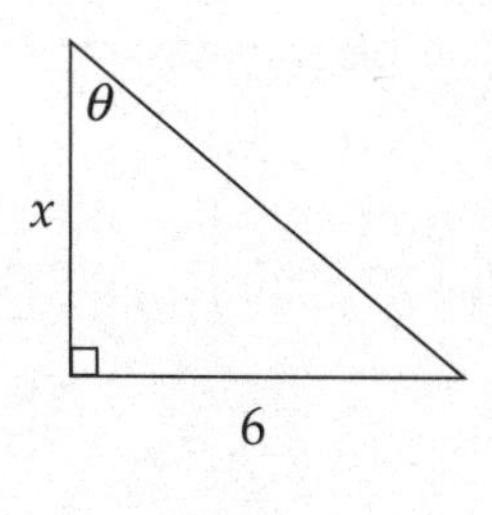

Figure 1

DIAGNOSTIC TEST ANSWER KEY

1. C
See "Right Triangles and SOHCAHTOA."

2. B
See "Level 1 Identities."

3. E
See "Level 2 Identities."

4. E
See "Graphing Trigonometric Functions."

5. C
See "Solving Trigonometric Equations."

6. D
See "Solving Triangles."

In the right triangle in Figure 1, if $\theta = 42°$, what is the value of x?

(A) 4.9 (B) 5.8 (C) 6.7 (D) 8.1 (E) 9.0

The 42° angle is opposite the given 6, and the side you're looking for is the adjacent side, so you can use the tangent to find *x*.

$$\tan = \frac{\text{opposite}}{\text{adjacent}}$$

$$\tan 42° = \frac{6}{x}$$

$$x = \frac{6}{\tan 42°}$$

Now, nobody expects you to know the tan of 42° from memory. This is one of those situations in which you have to use your calculator. Punch in "6/tan 42°" (make sure that your calculator is in degree mode), and you'll get something like 6.663675089, which is close to (C).

Level 1 Identities

You may come across questions like Example 2 on the Level 2 test.

Example 2

$$12\sqrt{3} - (8\cos x)\left(\frac{3\sqrt{3}}{2}\cos x\right) =$$

(A) $\sin^2 x$

(B) $12\sqrt{3}\sin^2 x$

(C) $12\sqrt{3} - 12\sqrt{3}\cos x$

(D) $12\sqrt{3}\cos^2 x$

(E) $12\sqrt{3} - \frac{19\sqrt{3}}{2}\cos x$

Start by multiplying to eliminate the parentheses:

$$12\sqrt{3} - (8\cos x)\left(\frac{3\sqrt{3}}{2}\cos x\right) = 12\sqrt{3} - \left(\frac{24\sqrt{3}}{2}\right)\cos^2 x$$

$$= 12\sqrt{3} - 12\sqrt{3}\cos^2 x$$

$$= 12\sqrt{3}(1 - \cos^2 x)$$

You always want to look for $\cos^2 x + \sin^2 x = 1$, which, rearranged, is $\sin^2 x = 1 - \cos^2 x$.

This relationship is really just a variation of the Pythagorean theorem. The test makers love this identity—it turns up a lot, so keep your eye out for it.

Here you can use the identity by substituting $\sin^2 x$ for $1 - \cos^2 x$:

$$12\sqrt{3}(1-\cos^2 x)$$
$$=12\sqrt{3}(\sin^2 x)$$

The answer is (B).

For Math 2 you have to know a lot more than just SOHCAHTOA and "$\sin^2 x + \cos^2 x = 1$." To begin with, you're supposed to know the other three trigonometric functions: cotangent, secant, and cosecant. Like the three basic functions (sine, cosine, and tangent), these other three functions can be defined in terms of "opposite," "adjacent," and "hypotenuse." They can also be defined as reciprocals of the three basic functions.

LEVEL 2 STRATEGY

Most angle measures on the Level 2 test are in degrees, but sometimes they're in radians, so read the questions carefully and switch your calculator to radian mode when appropriate.

Level 2 Identities

Besides questions that ask explicitly about such identities, there are other questions that become a lot easier and faster to answer when you know the right identities. That's the case with Example 3.

Example 3

If $\cos 2A = \frac{7}{19}$, what is the value of $\frac{1}{\cos^2 A - \sin^2 A}$?

(A) 0.18 (B) 0.37 (C) 0.74 (D) 1.36 (E) 2.71

Here's one way to answer this question. First find A :

$$\cos 2A = \frac{7}{19}$$
$$2A = \arccos\left(\frac{7}{19}\right)$$
$$2A \approx 68.38^\circ$$
$$A \approx 34.19^\circ$$

COTANGENT, SECANT, AND COSECANT

Cotangent = $\frac{\text{Adjacent}}{\text{opposite}} = \frac{1}{\text{Tangent}}$

Secant = $\frac{\text{Hypotenuse}}{\text{Adjacent}} = \frac{1}{\text{Cosine}}$

Cosecant = $\frac{\text{Hypotenuse}}{\text{opposite}} = \frac{1}{\text{Sine}}$

Then plug A = 34.19 into the expression you want to evaluate:

$$\begin{aligned}\frac{1}{\cos^2 A - \sin^2 A} &= \frac{1}{(\cos 34.19°)^2 - (\sin 34.19°)^2} \\ &\approx \frac{1}{(0.827)^2 - (0.562)^2} \\ &\approx \frac{1}{0.684 - 0.316} \\ &= \frac{1}{0.368} \\ &\approx 2.72\end{aligned}$$

The answer is (E), but the above method is much longer, more involved, and more open to calculator error than necessary. The best way to answer this question is to recognize the relationship between what's given and what's asked for. Answering the question involves just one quick calculation if you remember the relevant identity:

$$\cos 2A = \cos^2 A - \sin^2 A$$

So the given cos 2A = $\frac{7}{19}$ is the same as the denominator of the expression you're solving for:

$$\frac{1}{\cos^2 A - \sin^2 A} = \frac{1}{\cos 2A} = \frac{1}{\frac{7}{19}} = \frac{19}{7}$$

So the only calculating you need to do is this: 19 ÷ 7 ≈ 2.71.

So it can come in handy to know lots of trigonometric identities. You don't absolutely have to know the double-angle and half-angle identities—there's usually a way around them—but you'll be better equipped to work quickly and efficiently if you do know them.

Graphing Trigonometric Functions

Example 4 is about graphing a trigonometric function.

Example 4

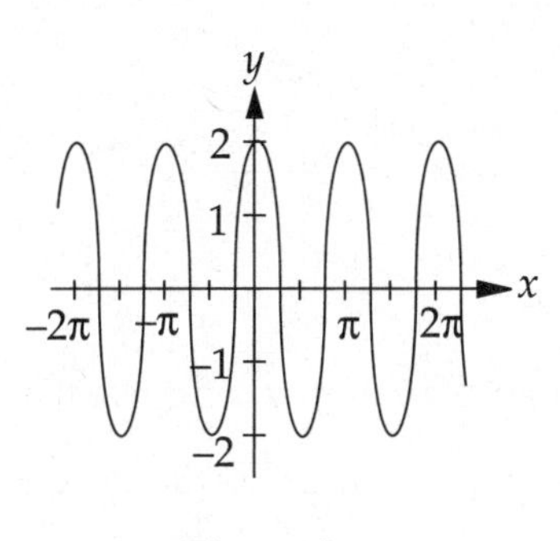

Figure 2

Which of the following is an equation of the graph shown in Figure 2?

(A) $y = \frac{1}{2}\sin 2x$

(B) $y = \frac{1}{2}\cos 2x$

(C) $y = 2\sin\left(\frac{x}{2}\right)$

(D) $y = 2\cos\left(\frac{x}{2}\right)$

(E) $y = 2\cos 2x$

To graph a trigonometric function, put the angles (usually in radians) on the *x*-axis and the results of applying the function on the *y*-axis. You can always use your graphing calculator (if you have one and know how to use it) to find the graph of a trigonometric function. That would be one way to do Example 4.

But ultimately you're better off if you can have a true understanding, a "feeling," for what the graphs of trigonometric functions look like. The way to do that is to start with a picture in your mind of what the graphs of the three basic functions look like.

LEVEL 2 IDENTITIES

Functions of Sums

$\sin(A + B) =$
$\sin A \cos B + \cos A \sin B$

$\cos(A + B) =$
$\cos A \cos B - \sin A \sin B$

$\tan(A + B) =$
$\frac{\tan A + \tan B}{1 - \tan A \tan B}$

Double-Angle Identities

$\sin 2x = 2\sin x \cos x$

$\cos 2x = \cos^2 x - \sin^2 x$
$= 1 - 2\sin^2 x$
$= 2\cos^2 x - 1$

$\tan 2x = \frac{2\tan x}{1 - \tan^2 x}$

Half-Angle Identities

$\sin\frac{1}{2}A = \pm\sqrt{\frac{1 - \cos A}{2}}$

$\cos\frac{1}{2}A = \pm\sqrt{\frac{1 + \cos A}{2}}$

$\tan\frac{1}{2}A = \pm\sqrt{\frac{1 - \cos A}{1 + \cos A}}$

RADIANS

π radians = 180 degrees

Basic trig function 1: The graph of $y = \sin x$ is a curve that goes through the origin because $\sin 0 = 0$ and rises to a crest at $\left(\frac{\pi}{2}, 1\right)$ because $\sin \frac{\pi}{2} = 1$:

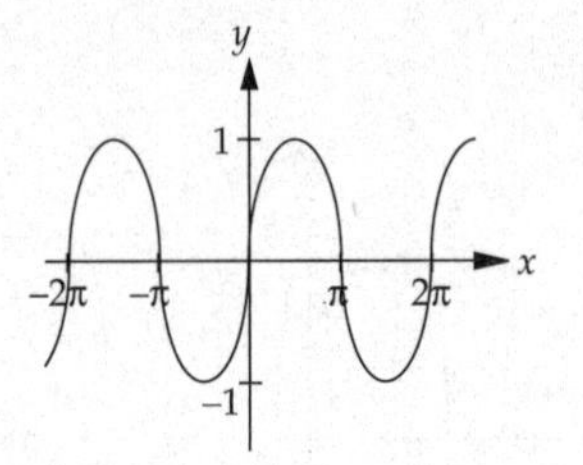

Basic trig function 2: The graph of $y = \cos x$ is a curve that looks just like the sine curve except it crosses the y-axis at (0, 1) because $\cos 0 = 1$, falls to cross the x-axis at $\left(\frac{\pi}{2}, 0\right)$ because $\cos \frac{\pi}{2} = 0$, and continues on down to its floor at (π, −1) because $\cos \pi = -1$:

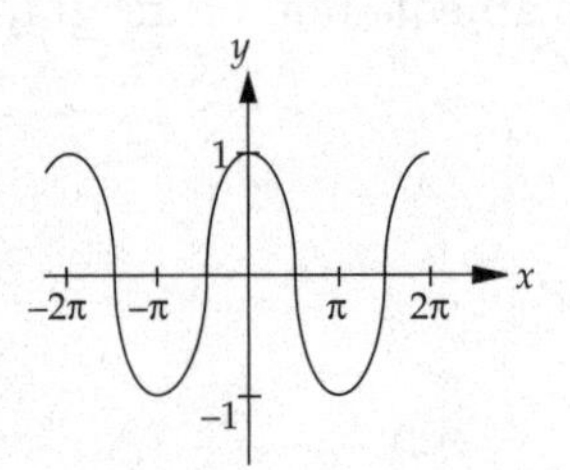

MATH 2 STRATEGY: MAKE A MENTAL PICTURE OF THE GRAPHS OF Y = SIN X, Y = COS X, AND Y = TAN X

Fix in your mind images of the sine, cosine, and tangent graphs, and you'll find that graphing trigonometric functions in general is easy and makes sense.

Basic trig function 3: The graph of $y = \tan x$ looks like this:

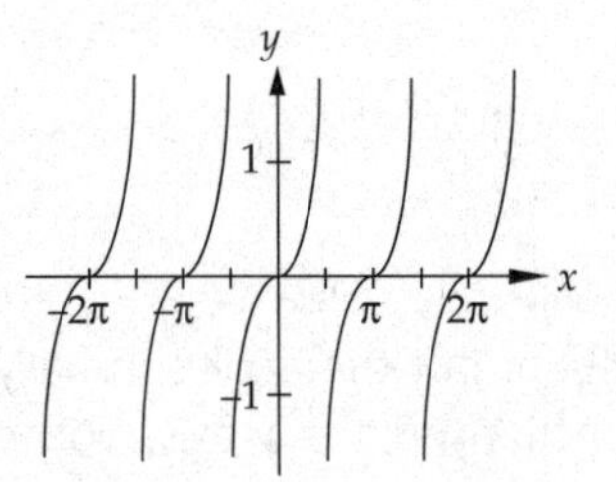

Many graphs of trigonometric functions can be variations on these.

The graph in Example 4, for instance, looks a lot like the cosine curve—it has a crest on the *y*-axis—but its maximum and minimum are ±2, and it falls to cross the *x*-axis at $\frac{\pi}{4}$ and bottoms out at $x = \frac{\pi}{2}$.

In other words, it's the cosine curve with twice the amplitude and half the period. The amplitude is affected by the number in front of the "cos." Twice the amplitude means that number is 2. The period is affected by the number in front of the *x*. Half the period means the coefficient of *x* is 2. So the equation is $y = 2\cos 2x$. And the answer is (E).

AMPLITUDE AND PERIOD

For functions in the form $y = a\sin bx$ or $y = a\cos bx$, the amplitude of the curve is *a*, and the period of the curve is $\frac{360}{b}$ degrees or $\frac{2\pi}{b}$ radians.

Of course, you can always do a question like Example 4 by picking a point or two. The point (0,2) satisfies equations (D) and (E) only.

Of those choices, (E) is the only one that works with the point $\left(\frac{\pi}{2}, -2\right)$.

Solving Trigonometric Equations

Here's a question that looks like a complicated trigonometry question but turns out not to be trigonometry at all.

PICK A POINT OR TWO

To identify the equation for a particular graph, pick a point or two from the graph and try the coordinates in the equations.

Example 5

If $6\sin^2\theta - \sin\theta = 1$ and $0 \leq \theta \leq \pi$, what is the value of $\sin\theta$?

(A) $\frac{1}{6}$ (B) $\frac{1}{3}$ (C) $\frac{1}{2}$ (D) 19 (E) 30

This sure looks like a trigonometry question. There are sines and thetas all over the place. But in fact this is just a plain old algebra question in disguise. You never need to find the sine or the arcsine of anything. You don't need to find θ. This is really just a quadratic equation where the unknown is written "$\sin\theta$" instead of the more usual "*x*." Perhaps you'll find the equation easier to deal with if you first replace "$\sin\theta$" with *x* :

DON'T BE FOOLED BY DISGUISES

Sometimes what looks like a trigonometry question turns out to be primarily an algebra question.

$$6\sin^2\theta - \sin\theta = 1$$
$$6x^2 - x = 1$$

Now solve for x as usual:

$$6x^2 - x = 1$$
$$6x^2 - x - 1 = 0$$
$$(2x-1)(3x+1) = 0$$
$$2x - 1 = 0 \text{ or } 3x + 1 = 0$$
$$x = \frac{1}{2} \text{ or } -\frac{1}{3}$$

It's given that $0 \le \theta \le \pi$. The sine is nonnegative, and the answer is (C), $\frac{1}{2}$.

Solving Triangles

You just saw an example that was an algebra question disguised as a trigonometry question. Example 6 is a trigonometry question disguised as a geometry question.

> **LAW OF SINES AND LAW OF COSINES**
>
> If you know any two angles and one side of a triangle, you can figure out the other sides by using the **Law of Sines**. For any triangle, the side lengths are proportional to the sines of the opposite angles:
>
> $$\frac{a}{\sin A} = \frac{b}{\sin B} = \frac{c}{\sin C}$$
>
> If you know two sides of a triangle and the angle between them, you can figure out the third side by using the **Law of Cosines**, which is a more general version of the Pythagorean theorem:
>
> $c^2 = a^2 + b^2 - 2ab \cos C$

Example 6

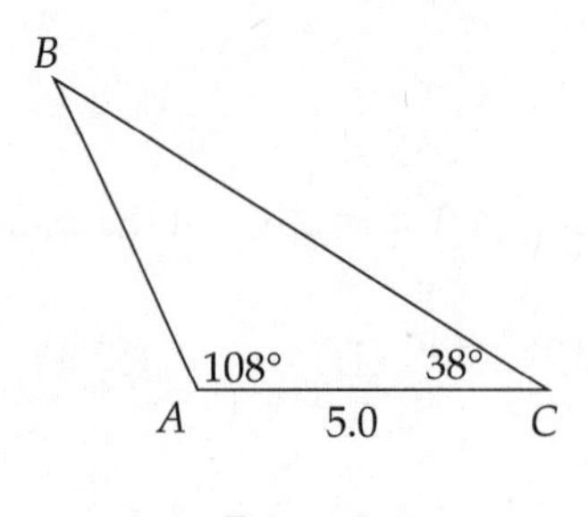

Figure 3

What is the length of side BC in Figure 3?

(A) 7.3 (B) 7.7 (C) 8.1 (D) 8.5 (E) 8.9

You might not think that this is a trigonometry question at first. There's no "sin," "cos," or any other explicit trig function mentioned in the stem. And the triangle's not a right triangle. But trig is the tool you have to use to answer this question. This is a case of "solving a triangle"; that is, finding the length of one or more sides.

With the Law of Sines and the Law of Cosines and a calculator, you can solve almost any triangle. If you know any two angles—which of course means that you know all three—and one side, you can use the Law of Sines to find the other two sides. If you know two sides and the angle between them—remember, it must be the angle between them—you can use the Law of Cosines to find the third side.

In Example 6, what you're given is two angles and a side, so you'll use the Law of Sines, which says simply that the sines are proportional to the opposite sides. Here the side you're looking for, *BC*, is opposite the 108°. The side you're given, *AC*, is opposite the unlabeled angle, which measures 180 – 108 – 38 = 34 degrees. Now you can set up the proportion.

$$\frac{BC}{\sin A} = \frac{AC}{\sin B}$$

$$\frac{BC}{\sin 108^\circ} = \frac{5.0}{\sin 34^\circ}$$

$$BC = \frac{5 \sin 108^\circ}{\sin 34^\circ} \approx 8.5$$

The answer is (D).

Now that you've reviewed the relevant trigonometry facts and formulas and learned a few essential Kaplan strategies, it's time to take another crack at some testlike trig questions. Try the Trigonometry Follow-Up Test.

THINGS TO REMEMBER:

SOHCAHTOA

$$\text{Sine} = \frac{\text{Opposite}}{\text{Hypotenuse}}$$

$$\text{Cosine} = \frac{\text{Adjacent}}{\text{Hypotenuse}}$$

$$\text{Tangent} = \frac{\text{Opposite}}{\text{Adjacent}}$$

Pythagorean Identity

$\sin^2 x + \cos^2 x = 1$

Cotangent, Secant, and Cosecant

Cotangent =

$$\frac{\text{Adjacent}}{\text{Opposite}} = \frac{1}{\text{Tangent}}$$

Secant =

$$\frac{\text{Hypotenuse}}{\text{Adjacent}} = \frac{1}{\text{Cosine}}$$

Cosecant =

$$\frac{\text{Hypotenuse}}{\text{Opposite}} = \frac{1}{\text{Sine}}$$

Level 2 Identities

Functions of Sums

$\sin(A + B) =$

$\sin A \cos B + \cos A \sin B$

$\cos(A + B) =$

$\cos A \cos B - \sin A \sin B$

$$\tan(A + B) = \frac{\tan A + \tan B}{1 - \tan A \tan B}$$

Double-Angle Identities

$\sin 2x = 2 \sin x \cos x$

$\cos 2x = \cos^2 x - \sin^2 x$

$= 1 - 2\sin^2 x$

$= 2\cos^2 x - 1$

$$\tan 2x = \frac{2\tan x}{1 - \tan^2 x}$$

Half-Angle Identities

$$\sin\frac{1}{2}A = \pm\sqrt{\frac{1 - \cos A}{2}}$$

$$\cos\frac{1}{2}A = \pm\sqrt{\frac{1 + \cos A}{2}}$$

$$\tan\frac{1}{2}A = \pm\sqrt{\frac{1 - \cos A}{1 + \cos A}}$$

Radians

π radians = 180 degrees

Amplitude and Period

For functions in the form $y = a \sin bx$ or $y = a \cos bx$, the amplitude of the curve is a and the period of the curve is $\frac{360}{b}$ degrees or $\frac{2\pi}{b}$.

Law of Sines and Law of Cosines

$$\frac{a}{\sin A} = \frac{b}{\sin B} = \frac{c}{\sin C}$$

$c^2 = a^2 + b^2 - 2ab \cos C$

TRIGONOMETRY FOLLOW-UP TEST

6 Questions (8 Minutes)

Directions: Solve problems 1–6. Fill in the oval corresponding to the best answer choice in the grid to the right of each question. (Answers and explanations begin on page 154.)

DO YOUR FIGURING HERE

1. In Figure 1, what is the value of θ?

   (A) 36.9°
   (B) 38.6°
   (C) 41.4°
   (D) 47.2°
   (E) 51.3°

   Ⓐ Ⓑ Ⓒ Ⓓ Ⓔ

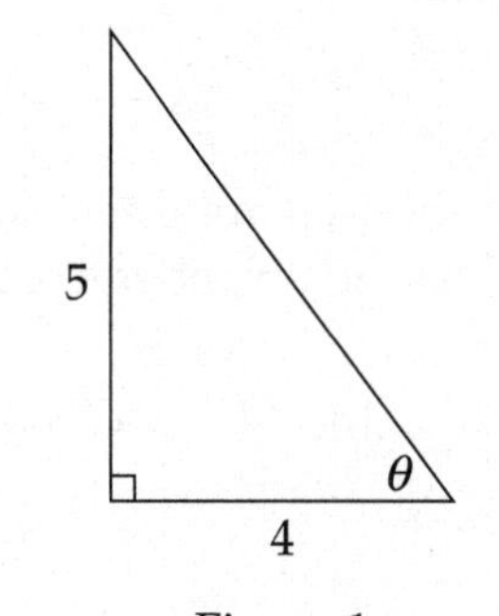

Figure 1

2. $\sin^3\theta + \sin\theta - \sin\theta\cos^2\theta =$

   (A) 0
   (B) $\sin\theta$
   (C) $\sin 2\theta$
   (D) $2\sin\theta$
   (E) $2\sin^3\theta$

   Ⓐ Ⓑ Ⓒ Ⓓ Ⓔ

3. If $\sin\theta = \frac{1}{2}\cos\theta$, and $0 \le \theta \le \frac{\pi}{2}$, the value of $\frac{1}{2}\sin\theta$ is

   (A) 0.22 (B) 0.25 (C) 0.45 (D) 0.50 (E) 0.75

   Ⓐ Ⓑ Ⓒ Ⓓ Ⓔ

4. If θ is an acute angle for which $\tan^2 \theta = 6\tan\theta - 9$, what is the degree measure of θ?

(A) 51.3
(B) 60.0
(C) 71.6
(D) 79.7
(E) 83.5

Ⓐ Ⓑ Ⓒ Ⓓ Ⓔ

5. What is the y-coordinate of the point at which the graph of $y = 2\sin x - \cos 2x$ intersects the y-axis?

(A) –2 (B) –1 (C) 0 (D) 1 (E) 2

Ⓐ Ⓑ Ⓒ Ⓓ Ⓔ

6. Figure 2 shows a regular pentagon with all of its vertices on the sides of a rectangle. If $BC = \frac{1}{4}$, what is the perimeter of the pentagon?

(A) 0.31
(B) 0.62
(C) 0.77
(D) 0.80
(E) 1.00

Ⓐ Ⓑ Ⓒ Ⓓ Ⓔ

DO YOUR FIGURING HERE

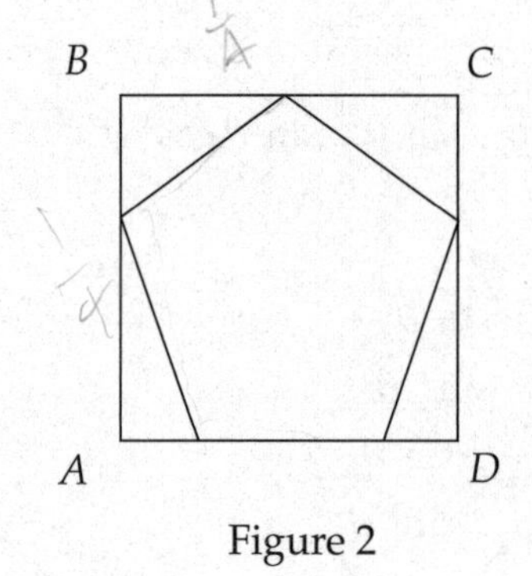

Figure 2

Turn the page
for answers and explanations
to the Follow-Up Test.

FOLLOW-UP TEST—ANSWERS AND EXPLANATIONS

1. E

Since you are given the opposite and adjacent sides, try using tangent.

$$\tan\theta = \frac{\text{opposite}}{\text{adjacent}} = \frac{5}{4} = 1.25$$

Now use the $\tan^{-1}$ function on your calculator (probably "2nd" and then "tan") to find

$$\tan^{-1}(1.25) = 51.34019175$$

This is closest to (E).

2. E

Whenever you have a complicated expression with trigonometric functions, try factoring. First factor out $\sin\theta$ from each term:

$$\sin^3\theta + \sin\theta - \sin\theta\cos^2\theta = (\sin\theta)(\sin^2\theta + 1 - \cos^2\theta)$$

Next look for identities or ways to use the definitions of the trig functions. In this case, you should notice that $1 - \cos^2\theta = (\sin^2\theta)$, so

$$(\sin\theta)(\sin^2\theta + 1 - \cos^2\theta)$$
$$= (\sin\theta)(\sin^2\theta + \sin^2\theta)$$
$$= (\sin\theta)(2\sin^2\theta) = 2\sin^3\theta$$

3. A

You could use the given equation $\sin\theta = \frac{1}{2}\cos\theta$ to find θ, but that's not really necessary. You could re-express the given equation putting it all in terms of $\sin\theta$:

$$\sin\theta = \frac{1}{2}\cos\theta$$
$$\sin\theta = \frac{1}{2}\sqrt{1-\sin^2\theta}$$
$$(\sin\theta)^2 = \left(\frac{1}{2}\sqrt{1-\sin^2\theta}\right)^2$$
$$\sin^2\theta = \frac{1}{4}(1-\sin^2\theta)$$
$$4\sin^2\theta = 1-\sin^2\theta$$
$$5\sin^2\theta = 1$$
$$\sin^2\theta = \frac{1}{5}$$
$$\sin\theta = \sqrt{\frac{1}{5}} \approx 0.45$$
$$\frac{1}{2}\sin\theta \approx 0.22$$

4. C

This is just a quadratic equation where the unknown is $\tan\theta$:

$$\tan^2\theta = 6\tan\theta - 9$$
$$\tan^2\theta - 6\tan\theta + 9 = 0$$
$$(\tan\theta - 3)^2 = 0$$
$$\tan\theta = 3$$
$$\theta = \arctan(3)$$
$$\approx 71.6^\circ$$

5. B

The graph intersects the y-axis at the point where $x = 0$:

$$\begin{aligned} y &= 2\sin x - \cos 2x \\ &= 2\sin(0) - \cos(0) \\ &= 2(0) - 1 \\ &= -1 \end{aligned}$$

While trigonometry problems are often considered to be more "difficult" than other algebraic problems, sometimes the problems themselves aren't actually that hard to solve. Keep this in mind when you see problems with sine and cosine.

6. C

Mark up the figure. The angles of the rectangle are right angles. To figure out the interior angles of the pentagon, use the formula for the degree measure of each interior angle of a regular polygon of n sides:

$$\begin{aligned} \text{Interior angle} &= \frac{(n-2)180}{n} \\ &= \frac{(5-2)180}{5} \\ &= 108 \end{aligned}$$

Label all the angles you can:

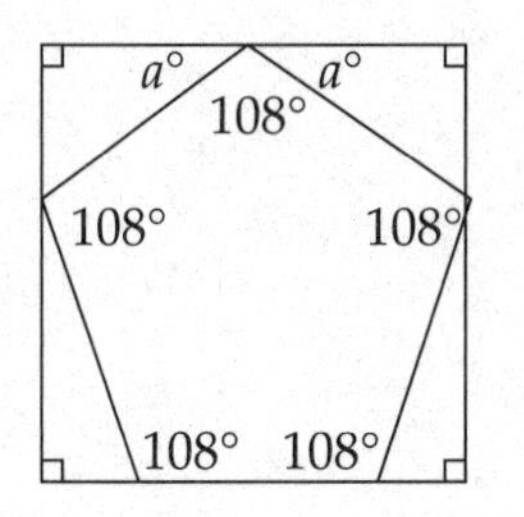

The angles marked $a°$ are equal and together with 108° add up to 180°:

$$\begin{aligned} 108 + 2a &= 180 \\ 2a &= 72 \\ a &= 36 \end{aligned}$$

So the right triangle in the upper left looks like this:

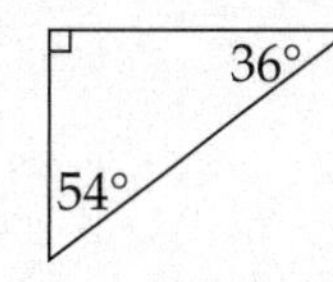

The leg on top (opposite the 54° angle) is half of BC, or $\frac{1}{8}$. The hypotenuse of this right triangle is one side of the pentagon whose perimeter you are looking for, so call the hypotenuse x:

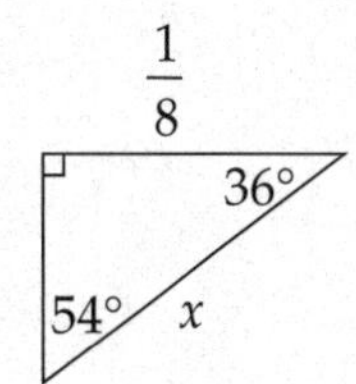

You have the leg that's *opposite* the 54° and you're looking for the *hypotenuse*, so use the *sine*:

$$\sin 54° = \frac{\frac{1}{8}}{x}$$

$$x = \frac{\frac{1}{8}}{\sin 54°} \approx 0.155$$

That's not the answer, however. That's just one side of the pentagon, and you want the perimeter, so multiply that side length by 5:

$$\text{Perimeter} \approx 5 \times 0.155 \approx 0.77$$

Chapter 9: **Functions**

- Substitution
- Minimums and maximums
- Graphing functions

Functions is a very large category on the Math 2 test. You need to be comfortable with the material in this chapter.

FUNCTIONS FACTS AND FORMULAS IN THIS CHAPTER

- Functions: Domain and Range
- Compound Functions
- Maximums and Minimums
- Undefined
- Inverse Functions

HOW TO USE THIS CHAPTER

Maybe you already know everything you need to know about functions. You can find out by taking the Functions Diagnostic Test. Check your answers using the answer key following the test. No matter how you score, don't worry! The answer key also shows where to find a detailed explanation for each question. The "Find Your Study Plan" section that follows the test will suggest next steps based on your performance on the Diagnostic.

Find Your Level

How you use this chapter really depends on how much time you have to prep. Find your level and pace below.

Standard Plan. Do everything in this chapter. It's all relevant to the Math 2 test.

> *Shortcut: Take the Functions Diagnostic Test and check your answers. The "Find Your Study Plan" section that follows the test will suggest next steps based on your Diagnostic Test score.*

Panic Plan. Look through the chapter quickly and make sure you're comfortable with the material. If you're not comfortable with functions, spend some time with this critical chapter.

FUNCTIONS DIAGNOSTIC TEST

DO YOUR FIGURING HERE

6 Questions (8 Minutes)

Directions: Solve problems 1–6 and choose the best answer from those given. Fill in the oval corresponding to the best answer choice in the grid to the right of each question. (Answers are on page 161.)

1. If $f(t) = t^2 + \frac{3}{2}t$, then $f(q - 1) =$

(A) $q^2 - \frac{3}{2}$

(B) $q^2 - \frac{1}{2}q$

(C) $q^2 - \frac{1}{2}q + \frac{1}{2}$

(D) $q^2 - \frac{1}{2}q - \frac{1}{2}$

(E) $q^2 - \frac{1}{2}q + 1$

Ⓐ Ⓑ Ⓒ Ⓓ Ⓔ

2. If $h(x) = \sqrt{x^2 + x - 4}$ and $g(x) = \sqrt{5x}$, what is the value of $h(g(6))$?

(A) 5.48 (B) 5.55 (C) 5.61 (D) 5.83 (E) 6.16

Ⓐ Ⓑ Ⓒ Ⓓ Ⓔ

3. What is the maximum value of $f(x) = 3 - (x + 2)^2$?

(A) −3 (B) −1 (C) 1 (D) 3 (E) 5

Ⓐ Ⓑ Ⓒ Ⓓ Ⓔ

DO YOUR FIGURING HERE

4. If $p(x) = |2 - x| + \frac{1}{2}$, which of the following could be the graph of $y = p(x)$?

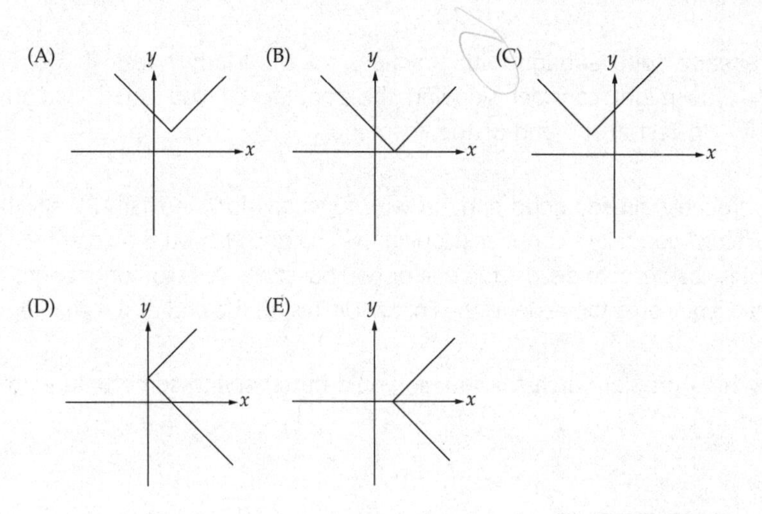

A B C D E

5. If $f(x) = \frac{\sqrt{x^2 - 4}}{x - 4}$, what are all the values of x for which $f(x)$ is defined?

(A) All real numbers except 4
(B) All real numbers except –2 and 2
(C) All real numbers greater than or equal to –2 and less than or equal to 2
(D) All real numbers less than or equal to –2 or greater than or equal to 2
(E) All real numbers less than or equal to –2 or greater than or equal to 2, except 4

A B C D E

6. If $f(x) = \frac{1}{3}x + 3$, then $f^{-1}(x) =$

(A) $-\frac{1}{3}x - 3$
(B) $-3x + \frac{1}{3}$
(C) $3x + \frac{1}{3}$
(D) $3x - 3$
(E) $3x - 9$

A B C D E

Find Your Study Plan

The answer key on the next page shows where in this chapter to find explanations for the questions you missed. Here's how you should proceed based on your Diagnostic Test score.

6: Superb! You're already good enough with functions for the Math 2 test. If you're taking a "shortcut," you might consider skipping this chapter. Or you could just go straight to the Follow-Up Test at the end of the chapter.

4–5: Good. You're probably already good enough with functions for the Math 2 test. If you're taking a "shortcut," you might consider skipping this chapter. But you should at least look at the parts of this chapter that discuss the one or two questions you did not get right. Then, if you have time, you could move on to the Follow-Up Test at the end of the chapter.

0–3: Functions may be a problem area for you, so you'd better spend some time with this chapter.

FUNCTIONS TEST TOPICS

We'll use the questions on the Functions Diagnostic Test to illustrate the conventions of functions.

DON'T BE AFRAID OF FUNCTIONS

There's nothing inherently difficult about functions. There are not a lot of formulas or theorems to remember. Getting comfortable with functions just means understanding the symbolism and conventions.

Who's Afraid of Functions?

Lots of students are afraid of functions. But there's nothing especially difficult about them. They just look scary. Once you "get" the conventions, you'll never be afraid of functions again. Here's a quick and painless review of the basic things you need to know about functions.

A function is a process that turns a number into another number. Squaring is an example of a function. For any number you can think of, there is a unique number that is its square. The conventional way of writing this function is:

$$f(x) = x^2$$

When you apply the function to some particular number, such as –5, you write it this way:

$$f(-5)$$

And to find the value of $f(-5)$, you plug $x = -5$ into the definition:

$$f(x) = x^2$$
$$f(-5) = (-5)^2 = 25$$

Substitution

The most straightforward functions questions you'll encounter on Math 2 are questions like Example 1 that ask you simply to apply a function to some number or expression.

DIAGNOSTIC TEST ANSWER KEY

1. D
See "Substitution."

2. C
See "Compound Functions."

3. D
See "Maximums and Minimums."

4. A
See "Graphing Functions."

5. E
See "Undefined."

6. E
See "Inverse Functions."

Example 1

If $f(x) = t^2 + \frac{3}{2}t$, then $f(q - 1) =$

(A) $q^2 - \frac{3}{2}$

(B) $q^2 - \frac{1}{2}q$

(C) $q^2 + \frac{1}{2}q + \frac{1}{2}$

(D) $q^2 - \frac{1}{2}q - \frac{1}{2}$

(E) $q^2 - \frac{1}{2}q + 1$

To find $f(q - 1)$, plug $t = q - 1$ into the definition. Substituting an algebraic expression for t is a little more complicated than substituting a number for t, but the idea's the same.

$$f(t) = t^2 + \frac{3}{2}t$$

$$f(q-1) = (q-1)^2 + \frac{3}{2}(q-1)$$

$$= q^2 - 2q + 1 + \frac{3}{2}q - \frac{3}{2}$$

$$= q^2 - \frac{1}{2}q - \frac{1}{2}$$

The answer is (D).

FUNCTIONS:

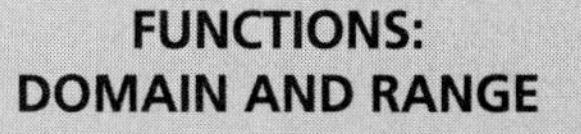

A **function** is a set of ordered pairs (x, y) such that for each value of x there is one and only one value of y.

The set of all allowable x values is called the **domain**. Note that, according to the SAT directions, the domain of any function f is assumed to be the set of all real numbers x for which $f(x)$ is a real number.

The corresponding set of all y values is called the **range**.

Compound Functions

The letter f is not the only letter used to designate a function, though it's the most popular. Second in popularity is the letter g, which is generally used in a question that includes two different functions. Take a look at Example 2.

COMPOUND FUNCTIONS

f(*g*(*x*)) means apply *g* first and then apply *f* to the result.

g(*f*(*x*)) means apply *f* first and then apply *g* to the result.

f(*x*)*g*(*x*) means apply *f* and *g* separately and then multiply the results.

Example 2

If $h(x)=\sqrt{x^2+x-4}$ and $g(x)=\sqrt{5x}$, what is the value of $h(g(6))$?

(A) 5.48 (B) 5.55 (C) 5.61 (D) 5.83 (E) 6.16

When one function is written inside another function's parentheses, apply the inside function first:

$$g(x)=\sqrt{5x}$$
$$g(6)=\sqrt{5(6)}=\sqrt{30}$$

Then apply the outside function to the result:

$$h(x)=\sqrt{x^2+x-4}$$
$$h(\sqrt{30})=\sqrt{(\sqrt{30})^2+\sqrt{(30)}-4}$$
$$=\sqrt{30+\sqrt{30}-4}$$
$$=\sqrt{26+\sqrt{30}}\approx 5.610456806$$

The answer is (C).

WATCH THOSE PARENTHESES

When it comes to functions, pay close attention to order and parentheses.

Notice that there's a difference between *h*(*g*(6)) and *g*(*h*(6)). In the latter, you're to apply the function *h* first and then the function *g* to the result. The order makes a difference: *h*(*g*(6)) ≈ 5.61, but *g*(*h*(6)) ≈ 5.55 (which is, of course, included as a distracter).

Notice also that there's a difference between *h*(*g*(*x*)) and *h*(*x*)*g*(*x*). In the former expression, a hierarchy is indicated: The function *g* is inside the parentheses of function *h*, so you apply *g* first, and then you apply *h* to the result. In the expression *h*(*x*)*g*(*x*), however, the two functions are written side by side, which indicates multiplication. You apply the function *h* to 6, and you apply the function *g* to 6, and then you multiply the results:

$$h(6)=\sqrt{6^2+6-4}=\sqrt{38}$$
$$g(6)=\sqrt{30}$$
$$h(6)g(6)=\sqrt{38}\times\sqrt{30}=\sqrt{1{,}140}\approx 33.76$$

So when it comes to functions, be sure to pay close attention to order and parentheses.

Maximums and Minimums

Another typical functions question is one that asks for a minimum or, as in Example 3, maximum value of a function.

Example 3

What is the maximum value of $f(x) = 3 - (x - 2)^2$?

(A) –3 (B) –1 (C) 1 (D) 3 (E) 5

If you have a graphing calculator (and know how to use it), you could graph the function and trace the graph to find the maximum. But it's really a lot easier if you conceptualize the situation. The expression $3 - (x - 2)^2$ will be at its maximum when the part being subtracted from the 3 is as small as it can be. That part after the minus sign, $(x - 2)^2$, is the square of something, so it can be no smaller than 0. When $x = 2$, $(x - 2)^2 = 0$, and the whole expression $3 - (x - 2)^2 = 3 - 0 = 3$. For any other value of x, the part after the minus sign will be greater than 0, and the whole expression will be less than 3. So 3 is the maximum value, and the answer is (D).

MAXIMUMS AND MINIMUMS

To find a maximum or minimum value of a function, look for parts of the expression—especially squares—that have upper or lower limits.

Graphing Functions

Like almost everything else with functions, graphing is no big deal once you understand the conventions. Example 4 provides a very good illustration.

Example 4

If $p(x) = |2 - x| + \frac{1}{2}$, which of the following could be the graph of $y = p(x)$?

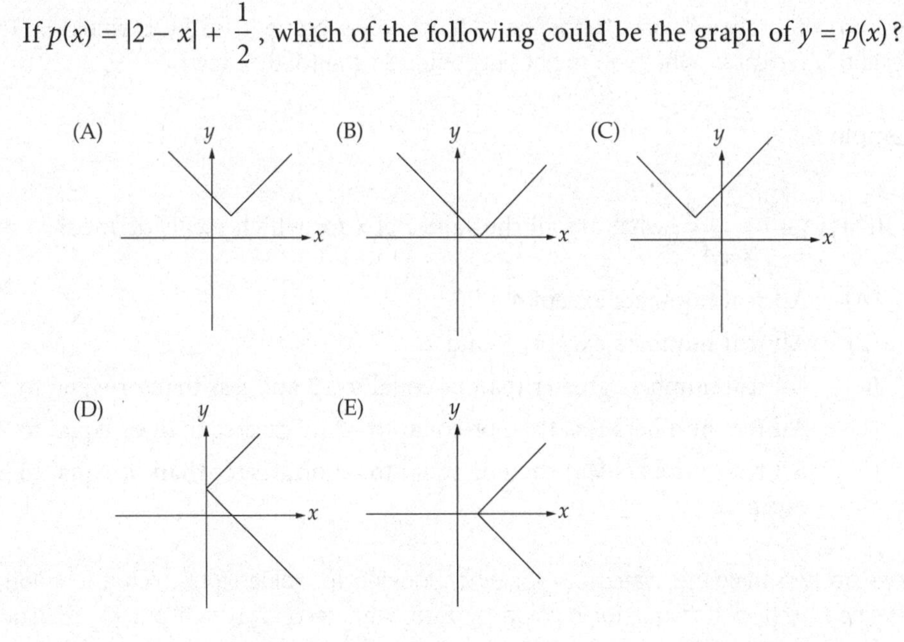

The question says $p(x)=|2-x|+\frac{1}{2}$ and asks for the graph of $y = p(x)$—meaning the graph of $y=|2-x|+\frac{1}{2}$.

The smallest that y can be is $\frac{1}{2}$, because the absolute value of anything is nonnegative. That narrows the choices to (A) and (C). Notice that (D) and (E) cannot be functions, because for some values of x they show two associated values of y. A function, by definition, yields no more than one y for any particular x.

To choose between (A) and (C), think about what value of x yields $y = \frac{1}{2}$.

$$
\begin{aligned}
|2-x|+\frac{1}{2}&=\frac{1}{2}\\
|2-x|&=0\\
2-x&=0\\
2&=x
\end{aligned}
$$

So the graph reaches its minimum point at $x = 2$, which means the answer must be (A).

Undefined

The issue of "defined" and "undefined" functions is brought up by Example 5. This question is typical of what you might encounter on the Math 2 test.

UNDEFINED

To find values of x for which a function is undefined, look for values that would make a denominator zero or that would make an expression under a radical negative.

Example 5

If $f(x)=\frac{\sqrt{x^2-4}}{x-4}$, what are all the values of x for which $f(x)$ is defined?

(A) All real numbers except 4
(B) All real numbers except –2 and 2
(C) All real numbers greater than or equal to –2 and less than or equal to 2
(D) All real numbers less than or equal to –2 or greater than or equal to 2
(E) All real numbers less than or equal to –2 or greater than or equal to 2, except 4

There are two things to watch out for when looking for values for which a function is undefined. First, watch out for division by zero. And second, watch out for negatives under radicals. In this case you have both.

The function is undefined when the denominator is zero:

$$f(x)=\frac{\sqrt{x^2-4}}{x-4}$$

denominator

$$x-4=0$$
$$x=4$$

The function is also undefined when the expression under the radical is negative:

$$f(x)=\frac{\sqrt{x^2-4}}{x-4}$$

expression under radical

$$x^2-4<0$$
$$x^2<4$$
$$-2<x<2$$

So the expression is defined for all x such that $x \le -2$ or $x \ge 2$, except $x = 4$. The answer is (E).

There is a third thing that can result in undefined values in a function: logarithms. Similar to square roots, the log of a negative number is undefined. Unlike roots, however, the log of 0 is *also* undefined. If you keep these three cases in mind, you can determine the total domain of a function with ease.

Inverse Functions

Perhaps the most advanced functions question you'll face on Math 2 will be one involving the inverse of a function—written $f^{-1}(x)$—as in Example 6.

Example 6

If $f(x)=\frac{1}{3}x+3$, then $f^{-1}(x)=$

(A) $-\frac{1}{3}x-3$

(B) $-3x+\frac{1}{3}$

(C) $3x+\frac{1}{3}$

(D) $3x-3$

(E) $3x-9$

Here's what you do to find the inverse of a function. First, put y in the place of $f(x)$ to make a more familiar-looking equation form:

$$f(x)=\frac{1}{3}x+3$$

$$y=\frac{1}{3}x+3$$

Second, solve the equation for x in terms of y:

$$y=\frac{1}{3}x+3$$

$$y-3=\frac{1}{3}x$$

$$3(y-3)=x$$

$$x=3y-9$$

Third, put $f^{-1}(x)$ in the place of x and put x in the place of y:

$$x=3y-9$$

$$f^{-1}(x)=3x-9$$

The answer is (E).

INVERSE FUNCTIONS

- To find the inverse of a function:
1. Replace $f(x)$ with y.
2. Solve the equation for x in terms of y.
3. Replace x with $f^{-1}(x)$ and replace y with x.
- Graphs of inverse functions are symmetric about the line $y = x$.
- The slopes of the lines of inverse functions are reciprocals.

You may also see a question on one or more of the following topics.

A **periodic function** is one that cycles over and over again, like the sine function.

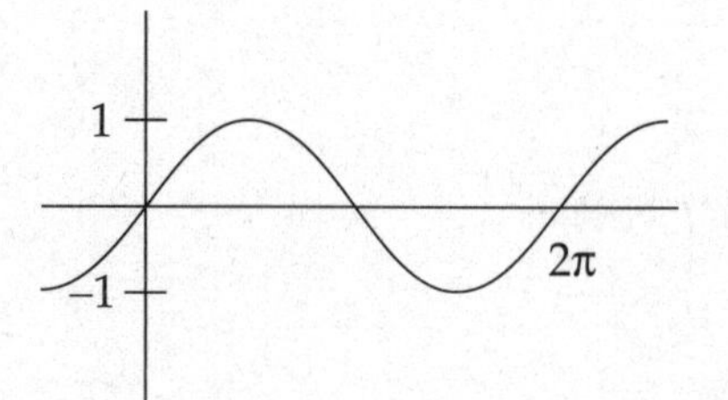

The fundamental period of a function is the length of a smallest complete cycle. For the sine function, the period is 2π. For large values, you can evaluate a function just by calculating the function at the remainder after dividing by the fundamental period.

For example, $\sin 9\pi = \sin \pi = 0$.

A **piecewise function** is a function that is defined in intervals; that is, for different values, you perform different calculations.

$$f(x) = \begin{cases} x^2 \text{ when } x < 4 \\ x \text{ when } 4 \le x < 25 \\ \sqrt{x} \text{ when } x \ge 25 \end{cases}$$

When evaluating piecewise functions, just be sure to make sure that you're calculating in the correct interval.

A **recursive function** is one that is defined by how the terms in the sequence are related to the preceding terms. For example, the function known as the factorial is a recursive function:

$$5! = 5 \times 4! = 5 \times 4 \times 3! = 5 \times 4 \times 3 \times 2! = 5 \times 4 \times 3 \times 2 \times 1$$

The Fibonacci sequence is also a recursive function, defined as follows:

$$f_0 = 0$$
$$f_1 = 1$$
$$f_n = f_{n-1} + f_{n-2}$$

On Test Day, you may be asked to find specific terms in a sequence using recursive definitions.

A **parametric function** is a function where two coordinates are defined through two separate nonparametric functions of a third value. Most often, parametric functions are used to describe position (in terms of *x* and *y*) in time (*t*).

Velocity $\vec{v}$ can be found using the formula $\left(\frac{x_2 - x_1}{t}, \frac{y_2 - y_1}{t}\right)$.

Speed $|\vec{v}|$ can be found using the formula $\sqrt{(\vec{v}_x)^2 + (\vec{v}_y)^2}$.

For example:

An object moving with constant velocity is at point G(4,2) when time $t = 0$ seconds and at point $H(-8,11)$ when $t = 4$ seconds.

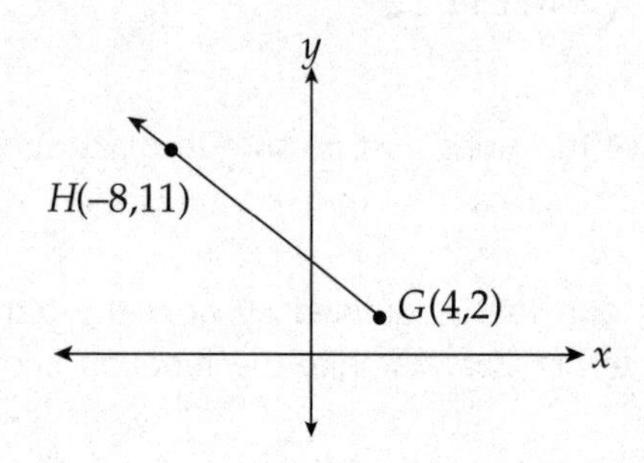

In 4 seconds, the object moves (–8,11) – (4,2) = (–12,9) units. In 1 second, it moves $\frac{1}{4}(-12,9) = \left(\frac{-12}{4}, \frac{9}{4}\right) = \left(-3, \frac{9}{4}\right)$, which is its velocity $\vec{v}$.

Its speed is $|\vec{v}| = \sqrt{(-3)^2 + \left(\frac{9}{4}\right)^2} + \sqrt{9 + \frac{81}{16}} = \sqrt{\frac{225}{16}} = 3.75$

The vector equation that describes the motion of this object is

$$(x, y) = (4,2) + t\left(-3, \frac{9}{4}\right)$$

$$x = 4 - 3t$$

$$y = 2 + \frac{9}{4}t$$

Now that you've had a good look at some typical (and a few unusual) Math 2 functions questions, it's time to try a few more on your own. See how well you can do with the questions in the Functions Follow-Up Test.

THINGS TO REMEMBER:

Functions: Domain and Range

A **function** is a set of ordered pairs (x,y) such that for each value of x there is one and only one value of y.

The set of all allowable x values is called the **domain**.
The corresponding set of all y values is called the **range**.

Compound Functions

$f(g(x))$ means apply g first and then apply f to the result.

$g(f(x))$ means apply f first and then apply g to the result.

$f(x)g(x)$ means apply f and g separately and then multiply the results.

Maximums and Minimums

To find a maximum or minimum value of a function, look for parts of the expression—especially squares—that have upper or lower limits.

Undefined

To find values of x for which a function is undefined, look for values that would make a denominator zero or that would make an expression under a radical negative.

Inverse Functions

To find the inverse of a function:

1. Replace $f(x)$ with y.
2. Solve the equation for x in terms of y.
3. Replace x with $f^{-1}(x)$ and replace y with x.

Graphs of inverse functions are symmetric about the line $y = x$.

The slopes of the lines of inverse functions are reciprocals.

FUNCTIONS FOLLOW-UP TEST

DO YOUR FIGURING HERE

6 Questions (8 Minutes)

Directions: Solve problems 1–6. Fill in the oval corresponding to the best answer choice in the grid to the right of each question. (Answers and explanations begin on page 173.)

1. If $f(x) = x^2 + 2x - 2$ and if $f(s - 1) = 1$, what is the smallest possible value of s?

(A) − 3
(B) − 2
(C) − 1
(D) 1
(E) 2

Ⓐ Ⓑ Ⓒ Ⓓ Ⓔ

2. If $f(x) = 2x^3$ and $g(x) = 3x$, what is the value of $g(f(-2)) - f(g(-2))$?

(A) −480
(B) −384
(C) 0
(D) 384
(E) 480

Ⓐ Ⓑ Ⓒ Ⓓ Ⓔ

3. For what value of x is $|16 - (x + 5)^2|$ at its minimum?

(A) −9
(B) −5
(C) 5
(D) 9
(E) 16

Ⓐ Ⓑ Ⓒ Ⓓ Ⓔ

4. The graph in Figure 1 could be the graph of which of the following functions?

(A) $f(x) = x^2 + 9$
(B) $f(x) = (x - 9)^2$
(C) $f(x) = 9 - x^2$
(D) $f(x) = -|x^2 + 9|$
(E) $f(x) = |-x^2 + 9|$

Ⓐ Ⓑ Ⓒ Ⓓ Ⓔ

5. If $f(x) = \dfrac{1}{\sqrt{1-x^2}}$, which of the following describes all the real values of x for which $f(x)$ is undefined?

(A) $x = -1$ or $x = 1$
(B) $x < -1$ or $x > 1$
(C) $x \leq -1$ or $x \geq 1$
(D) $-1 < x < 1$
(E) $-1 \leq x \leq 1$

Ⓐ Ⓑ Ⓒ Ⓓ Ⓔ

6. If $f(x)$ is a linear function and the slope of $y = f(x)$ is $\frac{1}{2}$, what is the slope of $y = f^{-1}(x)$?

(A) -2
(B) $-\frac{1}{2}$
(C) $\frac{1}{2}$
(D) 2
(E) It cannot be determined from the information given.

Ⓐ Ⓑ Ⓒ Ⓓ Ⓔ

DO YOUR FIGURING HERE

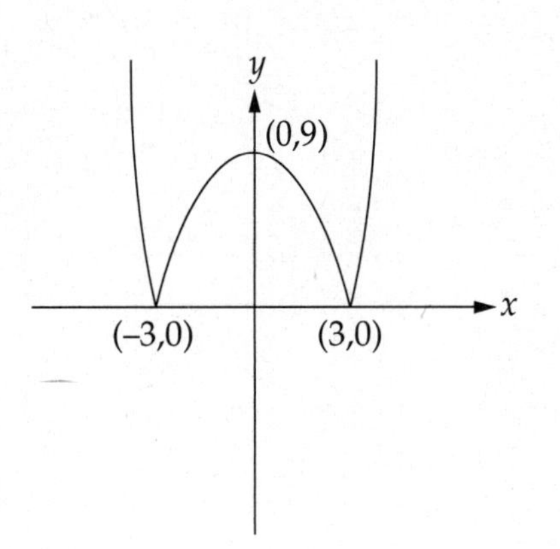

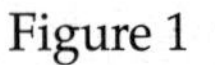

Figure 1

FOLLOW-UP TEST—ANSWERS AND EXPLANATIONS

1. B

Solution 1: Algebra

Plug $x = s - 1$ into the definition of the function:

$$f(x) = x^2 + 2x - 2$$
$$f(s - 1) = (s - 1)^2 + 2(s - 1) - 2$$
$$= s^2 - 2s + 1 + 2s - 2 - 2$$
$$= s^2 - 3$$

Now write an equation in which the result is equal to 1:

$$s^2 - 3 = 1$$
$$s^2 - 4 = 0$$
$$(s - 2)(s + 2) = 0$$
$$s = 2, s = -2$$

The smallest of these is –2, answer (B).

Solution 2: Less messy Algebra

First solve for x when the function is equal to 1:

$$f(x) = x^2 + 2x - 2 = 1$$
$$x^2 + 2x - 3 = 0$$
$$(x + 3)(x - 1) = 0$$
$$x = -3, x = 1$$

We know that $s - 1$ is taking the place of x, so we can write:

$$s - 1 = -3 \qquad s - 1 = 1$$
$$s = -2 \qquad s = 2$$

Solution 3: Backsolving

Plug in each of the answer choices, using $s - 1 = x$ to translate from s to x values, and see which is the smallest that works in the equation.

2. D

Perform the inside functions first:

$$f(-2) = 2(-2)^3 = 2(-8) = -16$$
$$g(-2) = 3(-2) = -6$$

Then perform the outside functions on the results:

$$g(f(-2)) = g(-16) = 3(-16) = -48$$
$$f(g(-2)) = f(-6) = 2(-6)^3 = 2(-216) = -432$$

Now you can find the difference:

$$g(f(-2)) - f(g(-2)) = -48 - (-432) = 384$$

3. A

The absolute value must be at least 0, so the expression must be at its minimum when it is equal to 0.

$$\left|16 - (x + 5)^2\right| = 0$$
$$16 - (x + 5)^2 = 0$$
$$(x + 5)^2 = 16$$
$$x + 5 = \pm 4$$
$$x = -9, x = -1$$

Of these, only –9 is an answer choice, so (A) is correct.

You can also backsolve fairly easily on this question—plug in each answer choice, and see which gives you the smallest value for the function.

4. E

Testing the point (3,0) in each of the answer choices, we find that this point is not on the graphs of choices (A), (B), and (D). Therefore, we can eliminate them. Looking at choices (C) and (E), for any $x > 3$, the graph of (C) is negative, so choice (C) can be eliminated. Therefore, choice (E) must be correct.

5. C

The function will be undefined for any values of x that make the denominator zero or that make the expression under the radical negative. The denominator is zero when

$$\sqrt{1-x^2}=0$$
$$1-x^2=0$$
$$x^2=1$$
$$x=\pm 1$$

The expression under the radical is negative when

$$1-x^2<0$$
$$-x^2<-1$$
$$x^2>1$$
$$x<-1 \text{ or } x>1$$

So the function is undefined when $x \le -1$ or when $x \ge 1$.

6. D

You might think this one can't be done because it does not tell you exactly what the function is that you want the inverse of.

It just says the slope is $\frac{1}{2}$, but, in fact, it doesn't actually ask for the inverse function; it just asks for the slope.

You can spot the correct answer immediately if you remember how the slopes of inverse functions are related: They're reciprocals.

You can still find the answer without remembering that relationship. You know the slope of $f(x)$ is $\frac{1}{2}$.

You don't know the y-intercept, so just call it b :

$$f(x)=\frac{1}{2}x+b$$

Find the inverse function:

$$f(x)=\frac{1}{2}x+b$$
$$y=\frac{1}{2}x+b$$
$$y-b=\frac{1}{2}x$$
$$2y-2b=x$$
$$x=2y-2b$$
$$f^{-1}(x)=2x-2b$$

The slope is the coefficient of x, which is 2.

Chapter 10: **Miscellaneous Topics**

- Symbols, definitions, and instructions
- Imaginary and complex numbers
- Logic
- Percents, averages, and rates
- Permutations and combinations
- Probability
- Logarithms
- Sequences
- Series, vectors, and regression

This chapter covers a dozen largely unrelated math topics that are regularly tested on the Math 2 test. You will not see all the topics in this chapter on any one edition of the test, but you can be sure that some will appear.

MISCELLANEOUS FACTS AND FORMULAS IN THIS CHAPTER

- Rules of Imaginary Numbers
- Contrapositive
- Percent Increase and Decrease Formulas
- Average Rate Formula
- Average, Median, and Mode
- Permutations and Combinations Formulas
- Probability Formulas
- Rules of Logarithms
- Limit of an Algebraic Fraction
- Sequences Formulas

HOW TO USE THIS CHAPTER

Maybe you already know everything you need to know about these math topics. You can find out by taking the Miscellaneous Topics Diagnostic Test. Check your answers using the answer key following the test. No matter how you score, don't worry! The answer key also shows where to find a detailed explanation for each question. The "Find Your Study Plan" section that follows the test will suggest next steps based on your performance on the Diagnostic.

Find Your Level

How you use this chapter really depends on how much time you have to prep. Find your level and pace on the following page.

Standard Plan. Do everything in this chapter. It's all relevant to the Math 2 test.

Shortcut. Take the Miscellaneous Topics Diagnostic Test and check your answers. The "Find Your Study Plan" section that follows the test will suggest next steps based on your Diagnostic Test score.

Panic Plan. Make sure you're comfortable with the fundamental material in previous chapters before you spend any time on these miscellaneous topics.

MISCELLANEOUS TOPICS DIAGNOSTIC TEST

DO YOUR FIGURING HERE

12 Questions (15 Minutes)

Solve problems 1–12. Fill in the oval corresponding to the best answer choice in the grid to the right of each question. (Answers are on page 182.)

1. For all nonzero numbers x and y, the operation ∇ is defined by the equation $x \nabla y = \frac{|x|}{x^2} + \frac{y^2}{|y|}$ when $x > y$ and by the equation $x \nabla y = \frac{|x|}{y^2} - \frac{x^2}{|y|}$ when $x \leq y$. If $x \nabla y < 0$, then which of the following could be true?

   I. $x^3 = y^3$
   II. $(y + x)(y - x) > 0$
   III. $x - y > 0$

   (A) I only
   (B) II only
   (C) III only
   (D) I and II
   (E) II and III

   Ⓐ Ⓑ Ⓒ Ⓓ Ⓔ

2. If $i = \sqrt{-1}$, then which of the following has the greatest value?

   (A) $i^4 + i^3 + i^2 + i$
   (B) $i^8 + i^6 + i^4 + i^2$
   (C) $i^{12} + i^9 + i^6 + i^3$
   (D) $i^{16} + i^{12} + i^8 + i^4$
   (E) $i^{20} + i^{15} + i^{10} + i^5$

   Ⓐ Ⓑ Ⓒ Ⓓ Ⓔ

3. "If a zic is a zac, then a zic is not a zoc."

   If the statement above is true, then which of the following statements must also be true?

   (A) "If a zic is a zoc, then a zic is not a zac."
   (B) "If a zic is not a zac, then a zic is a zoc."
   (C) "If a zic is not a zoc, then a zic is a zac."
   (D) "If a zic is not a zac, then a zoc is not a zac."
   (E) "If a zoc is a zic, then a zac is a zic."

   Ⓐ Ⓑ Ⓒ Ⓓ Ⓔ

DO YOUR FIGURING HERE

4. A delicatessen charges for a certain type of salad at a rate of 2 dollars for the first half pound and 75 cents for each additional half pound. Which of the following expressions represents the total charge, in cents, for p pounds of the salad, where p is a positive integer?

   (A) $25(6p + 8)$
   (B) $25(6p + 5)$
   (C) $25(3p + 5)$
   (D) $1.25 + 1.5p$
   (E) $2 + 0.75(p - 1)$

   Ⓐ Ⓑ Ⓒ Ⓓ Ⓔ

5. During a weight-lifting routine, an athlete calculates the total amount of weight moved while performing a certain exercise by multiplying the amount of weight per lift by the number of lifts. On Wednesday, the weight per lift was 20 percent greater than the weight per lift on Monday, while the number of lifts was 25 percent less than the number of lifts performed on Monday. On Friday, the weight per lift was 25 percent greater than the weight per lift on Wednesday, while the number of lifts was 40 percent less than the number of lifts performed on Wednesday. The total weight moved by the athlete while performing the exercise on Monday was approximately what percent greater than the total weight she moved while performing the exercise on Friday?

   (A) 15% (B) 33% (C) 48% (D) 50% (E) 100%

   Ⓐ Ⓑ Ⓒ Ⓓ Ⓔ

6. Last year a manufacturer had a sales total for January that was 50 percent greater than the average (arithmetic mean) of the monthly sales totals for February through December. The sales total for January was what fraction of the sales total for the year?

(A) $\frac{1}{11}$ (B) $\frac{1}{8}$ (C) $\frac{3}{25}$ (D) $\frac{3}{22}$ (E) $\frac{3}{5}$

Ⓐ Ⓑ Ⓒ Ⓓ Ⓔ

7. Kayla drove from Bayside to Chatham at a constant speed of 21 miles per hour and then returned along the same route from Chatham to Bayside. If her average speed for the entire journey was 26.25 miles per hour, at what average speed, in miles per hour, did Kayla return from Chatham to Bayside?

(A) 28
(B) 31
(C) 31.5
(D) 35
(E) It cannot be determined from the information given.

Ⓐ Ⓑ Ⓒ Ⓓ Ⓔ

8. Of the 12 members of a high school drama club, 8 are seniors. The club plans to establish an 8-member committee to interview potential club members. If exactly 6 members of the committee must be seniors, how many committees are possible?

(A) 21
(B) 34
(C) 168
(D) 336
(E) 495

Ⓐ Ⓑ Ⓒ Ⓓ Ⓔ

DO YOUR FIGURING HERE

DO YOUR FIGURING HERE

9. Two fair dice are tossed simultaneously. What is the probability that the product of the numbers that land showing is at least 25?

(A) $\frac{1}{36}$ (B) $\frac{1}{12}$ (C) $\frac{1}{9}$ (D) $\frac{1}{4}$ (E) $\frac{3}{2}$

Ⓐ Ⓑ Ⓒ Ⓓ Ⓔ

10. $\log_3 \sqrt[9]{3} =$

(A) $\frac{1}{9}$ (B) $\frac{3}{2}$ (C) $\frac{1}{2}$ (D) 1 (E) 3

Ⓐ Ⓑ Ⓒ Ⓓ Ⓔ

11. $\lim\limits_{x \to 3} \frac{x^3 + x^2 - 12x}{x^2 - 9}$

(A) -7

(B) $-\frac{7}{2}$

(C) 0

(D) $\frac{7}{2}$

(E) The limit does not exist.

Ⓐ Ⓑ Ⓒ Ⓓ Ⓔ

12. If the first term in a geometric sequence is 3, and if the third term is 48, what is the 11th term?

(A) 228

(B) 528

(C) 110,592

(D) 3,145,728

(E) 12,582,912

Ⓐ Ⓑ Ⓒ Ⓓ Ⓔ

Find Your Study Plan

The answer key is on the next page. It doesn't matter much how you did on the Miscellaneous Topics Diagnostic Test as a whole, because the topics are so varied. What matters is exactly what questions you had trouble with. No matter what your Diagnostic Test score is, or what level test you're preparing for, your next step is to read and study those parts of this chapter that address the particular diagnostic questions you got wrong. The answer key shows where in this chapter to find explanations for the questions you missed.

MISCELLANEOUS TEST TOPICS

We'll use the questions on the Miscellaneous Topics Diagnostic Test to point out more topics you'll need to be prepared to deal with on the Math 2 test.

Symbolism, Definitions, and Instructions

One miscellaneous question type that turns up with great regularity is the type that presents you with a new symbol representing an unfamiliar operation, or a new word with an unfamiliar definition, or a new procedure with unfamiliar steps. In Example 1 you have a new symbol and operation:

DON'T FREAK OUT.

If you see something—a symbol, a word—you've never seen before, chances are the test makers just made it up to test your ability to stay cool in the face of something new and unfamiliar.

Example 1

For all nonzero numbers x and y, the operation ∇ is defined by the equation $x \nabla y = \frac{|x|}{x^2} + \frac{y^2}{|y|}$ when $x > y$ and by the equation $x \nabla y = \frac{|x|}{y^2} - \frac{x^2}{|y|}$ when $x \leq y$. If $x \nabla y < 0$, then which of the following could be true?

I. $x^3 = y^3$
II. $(y + x)(y - x) > 0$
III. $x - y > 0$

(A) I only
(B) II only
(C) III only
(D) I and II
(E) II and III

Symbolism questions often look intimidating at first, but you'll find that, on closer inspection, they merely teach you some invented rule and then ask you to show that you understand the rule. Of course, on a test as challenging as the Math 2, symbolism questions often include a twist or two. Here, for example, you're told that ∇ means one thing whenever $x > y$, but something different whenever $x \leq y$. And you're expected to juggle these definitions in the context of a Roman Numeral question.

DIAGNOSTIC TEST ANSWER KEY

1. D
See "Symbolism, Definitions, and Instructions."

2. D
See "Imaginary and Complex Numbers."

3. A
See "Logic."

4. B
See "Translating from English into Algebra."

5. C
See "Percent Increase and Decrease."

6. C
See "Averages."

7. D
See "Distance, Rate, and Time."

8. C
See "Permutations and Combinations."

9. C
See "Probability."

10. A
See "Logarithms."

11. D
See "Limits."

12. D
See "Sequences."

Take one issue at a time. You know that $x \nabla y < 0$. What does that tell you about the relationship between x and y themselves? Could x be greater than y? No. When x is greater than y, $x \nabla y = \frac{|x|}{x^2} + \frac{y^2}{|y|}$. But absolute value is always nonnegative and (because x and y are nonzero) squares must be positive in this question, so $\frac{|x|}{x^2} + \frac{y^2}{|y|}$ must be positive. Because, again, $x \nabla y < 0$, you must be dealing here with the other definition: $\frac{|x|}{y^2} - \frac{x^2}{|y|}$. And that's the definition that holds if and only if $x \leq y$.

Having invested some time thinking about the meaning of the question stem, notice now how breezily you move through the statements. Which one(s) have $x \leq y$? In I, in order for $x^3 = y^3$, x and y themselves must be equal. (The same would be true of x and y raised to any odd power, though not to even powers, in which case x could equal $\pm y$.) So given $x \leq y$, "could it be true" that $x = y$? Yes. So I must be included in the answer: eliminate (B), (C), and (E).

Consider II. What do you think of when you see $(y + x)(y - x)$? This expression is another way of writing the classic factorable $y^2 - x^2$. So $y^2 - x^2 > 0$, which means that $y^2 > x^2$. If $x \leq y$, "could it be true" that $y^2 > x^2$? Sure—as Picking Numbers shows: Say $x = 1$ and $y = 2$. So II must also be included in the answer, which must be (D).

Only for the sake of rounding out your knowledge, consider III. If $x \leq y$, "could it be true" that $x - y > 0$? If $x - y > 0$, then $x > y$. If $x \leq y$, it could not be true that $x > y$.

The answer is (D).

Imaginary and Complex Numbers

There's a chance you will encounter a question that begins, "If $i^2 = -1$...," like the example below.

Example 2

If $i = \sqrt{-1}$, then which of the following has the greatest value?

(A) $i^4 + i^3 + i^2 + i$
(B) $i^8 + i^6 + i^4 + i^2$
(C) $i^{12} + i^9 + i^6 + i^3$
(D) $i^{16} + i^{12} + i^8 + i^4$
(E) $i^{20} + i^{15} + i^{10} + i^5$

Imaginary numbers: To answer an imaginary numbers question like this, you have to know how to use your i's. When you're adding or subtracting, i's act a lot like variables:

RULES OF IMAGINARY NUMBERS

$ai + bi = (a + b)i$
$ai - bi = (a - b)i$
$(ai)(bi) = -ab$
$i^2 = -1$
$i^3 = -i$
$i^4 = 1$

$$i+i=2i$$
$$i-i=0$$
$$2i+3i=5i$$
$$2i-3i=-i$$

But when you're multiplying, or raising to a power, don't forget to take that extra step of changing i^2 to –1:

$$i\times i=i^2=-1$$
$$(2i)(3i)=6i^2=6(-1)=-6$$
$$(3i)^2=(3i)(3i)=9i^2=9(-1)=-9$$

COMPLEX NUMBERS

A number in the form $a + bi$, where a and b are real numbers, is called a complex number.

Notice that when you raise i to successive powers, a pattern develops:

$i^1=i$	$i^5=i$	$i^9=i$
$i^2=-1$	$i^6=-1$	$i^{10}=-1$
$i^3=-i$	$i^7=-i$	$i^{11}=-i$
$i^4=1$	$i^8=1$	$i^{12}=1$

In light of this pattern, consider the choices in Example 2. The exponents in the choices are descending multiples of 1, 2, 3, 4, and 5, respectively. Only in (D)—exponent-multiples of 4—do you add overall values of 1 + 1 + 1 + 1 to get 4. In every other choice, you add combinations of i, –1, $-i$, and 1 in varying orders. But no matter in what order you add the terms, the sum is zero.

The answer is (D).

To multiply complex numbers, use FOIL. (Review chapter 4 if you forgot what FOIL stands for. It's a way of remembering the order for multiplying binomials: first, outer, inner, last.) Just remember again to take that extra step of changing i^2 to –1:

$$(2-3i)(2-3i)$$
$$=(2)(2)+(2)(-3i)+(-3i)(2)+(-3i)(-3i)$$
$$=4-6i-6i+9i^2$$
$$=4-12i-9$$
$$=-5-12i$$

Logic

If you encounter a logic question on Math 2, it is likely that about all you'll have to know is the so-called contrapositive. That's what Example 3 is getting at.

> **CONTRAPOSITIVE**
>
> "If *p*, then *q*," is logically equivalent to "If not *q*, then not *p*."

Example 3

"If a zic is a zac, then a zic is not a zoc."

If the statement above is true, then which of the following statements must also be true?

(A) "If a zic is a zoc, then a zic is not a zac."
(B) "If a zic is not a zac, then a zic is a zoc."
(C) "If a zic is not a zoc, then a zic is a zac."
(D) "If a zic is not a zac, then a zoc is not a zac."
(E) "If a zoc is a zic, then a zac is a zic."

This example uses nonsense words to test your understanding of a basic rule of logic: Given a true statement, the statement's converse is not necessarily true, the statement's inverse is not necessarily true, but the statement's contrapositive must be true. That sounds like a mouthful; as you think about what follows, concern yourself not with fancy terminology such as *inverse* and *converse*—terms that won't appear on the test—but with the meanings of these terms.

The following statement is an *if-then* claim: If A, then B. The converse is what you get when you flip the *if* and the *then*: If B, then A. The inverse is what you get when you negate the *if* and the *then*: If not A, then not B. Finally, the contrapositive is what you get when you *both* flip *and* negate the A and the B: If not B, then not A. Again, given a true statement, the statement's contrapositive must follow.

Consider an example: If I am in Iowa, then I am in the United States. The converse of this statement is: If I am in the United States, then I am in Iowa. Is it *possibly* true that a person in the United States is in Iowa? Yes. Is it *necessarily* true? No. The inverse of the original statement is, If I am not in Iowa, then I am not in the United States. Is it *possibly* true that a person not in Iowa is not in the United States? Yes. Is it *necessarily* true? No. Now consider the contrapositive of the original statement: If I am not in the United States, then I am not in Iowa. Not only is it *possible* that a person not in the United States not be in Iowa, but it is *necessarily* true that someone not in the United States is not in Iowa.

Finally, focus on Example 3, which asks for what *must* be true. Form the contrapositive by both flipping and negating the *if* and the *then*. Flipping them produces, If a zic is not a zoc, then a zic is a zac. Now negate the terms: If a zic is a zoc, then a zic is not a zac.

The answer is (A).

Translating from English into Algebra

Solving a word problem means taking a situation that is described verbally and turning it into one that is described mathematically. It means translating from English into algebra, which is exactly what Example 4 asks you to do.

Example 4

A delicatessen charges for a certain type of salad at a rate of 2 dollars for the first half pound and 75 cents for each additional half pound. Which of the following expressions represents the total charge, in cents, for p pounds of the salad, where p is a positive integer?

(A) $25(6p + 8)$
(B) $25(6p + 5)$
(C) $25(3p + 5)$
(D) $1.25 + 1.5p$
(E) $2 + 0.75(p - 1)$

With intricate word problems such as this, invest time in reading the question stem closely, and perhaps even in reading it twice, to ensure that you understand not only the general situation and what's asked for, but also nuances. In this case, for example, notice that you're asked for the answer in cents, so begin the translation process by expressing the 2-dollar initial charge as 200 (cents). Note that this amount pays for the first half pound.

Now ask yourself: What is the weight of the entire purchase? It's p pounds. But consider that you're better off thinking of the same weight as $2p$ half pounds, because half pounds are the increments in which the salad is purchased. You then start to see how the money in question is spent: Adding 200 cents for the first half pound plus 75 cents for each additional half pound should produce the answer.

How many additional half pounds remain to be paid for once 200 cents have been paid for the first half pound? If the purchase weighs a total of $2p$ half pounds, and if one of the half pounds has already been paid for, then $2p - 1$ half pounds remain to be paid for at 75 cents each:

$$200 + 75(2p - 1)$$
$$200 + 150p - 75$$
$$125 + 150p$$
$$25(5 + 6p)$$

If you work better with thinking that's less abstract and more "on the ground," you could have Picked Numbers here. The presence of a variable in the choices was a tip-off to do so.

The answer is (B).

Percent Increase and Decrease

Some word problems present classic situations for which there are standard approaches, or even formulas. Example 5, for instance, requires that you know how to find a percent increase.

PERCENT INCREASE AND DECREASE FORMULAS

Percent increase = $\frac{\text{Amount of increase}}{\text{Original Amount}} \times 100\%$

Percent decrease = $\frac{\text{Amount of decrease}}{\text{Original Amount}} \times 100\%$

Example 5

During a weight-lifting routine, an athlete calculates the total amount of weight moved while performing a certain exercise by multiplying the amount of weight per lift by the number of lifts. On Wednesday, the weight per lift was 20 percent greater than the weight per lift on Monday, while the number of lifts was 25 percent fewer than the number of lifts performed on Monday. On Friday, the weight per lift was 25 percent greater than the weight per lift on Wednesday, while the number of lifts was 40 percent fewer than the number of lifts performed on Wednesday. The total weight moved by the athlete while performing the exercise on Monday was approximately what percent greater than the total weight she moved while performing the exercise on Friday?

(A) 15% (B) 33% (C) 48% (D) 50% (E) 100%

Wow—that's a lot of words. But the question becomes easily manageable if you structure and chart its data in an orderly way. Use a grid like this:

	Monday	Wednesday	Friday
weight per lift			
number of lifts			
total weight moved			

Believe it or not, the establishment of an organizing structure such as this is probably the single most important step in the solution. From here on in, all you need do is put the right numbers in the right places, and calculate correctly.

On Monday, the athlete lifted some amount of weight—call it w—some number of times—call it t. The total weight moved, then, is wt:

	Monday	Wednesday	Friday
weight per lift	w		
number of lifts	t		
total weight moved	wt		

On Wednesday, the weight per lift increased by 20 percent—that is, $1.0w + 0.2w$—while the number of lifts decreased by 25 percent—that is, $1.0t - 0.25t$:

	Monday	Wednesday	Friday
weight per lift	w	$1.2w$	
number of lifts	t	$0.75t$	
total weight moved	wt		

The weight per lift on Friday was 25 percent more than the weight per lift on Wednesday, while the number of lifts on Friday was 40 percent fewer than the number of lifts on

Wednesday. In other words, $1.2w$ was increased by $(0.25)(1.2w)$, so $1.2w + 0.3w = 1.5w$, and $0.75t$ was decreased by $(0.4)(0.75t)$, so $0.75t - 0.3t = 0.45t$:

	Monday	Wednesday	Friday
weight per lift	w	$1.2w$	$1.5w$
number of lifts	t	$0.75t$	$0.45t$
total weight moved	wt		

Using your calculator, compute the total weight moved on Friday: $(1.5w)(0.45t) = 0.675wt$.

Finally, focus on the exact wording of what's asked: "The total weight... Monday...approximately what percent greater than...total weight on Friday?" The question asks for the increase from Friday's total to Monday's total. (Don't get confused by the fact that Monday comes before Friday; the logic of what's asked here makes Friday's amount the starting point and requires you to express Monday's amount as an increase over Friday's.)

Use the percent change formula:

$$\text{Percent increase} = \frac{\text{Monday minus Friday}}{\text{Friday}} = \frac{1.000wt - 0.675wt}{0.675wt} = \frac{0.325wt}{0.675wt} = \frac{0.325}{0.675}.$$

On your calculator, divide 0.325 by 0.675. You'll get approximately 0.481, which is 48.1%. this is closest to choice (C).

Note that because this is a percent question, it's a strong candidate for Picking Numbers. Had you tried to do so, making w and t each 100, for example, you would have moved through the question quickly and easily. Remember that, in the abstract, neither Picking Numbers nor algebra is better than the other. Whatever gets you to the answer faster is better.

The answer is (C).

AVERAGE, MEDIAN, AND MODE

Average (arithmetic mean)

$$= \frac{\text{Sum of the terms}}{\text{Number of terms}}$$

Median = middle value (or average of two middle values)

Mode = most frequent value

Averages

Another classic situation with a standard approach is an averages question like Example 6.

Example 6

Last year a manufacturer had a sales total for January that was 50 percent greater than the average (arithmetic mean) of the monthly sales totals for February through December. The sales total for January was what fraction of the sales total for the year?

(A) $\frac{1}{11}$ (B) $\frac{1}{8}$ (C) $\frac{3}{25}$ (D) $\frac{3}{22}$ (E) $\frac{3}{5}$

> **USE THE SUM**
>
> The key to many averages questions is to use this variant of the Average formula:
>
> Sum of the terms = (Average) × (Number of terms)

Call the average monthly sales for February through December $2x$. Then the monthly sales for January were $3x$. Total annual sales equal the sum of monthly sales for January and *each* of the 11 months from February through December, inclusive. In other words, total annual sales equal $3x + 2x(11) = 25x$. And of that $25x$, what portion is January? Again—$3x$. So January's total sales as a fraction of total annual sales is $\frac{3x}{25x} = \frac{3}{25}$.

The answer is (C).

Distance, Rate, and Time

Yet another classic situation with a standard approach is the distance-rate-and-time question. Example 7 is an interesting variation.

> **AVOID THE SPEED TRAP**
>
> When a problem asks for average rate or average speed, don't just average the given rates or speeds. It's not that simple. Use the Average Rate formula.

Example 7

Kayla drove from Bayside to Chatham at a constant speed of 21 miles per hour and then returned along the same route from Chatham to Bayside. If her average speed for the entire journey was 26.25 miles per hour, at what average speed, in miles per hour, did Kayla return from Chatham to Bayside?

(A) 28
(B) 31
(C) 31.5
(D) 35
(E) It cannot be determined from the information given.

Here's another great example of how structuring the data in a question is often the most important step in the solution; once you've imposed order on the situation described in a question stem, actually getting to the answer is often merely a matter of putting the right numbers in the right places, and calculating with care.

The question requires two formulas. First is the rate formula, which is applied to two different legs of a journey:

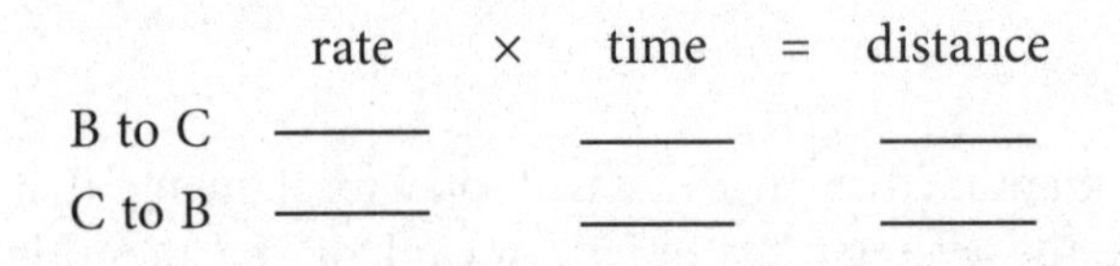

	rate	×	time	=	distance
B to C	____		____		____
C to B	____		____		____

> **AVERAGE RATE FORMULAS**

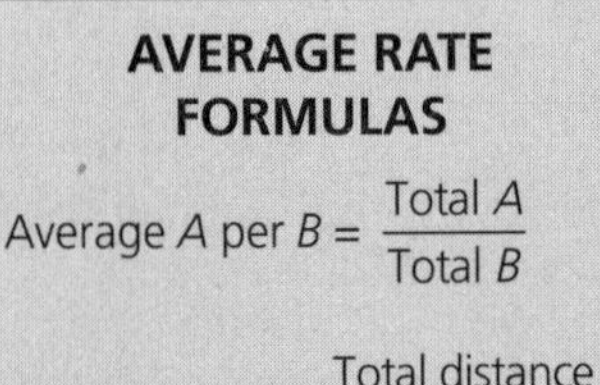

$$\text{Average } A \text{ per } B = \frac{\text{Total } A}{\text{Total } B}$$

$$\text{Average speed} = \frac{\text{Total distance}}{\text{Total time}}$$

Second is the average speed formula: Average speed $= \frac{\text{Total distance}}{\text{Total time}}$. First consider which slots you can fill in Rate × Time = Distance. Notice that you're given a rate for the trip from B to C, and you know that the distance going—which you can call d—is the same as the distance returning. Call the rate from C back to B r. Notice that now is *not* the time to do much with the 26.25. One of the most important lessons to learn about the Math 2 test is *when* to use which numbers; 26.25 will be important later:

	rate	×	time	=	distance
B to C	21		______		d
C to B	r		______		d

Given these values and variables, you can express the time going and the time returning:

	rate	×	time	=	distance
B to C	21		$\frac{d}{21}$		d
C to B	r		$\frac{d}{r}$		d

Now use the average speed formula. Notice the algebra that follows may be a little *mechanically* intricate, but it is not conceptually overwhelming: $\frac{\text{total distance}}{\text{total time}} =$

$$= \frac{2d}{\frac{d}{21}+\frac{d}{r}}$$

$$= \frac{2d}{\frac{dr}{21r}+\frac{21d}{21r}}$$

$$= \frac{2d}{\frac{dr+21d}{21r}}$$

$$= \frac{2d}{\frac{d(r+21)}{21r}}$$

$$= 2d \times \frac{21r}{d(r+21)}$$

$$= \frac{42dr}{d(r+21)}$$

$$= \frac{42r}{r+21}$$

Now is the time to bring in the 26.25, which you're given to be the average speed for the entire journey. Use it to solve for r, the answer:

$$26.25 = \frac{42r}{r+21}$$

$$26.25r + (26.25)(21) = 42r$$

$$26.25r + 551.25 = 42r$$

$$551.25 = 15.75r$$

$$\frac{551.25}{15.75} = r$$

$$35 = r$$

The answer is (D).

Permutations and Combinations

To be successful with combinations and permutations questions like Example 8, you have to remember the relevant formulas.

PERMUTATIONS AND COMBINATIONS FORMULAS

The number of permutations of n distinct objects is:

$n! = n(n-1)(n-2)\cdots(3)(2)(1)$

The number of permutations of n objects, a of which are indistinguishable, and b of which are indistinguishable, etc., is:

$$\frac{n!}{a!b!}$$

The number of permutations of n objects, taken r at a time, is:

$${}_nP_r = \frac{n!}{(n-r)!}$$

The number of combinations of n objects, taken r at a time, is:

$${}_nC_r = \frac{n!}{r!(n-r)!}$$

Example 8

8. Of the 12 members of a high school drama club, 8 are seniors. The club plans to establish an 8-member committee to interview potential club members. If exactly 6 members of the committee must be seniors, how many committees are possible?

(A) 21
(B) 34
(C) 168
(D) 336
(E) 495

That the question asks for "possible" groupings suggests that you should expect it to hinge on combinations. Of the 12 members, if 8 are seniors, 4 must be non-seniors. The question requires that 6 of the 8 seniors, and therefore 2 of the 4 non-seniors, be chosen. First find the number of possible combinations of seniors on the committee; then do the same for non-seniors. The answer is the product of the two results. (Note: It's the *product*, not the sum—which is the trap in (B).)

A group of 8 seniors can be combined into how many different subgroups of 6? The answer is 28:

$$\frac{n!}{k!(n-k)!} = \frac{8!}{6!2!} = \frac{8\times7\times6\times5\times4\times3\times2\times1}{6\times5\times4\times3\times2\times1\times2\times1} = \frac{8\times7}{2\times1} = 28$$

A group of 4 non-seniors can be combined into how many different subgroups of 2? The answer is 6:

$$\frac{n!}{k!(n-k)!} = \frac{4!}{2!2!} = \frac{4\times3\times2\times1}{2\times1\times2\times1} = 3 \times 2 = 6$$

Any one of the 28 subgroups of seniors can be combined with any one of the 6 subgroups of non-seniors, so the total number of combinations of the seniors and non-seniors is 28 × 6 = 168.

The answer is (C).

If you're good at memorizing formulas—if you can master the permutations and combinations formulas without too much trouble—go ahead and do it. But all these formulas probably won't get you more than one right answer on the Math 2 test, so they don't deserve a whole lot of time and effort.

Probability

To answer a probability question, you need to know not only how to use the general probability formula, but also how to deal with multiple probabilities and probabilities that events will not occur. You have a little bit of everything in Example 9.

> **PROBABILITY FORMULAS**
>
> Probability = $\frac{\text{\# of favorable outcomes}}{\text{Total \# of possible outcomes}}$
>
> If the probability that an event will occur is a, then the probability that the event will not occur is $1 - a$.
>
> If the probability that one event will occur is a and the independent probability that another event will occur is b, then the probability that both events will occur is ab.

Example 9

Two fair dice are tossed simultaneously. What is the probability that the product of the numbers that land showing is at least 25?

(A) $\frac{1}{36}$ (B) $\frac{1}{12}$ (C) $\frac{1}{9}$ (D) $\frac{1}{4}$ (E) $\frac{3}{2}$

It's unlikely that a probability question on Math 2 will hinge only on applying the basic probability formula. Instead, a question is much more likely to call on you to figure out the probability of two independent events, or to determine the likelihood that an event will not occur by subtracting the probability that it will occur from 1. In this example, you're asked to figure the probability of an event that could occur in more than one way. To do so, follow four simple steps.

First, establish the sequence of events. Here you have two tosses of dice:

Toss 1 Toss 2

Second, write out the different ways in which what you want to occur could occur. For example, to yield a product of at least 25, the dice could land showing a five (5) and a six (6). Note this and all the other ways of getting at least 25:

	Toss 1	Toss 2
One way:	5	5
Another way:	5	6
Another way:	6	5
Another way:	6	6

Third, replace each number with its simple independent probability. In other words, think of a single, simple toss of one die. What's the probability of getting 5? One in six—so replace every 5 with $\frac{1}{6}$. And on a single toss of a die, what's the probability of getting 6? Again, one in six—so replace every 6 with $\frac{1}{6}$:

	Toss 1	Toss 2
One way:	$\frac{1}{6}$	$\frac{1}{6}$
Another way:	$\frac{1}{6}$	$\frac{1}{6}$
Another way:	$\frac{1}{6}$	$\frac{1}{6}$
Another way:	$\frac{1}{6}$	$\frac{1}{6}$

Finally, because the probability of an event that could occur in more than one way is simply the sum of the probabilities of the various ways in which the event could occur, multiply across and add down:

	Toss 1	Toss 2	
One way:	$\frac{1}{6}$	$\frac{1}{6}$	$= \frac{1}{36}$
Another way:	$\frac{1}{6}$	$\frac{1}{6}$	$= \frac{1}{36}$
Another way:	$\frac{1}{6}$	$\frac{1}{6}$	$= \frac{1}{36}$
Another way:	$\frac{1}{6}$	$\frac{1}{6}$	$= \frac{1}{36}$
			$\frac{4}{36} = \frac{1}{9}$

The answer is (C).

Logarithms

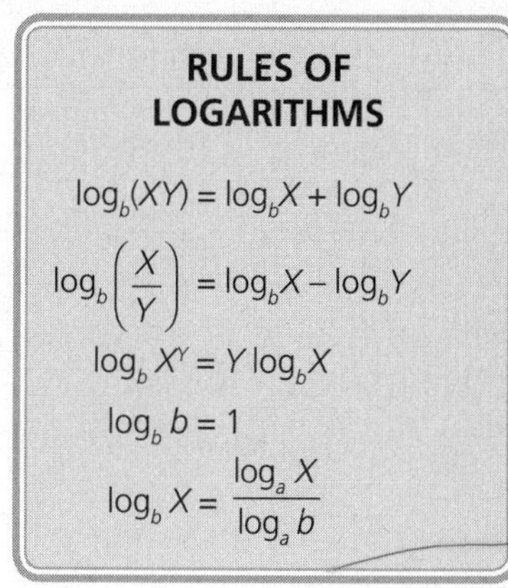

Maybe you'll see a logarithms question like Example 10 on your test.

Example 10

$$\log_3 \sqrt[9]{3} =$$

(A) $\frac{1}{9}$ (B) $\frac{3}{2}$ (C) $\frac{1}{2}$ (D) 1 (E) 3

Logarithms are not hard once you're familiar and experienced with them and know all the rules. If you're a long way from fully understanding logs, don't worry about them for this test. Logs will probably account for no more than one question. It's not worth spending a lot of time and effort figuring out logs when they appear so infrequently on the test.

If you understand what a log means, then Example 10 is not very difficult. The log to the base 3 of $\sqrt[9]{3}$ is the exponent that, attached to the base 3, will give you $\sqrt[9]{3}$:

$$\log_3 \sqrt[9]{3} = x$$

$$3^x = \sqrt[9]{3}$$

What you're doing is solving an equation with the variable in an exponent. To do that, re-express both sides of the equation in terms of the same base.

$$3^x = \sqrt[9]{3}$$

$$3^x = 3^{\frac{1}{9}}$$

$$x = \frac{1}{9}$$

The answer is (A).

Learn the rules of logarithms if you want to be sure to get a possible logarithms question right. But don't be afraid to ignore logs and to skip the logs question if you get one.

Limits

If you see a limits question like Example 11, forget about abstract theorems and calculus. There's a standard approach to a question like Example 11.

LIMIT OF AN ALGEBRAIC FRACTION

To find a limit of an algebraic fraction, first factor the numerator and denominator and cancel any factors they have in common. Then plug in and evaluate.

Example 11

$$\lim_{x \to 3} \frac{x^3 + x^2 - 12x}{x^2 - 9} =$$

(A) -7

(B) $-\frac{7}{2}$

(C) 0

(D) $\frac{7}{2}$

(E) The limit does not exist.

If you just try to plug $x = 3$ into the expression and evaluate it, you'll end up with zero over zero, which is undefined. The trick is first to factor the numerator and denominator and then to cancel factors they have in common:

$$\frac{x^3 + x^2 - 12x}{x^2 - 9} = \frac{x(x^2 + x - 12)}{(x+3)(x-3)}$$
$$= \frac{x(x+4)(x-3)}{(x-3)(x+3)}$$
$$= \frac{x(x+4)}{x+3}$$

Now plug in $x = 3$:

$$\frac{x(x+4)}{x+3} = \frac{(3)(7)}{6} = \frac{21}{6} = \frac{7}{2}$$

The answer is (D).

SEQUENCES FORMULAS

Arithmetic Sequences:

If d is the difference from one term to the next, a_1 is the first term, a_n is the nth term, and S_n is the sum of the first n terms:

$$a_n = a_1 + (n-1)d$$
$$S_n = n\left(\frac{a_1 + a_n}{2}\right)$$
$$= \frac{n}{2}[2a_1 + (n-1)d]$$

Geometric Sequences:

If r is the ratio between consecutive terms, a_1 is the first term, a_n is the nth term, and S_n is the sum of the first n terms:

$$a_n = a_1 r^{n-1}$$
$$S_n = \frac{a_1 - a_1 r^n}{1 - r}$$

To find the sum S_∞ of an infinite series when $-1 < r < 1$:

$$S_\infty = \frac{a_1}{1 - r}$$

Sequences

Our final commonly seen miscellaneous topic is typified by Example 12. It's a topic that entails a lot of formulas.

Example 12

If the first term in a geometric sequence is 3, and if the third term is 48, what is the 11th term?

(A) 228

(B) 528

(C) 110,592

(D) 3,145,728

(E) 12,582,912

This one's a snap if you know the formula. To find the *n*th term of a geometric sequence, you need to know the first term (here given as 3), and you need to know the ratio *r* between consecutive terms. You're not given consecutive terms but rather the first and the third. The second term would be the first term times *r*, and the third term would be the second term times *r*. Therefore,

$$a_1 \times r \times r = a_3$$
$$a_1 r^2 = a_3$$
$$3r^2 = 48$$
$$r^2 = 16$$
$$r = \pm 4$$

Notice that the given information allows for two possible values for *r*, +4 and −4. The sequence might be this:

$$3, 12, 48, 192, \ldots$$

Or it might be this:

$$3, -12, 48, -192, \ldots$$

The only difference in the latter is that the second, fourth, and all other even-numbered terms are negative. You're looking for the eleventh term, which is an odd-numbered term, so it will be the same number in either case. For simplicity's sake, use $r = 4$ in the formula:

$$a_n = a_1 r^{n-1}$$
$$= (3)(4^{10})$$
$$= 3{,}145{,}728$$

The answer is (D).

If you don't already at least half-know these formulas, they're probably not worth worrying about. But if you can remember them without too much trouble, one of them might come in handy on Test Day.

Finally, here are a few other topics that occasionally appear in the test.

Series

You've already seen arithmetic (adding/subtracting) and geometric (multiplying/dividing) sequences, where the question usually asks you for the *n*th term. Some questions on the test may ask you to work with a series, which results in a sum of the terms of a sequence. The formula for the sum of the first *n* terms of an arithmetic series is

$$S_n = \frac{n(a_1 + a_n)}{2}$$

where a_1 is the first term and a_n is the *n*th term. Here's an example:

If the first term in an arithmetic series is 3, and the 10th term is 38, what is the sum of the first 10 terms?

Just plug into the formula the information from the question: n = 10, a_1 = 3, and a_n = a_{10} = 38.

$$S_n = \frac{n(a_1 + a_n)}{2} = \frac{10(3+38)}{2} = \frac{10(41)}{2} = \frac{410}{2} = 205$$

The formula for the sum of the first n terms of a geometric series is

$$S_n = \frac{a_1(1-r^n)}{(1-r)}$$

where a term should be a_1 is the first term and r is the common ratio.

There is another way of expressing sums of terms. The Σ symbol is used to indicate summation, as in the following example:

$$\begin{aligned}\sum_{n=0}^{4}(2 + 3^n) &= (2 + 3^0) + (2 + 3^1) + (2 + 3^2) + (2 + 3^3) + (2 + 3^4)\\ &= (2 + 1) + (2 + 3) + (2 + 9) + (2 + 27) + (2 + 81)\\ &= 3 + 5 + 11 + 29 + 83\\ &= 131\end{aligned}$$

The Σ symbol asks you to evaluate the function $2 + 3^n$ for all of the values of n from 0 through 4 inclusive and then sum their values.

Vectors

Some properties, such as force, velocity, or acceleration, need more than just a number to describe them. We must know direction as well as size or magnitude in order to work with them. Those quantities that are described by both direction and magnitude are called vectors. Since vectors have information about both magnitude and direction, we draw them with arrows.

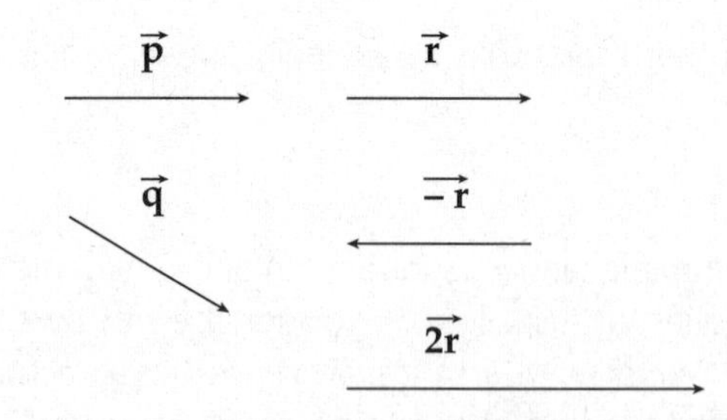

Two vectors are equal if and only if they are equal in both magnitude and direction. So $\vec{\mathbf{p}}$ is not equal to $\vec{\mathbf{q}}$, although they are the same length, because they are in different directions.

In plane geometry, two lines are perpendicular if their slopes are negative reciprocals of each other. In vector geometry, two vectors are perpendicular if and only if their dot product $\vec{r} \bullet \vec{s} = (x_1x_2) + (y_1y_2) + (z_1z_2)$ is zero.

If $\vec{\mathbf{r}}$ is a vector, then $-\vec{\mathbf{r}}$ is defined as having the same magnitude but the reverse direction to $\vec{\mathbf{r}}$. Subtracting $\vec{\mathbf{r}}$ is the same as adding $-\vec{\mathbf{r}}$.

Multiplying a vector by a number just has the effect of changing its scale. So, for example, $2\vec{\mathbf{r}}$ would be twice as long as $\vec{\mathbf{r}}$ and also in the same direction as $\vec{\mathbf{r}}$.

When two vectors are continuous (like a traveling vehicle that makes a turn), you can use the triangle law of addition and treat the two vectors as though they were sides of a triangle.

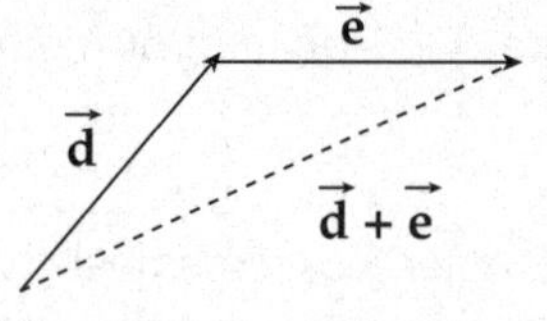

If two vectors are not continuous (like two forces pulling on an object), you must use the parallelogram law of addition.

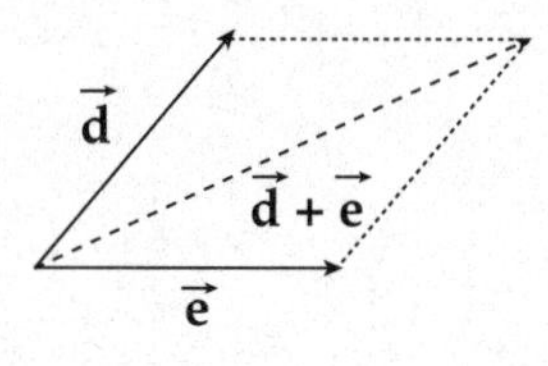

Here's an example:

Two forces pull on an object with forces of 6 N north and 8 N east. What is the resultant force and direction?

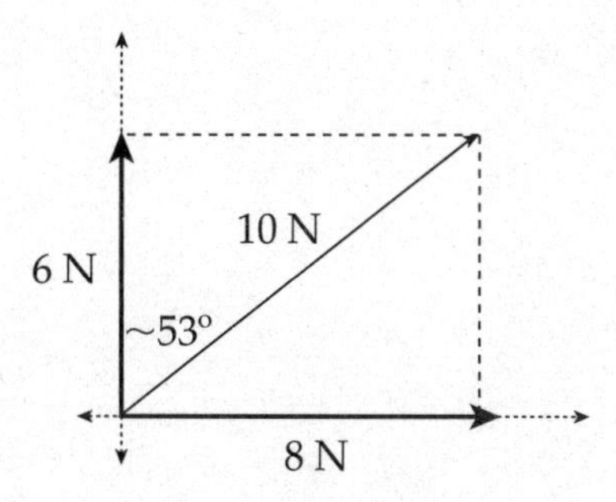

By treating the forces like sides of a parallelogram, you can find the resultant force.

Regression

Some questions on Test Day might involve scatterplots and regression. In order to interpret a scatterplot, you may need to make a guess about the correlation coefficient, which determines how closely the data points tend to come to the line. The closer that the absolute value of the correlation coefficient is to 1, the stronger the regression line is.

Scatterplots with a linear regression demonstrate a constant increase or decrease in values. In other words, the line that best fits the data will be a straight line, and its equation will be in the familiar form $y = mx + b$.

If the data suggest a gradual increase or decrease, the line that best describes the data will be a parabola, and the data can be summarized through quadratic regression. The least-squares curve will be in the form $y = ax^2 + bx + c$.

When there is a sharp increase or decrease in values, an exponential regression may be the best way to describe the data. The equation of the best-fit curve will be in the form $y = ab^x$.

There's a lot of material in this chapter, but remember that it's just the tip of the content pyramid. You can use the Miscellaneous Topics Follow-Up Test to see how much you've picked up from this chapter, but keep it in perspective. Whichever test you're taking, it's a higher priority to master the material in the chapters preceding this one.

THINGS TO REMEMBER:

Rules of Imaginary Numbers

$$ai + bi = (a + b)i$$
$$ai - bi = (a - b)i$$
$$(ai)(bi) = -ab$$
$$i^2 = -1$$
$$i^3 = -i$$
$$i^4 = 1$$

Complex numbers

A number in the form $a + bi$, where a and b are real numbers, is called a complex number.

Contrapositive

"If p, then q," is logically equivalent to "If not q, then not p."

Percent Increase and Decrease

$$\text{Percent increase} = \frac{\text{Amount of increase}}{\text{Original Amount}} \times 100\%$$

$$\text{Percent decrease} = \frac{\text{Amount of decrease}}{\text{Original Amount}} \times 100\%$$

Average Rate Formula

$$\text{Average } A \text{ per } B = \frac{\text{Total } A}{\text{Total } B}$$

$$\text{Average speed} = \frac{\text{Total Distance}}{\text{Total Time}}$$

Average, Median, and Mode

$$\text{Average (arithmetic mean)} = \frac{\text{Sum of the terms}}{\text{Number of terms}}$$

Median = middle value (or average of two middle values)

Mode = most frequent value

Sum of the terms = (Average) × (Number of terms)

Permutations and Combinations

The number of permutations of n distinct objects is

$$n! = n(n - 1)(n - 2) \cdots (3)(2)(1)$$

The number of permutations of n objects, a of which are indistinguishable, and b of which are indistinguishable, etc., is

$$\frac{n!}{a!b!\ldots}$$

The number of permutations of n objects, taken r at a time, is

$${}_nP_r = \frac{n!}{(n-r)!}$$

The number of combinations of n objects, taken r at a time, is

$${}_nC_r = \frac{n!}{r!(n-r)!}$$

Probability

$$\text{Probability} = \frac{\text{\# of favorable outcomes}}{\text{Total \# of possible outcomes}}$$

If the probability that an event will occur is a, then the probability that the event will not occur is $1 - a$.

If the probability that one event will occur is a and the independent probability that another event will occur is b, then the probability that both events will occur is ab.

Rules of Logarithms

$\log_b(XY) = \log_b X + \log_b Y$

$\log_b\left(\frac{X}{Y}\right) = \log_b X - \log_b Y$

$\log_b X^Y = Y\log_b X$

$\log_b b = 1$

$\log_b X = \frac{\log_a X}{\log_a b}$

Limit of an Algebraic Fraction

To find a limit of an algebraic fraction, first factor the numerator and denominator and cancel any factors they have in common. Then plug in and evaluate.

Sequences Formulas

Arithmetic Sequences:

$a_n = a_1 + (n - 1)d$

$S_n = n\left(\frac{a_1 + a_n}{2}\right)$

$= \frac{n}{2}\ [2a_1 + (n - 1)d]$

Geometric Sequences:

$a_n = a_1 r^{n-1}$

$S_n = \frac{a_1 - a_1 r^n}{1 - r}$

To find the sum S_∞ of an infinite series when $-1 < r < 1$:

$S_\infty = \frac{a_1}{1 - r}$

MISCELLANEOUS TOPICS FOLLOW-UP TEST

DO YOUR FIGURING HERE

12 Questions (15 Minutes)

Solve problems 1–12. Fill in the oval corresponding to the best answer choice in the grid to the right of each question. (Answers and explanations begin on page 206.)

1. The "panvoid" of a function is defined as the sum of the integers that do not fall within the domain of the function. All of the following functions have equal panvoids EXCEPT

(A) $f(x) = \frac{3x}{x^2 - 4}$

(B) $f(x) = \frac{3 - x}{x^3 - x}$

(C) $f(x) = \frac{3x}{3x^2 - 27}$

(D) $f(x) = \frac{x + 2}{x}$

(E) $f(x) = \frac{2 - x}{x^2 - x}$

Ⓐ Ⓑ Ⓒ Ⓓ Ⓔ

2. If $i^2 = -1$, $4 + i$, and $4 - i$ are roots of which of the following equations?

(A) $x^2 + 8x - 17 = 0$
(B) $x^2 - 8x + 17 = 0$
(C) $x^2 - 8x - 17 = 0$
(D) $x^2 + 10x - 8 = 0$
(E) $x^2 - 8x + 8 = 0$

Ⓐ Ⓑ Ⓒ Ⓓ Ⓔ

DO YOUR FIGURING HERE

3. If all Martians vacation on Venus, then

(A) a being that does not vacation on Venus is not a Martian.
(B) anyone who vacations on Venus is a Martian.
(C) no beings from Pluto vacation on Venus.
(D) all beings from Venus vacation on Mars.
(E) no beings from Pluto vacation on Saturn.

Ⓐ Ⓑ Ⓒ Ⓓ Ⓔ

4. Today Anselm is three times as old as his brother Bartholomew, and Bartholomew is 4 years younger than his sister Catherine. If Anselm, Bartholomew, and Catherine are all alive 5 years from today, which of the following must be true on that day?

I. Anselm is three times as old as Bartholomew.
II. Catherine is 4 years older than Bartholomew.
III. Anselm is older than Catherine.

(A) I only
(B) II only
(C) III only
(D) I and II
(E) II and III

Ⓐ Ⓑ Ⓒ Ⓓ Ⓔ

5. As part of a laboratory experiment, the number of times a light is flashed on a certain culture of bacteria is reduced 20 percent each day. If the light flashed 704 times on the fourth day of the experiment, then the number of times it flashed on the first day is closest to

(A) 880
(B) 1,217
(C) 1,267
(D) 1,375
(E) 1,460

Ⓐ Ⓑ Ⓒ Ⓓ Ⓔ

DO YOUR FIGURING HERE

6. The average (arithmetic mean) of six numbers is 3. If the average of the least and the greatest of these numbers is 5, then the other four numbers could be any of the following EXCEPT

(A) −4, 0, 5, 7
(B) 0, 0, 0, 8
(C) $\frac{1}{2}$, $\frac{3}{2}$, 1, 5
(D) 1, 1, 1, 1
(E) 1, 2, 2, 3

Ⓐ Ⓑ Ⓒ Ⓓ Ⓔ

7. On the second leg of a certain journey, a person traveled at an average speed 50 percent greater than the average speed at which the person traveled the first leg of the journey, and for an amount of time 50 percent greater than the amount of time spent on the first leg of the journey. The person's average speed for the entire journey was what percent greater than the person's average speed on the first leg of the journey?

(A) 22.5%
(B) 25%
(C) 30%
(D) 50%
(E) 100%

Ⓐ Ⓑ Ⓒ Ⓓ Ⓔ

8. Ten points lie on the circumference of a circle. Which of the following is the value that results when the number of triangles that can be created by connecting these points is subtracted from the number of heptagons (seven-sided polygons) that can be created by connected these points?

(A) 210 (B) 35 (C) 21 (D) 4 (E) 0

Ⓐ Ⓑ Ⓒ Ⓓ Ⓔ

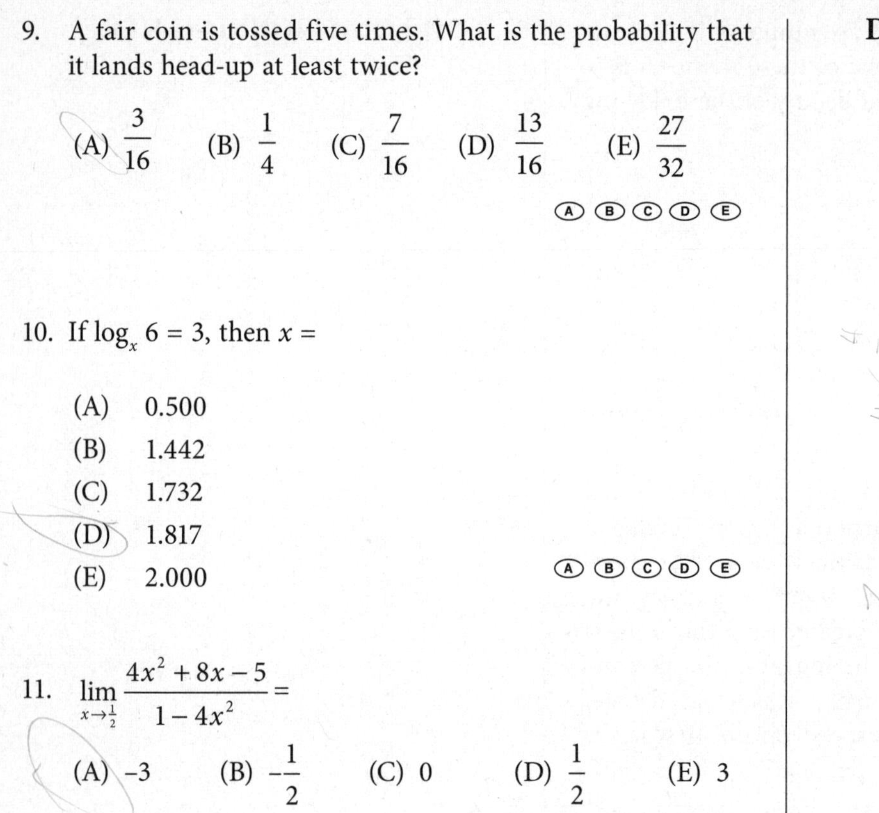

DO YOUR FIGURING HERE

9. A fair coin is tossed five times. What is the probability that it lands head-up at least twice?

(A) $\frac{3}{16}$ (B) $\frac{1}{4}$ (C) $\frac{7}{16}$ (D) $\frac{13}{16}$ (E) $\frac{27}{32}$

Ⓐ Ⓑ Ⓒ Ⓓ Ⓔ

10. If $\log_x 6 = 3$, then $x =$

(A) 0.500
(B) 1.442
(C) 1.732
(D) 1.817
(E) 2.000

Ⓐ Ⓑ Ⓒ Ⓓ Ⓔ

11. $\lim_{x \to \frac{1}{2}} \frac{4x^2 + 8x - 5}{1 - 4x^2} =$

(A) −3 (B) $-\frac{1}{2}$ (C) 0 (D) $\frac{1}{2}$ (E) 3

Ⓐ Ⓑ Ⓒ Ⓓ Ⓔ

12. What is the sum of the infinite geometric series

$2 + \left(-\frac{1}{2}\right) + \left(\frac{1}{8}\right) + \left(-\frac{1}{32}\right) + \ldots?$

(A) $1\frac{3}{8}$
(B) $1\frac{2}{5}$
(C) $1\frac{1}{2}$
(D) $1\frac{3}{5}$
(E) $1\frac{5}{8}$

Ⓐ Ⓑ Ⓒ Ⓓ Ⓔ

Turn the page
for answers and explanations
to the Follow-Up Test.

FOLLOW-UP TEST—ANSWERS AND EXPLANATIONS

1. E

This complex-looking question becomes easily manageable once you spend a few moments asking yourself some questions about the make-believe definition given in the question stem. When would an integer *not* fall within the domain of a function? When it would render the function undefined. Noticing that in each choice the function contains a fraction, you would then ask, what would make a fraction undefined? A value of zero in the denominator. So in each choice, get the forbidden values by setting the denominator equal to zero.

In (A), $x^2 - 4 \neq 0$, so $x \neq \pm 2$. The forbidden values, then, are ± 2, and the panvoid of the function is $2 + (-2) = 0$. In (B) $x^3 - x \neq 0$, so $x(x^2 - 1) \neq 0$, meaning that $x(x + 1)(x - 1) \neq 0$, and therefore that $x = 0$, $x = -1$, and $x = 1$ fall outside the domain of the function. The panvoid of the function, then, is $0 + (-1) + 1 = 0$. Notice that the function in (C) is not in simplest form: $\frac{3x}{3x^2 - 27} = \frac{3x}{3(x^2 - 9)} = \frac{x}{x^2 - 9} = \frac{x}{(x+3)(x-3)}$. If $(x + 3)(x - 3) \neq 0$, then $x \neq \pm 3$, and the panvoid of the function is $-3 + 3 = 0$. In (D), $x \neq 0$, so the panvoid of the function is 0. The answer must be (E): If $x^2 - x \neq 0$, then $x(x - 1) \neq 0$, so $x \neq 0$ and $x \neq 1$. The panvoid of (E) is therefore $0 + 1 = 1$.

2. B

Think about the general form of a quadratic equation: $ax^2 + bx + c = 0$. If the roots of the equation are the complex numbers $4 \pm i$, then in the quadratic formula $x = \frac{-b \pm \sqrt{b^2 - 4ac}}{2a}$, the real part of $4 \pm i$ equals $\frac{-b}{2a}$, and the imaginary part of $4 \pm i$ equals $\frac{\sqrt{b^2 - 4ac}}{2a}$. That is, $4 = \frac{-b}{2a}$ and $i = \frac{\sqrt{b^2 - 4ac}}{2a}$. You can express each of these equations a bit more simply, since the *a*-value in every choice is 1. So $4 = \frac{-b}{2}$ and $i = \frac{\sqrt{b^2 - 4c}}{2}$. If $4 = \frac{-b}{2}$, then $b = -8$. If $i = \frac{\sqrt{(-8)^2 - 4c}}{2}$, then $2i = \sqrt{(-8)^2 - 4c}$ and $4i^2 = 64 - 4c$.

Finally, $4c = 64 - 4i^2$ and $i^2 = -1$, so $4c = 64 + 4$, $4c = 68$, and $c = 17$. Collecting these values of *a*, *b*, and *c*, $ax^2 + bx + c = 0$ is $x^2 - 8x + 17 = 0$.

Another way to approach this problem is to generate the expression from the factors. If $4 + i$ and $4 - i$ are roots, then $[x - (4 + i)]$ and $[x - (4 - i)]$ are factors. Multiplying these together will result in the full expression:

$$[x - (4+i)][(x - (4-i)]$$
$$= x^2 - (4-i)x - (4+i)x + (4-i)(4+i)$$
$$= x^2 - 4x + ix - 4x - ix + 16 + 4i - 4i - i^2$$

Notice all the terms with just i cancel out, leaving $x^2 - 8x + 16 - i^2$. But since $i^2 = -1$, this becomes $x^2 - 8x + 16 - (-1) = x^2 - 8x + 17$.

3. A

The statement "all Martians vacation on Venus," is logically equivalent to the statement "If Martian, then vacation on Venus." The contrapositive of this statement is "If no vacation on Venus, then not a Martian"—which is just about identical to (A).

4. B

Translate carefully. If "Anselm is three times as old as Bartholomew," then $A = 3B$. And if "Bartholomew is 4 years younger than Catherine," then $B = C - 4$. Consider the statements one by one, Picking Numbers as you go. Must it be true, as I has it, that in five years, $A = 3B$? No. Say $A = 3$ and $B = 1$. Then in five years, $A = 8$ and $B = 6$, and $8 \neq 3(6)$. Eliminate (A) and (D). Must it be true, as II has it, that in five years, $C = B + 4$? It must, as any numbers you pick will demonstrate. Siblings who are four years apart now will always be four years apart. Eliminate (C). Must it be true, as III has it, that $A > C$? No, it *could* be true—if, for example $A = 90$, then $B = 30$ and $C = 34$. But it doesn't *have* to be true—as is shown, for example, when $A = 3$; then $B = 1$ and $C = 5$. In five years, $A = 8$ and $C = 10$. Eliminate (E).

5. D

The key insight here is to notice that when 20% of the flashes are removed, 80% remain. Call the number of flashes on the first day F. Then the number of flashes on the second day is $F - \frac{1}{5}F = \frac{4}{5}F$. On the third day, the number is 80% or $\frac{4}{5}$ of $\frac{4}{5}$ of F—which you can express more simply as $\left(\frac{4}{5}\right)^2 F$. Similarly, on the fourth day, the number of flashes is $\left(\frac{4}{5}\right)^3 F$. You're given that the number of flashes on the fourth day is 704, so $704 = \left(\frac{4}{5}\right)^3 F = \frac{4^3}{5^3}F = \frac{64}{125}F = \frac{64F}{125}$. If $704 = \frac{64F}{125}$, then $F = 704\left(\frac{125}{64}\right)$. Use your calculator: $F = 1{,}375$.

6. D

When you're given two parts of a three-part formula, determining the third part is almost always a smart move. Begin with Average $= \frac{\text{Sum of terms}}{\text{Number of terms}}$. Then $3 = \frac{\text{least} + a + b + c + d + \text{greatest}}{6}$, and least + a + b + c + d + greatest = 18. If the average of least and greatest is 5, then least + greatest = 10, so $10 + a + b + c + d = 18$ and $a + b + c + d = 8$. Only in (D) do the numbers not sum to 8.

7. C

An intended part of the challenge of this question is the quantity and complexity of the data it puts in front of you. Organize and relate these data by placing them in a structure such as this one:

	rate	×	time	=	distance
Leg 1	____		____		____
Leg 2	____		____		____

Call the rate of Leg 1 r and the time of Leg 1 t. Then,

	rate	×	time	=	distance
Leg 1	r		t		rt
Leg 2	$1.5r$		$1.5t$		$2.25rt$

Establish the average speed for the entire journey: average speed $= \frac{\text{Total distance}}{\text{Total time}} = \frac{3.25rt}{2.5t} = \frac{325r}{250} = 1.3r$. Finally, focus on the exact wording of what you're asked: the percent by which $1.3r$ is greater than r. You could use the percent change formula, but no need: $1.3r$ is what percent greater than $1.0r$? It is 0.3, or 30%.

Picking Numbers is also a great strategy for this question. Letting $r = 10$ and $t = 10$ enables you to slice through the intricacies of the question in seconds!

8. E

An understandable but fruitless first reaction to this question is to draw a circle and then some triangles within it; before long, the lines become an indecipherable mass. There must be another solution. Think a bit more abstractly about what's happening mathematically when you take ten points and make triangles: you're essentially chunking a group of 10 into subgroups of 3. Similarly, when you make heptagons in this circle, you chunk subgroups of 7 from a group of 10. So the number of triangles is

$$\frac{n!}{k!(n-k)!}$$

$$= \frac{10!}{3!7!}$$

$$= \frac{10\times9\times8\times7\times6\times5\times4\times3\times2\times1}{3\times2\times1\times7\times6\times5\times4\times3\times2\times1}$$

$$= \frac{10\times9\times8}{3\times2\times1}$$

$$= 10 \times 3 \times 4 = 120$$

And the number of heptagons is also $\frac{10!}{7!3!}$. No need to work out the value of this expression; it must have a value of 120, because it's equivalent to the expression representing the number of triangles. That the 3! and the 7! have been transposed makes no difference, since the order in which multiplication occurs is irrelevant. So the number of heptagons minus the number of triangles is $120 - 120 = 0$.

9. D

First think about this situation from the perspective, not of formulas you must know for Math 2, but of simple common sense. Imagine yourself actually tossing a coin five times. Does it seem relatively likely or unlikely that you'll get heads twice or more? It's relatively likely—and that tells you something about the smart way to handle this question. When simple reflection—plain old common sense—makes you think, "The event they're asking for has a pretty high probability of occurring," let your reaction be, "It'd probably be easier to figure out the probability of its *not* occurring and just subtract that probability from 1." In other words,

$$1 - \text{Prob}_{0 \text{ heads or } 1 \text{ heads}} = \text{Prob}_{2,\,3,\,4, \text{ or } 5 \text{ heads}}$$

Now focus on the probability of getting zero or one heads, following the four steps discussed in the body of this chapter. Recall that step one was to establish the sequence of events—in this case, a series of tosses:

Toss 1	Toss 2	Toss 3	Toss 4	Toss 5

Second, write out the different ways in which what you want to occur could occur. Suppose that H means heads and T means tails. Keep in mind that, at this point, what you want to occur is zero heads or one heads (H):

	Toss 1	Toss 2	Toss 3	Toss 4	Toss 5
One way:	T	T	T	T	T
Another way:	H	T	T	T	T
Another way:	T	H	T	T	T
Another way:	T	T	H	T	T
Another way:	T	T	T	H	T
Another way:	T	T	T	T	H

Third, replace each letter with the simple independent probability it represents by thinking about one toss of one coin. What's the probability of getting T? It's $\frac{1}{2}$—so replace every T with $\frac{1}{2}$. What's the probability of getting H? It's also $\frac{1}{2}$—so replace every H with $\frac{1}{2}$:

	Toss 1	Toss 2	Toss 3	Toss 4	Toss 5
One way:	$\frac{1}{2}$	$\frac{1}{2}$	$\frac{1}{2}$	$\frac{1}{2}$	$\frac{1}{2}$
Another way:	$\frac{1}{2}$	$\frac{1}{2}$	$\frac{1}{2}$	$\frac{1}{2}$	$\frac{1}{2}$
Another way:	$\frac{1}{2}$	$\frac{1}{2}$	$\frac{1}{2}$	$\frac{1}{2}$	$\frac{1}{2}$
Another way:	$\frac{1}{2}$	$\frac{1}{2}$	$\frac{1}{2}$	$\frac{1}{2}$	$\frac{1}{2}$
Another way:	$\frac{1}{2}$	$\frac{1}{2}$	$\frac{1}{2}$	$\frac{1}{2}$	$\frac{1}{2}$
Another way:	$\frac{1}{2}$	$\frac{1}{2}$	$\frac{1}{2}$	$\frac{1}{2}$	$\frac{1}{2}$

Finally, because the probability of an event that could occur in more than one way is simply the sum of the probabilities of the various ways in which the event could occur, multiply across and add down:

	Toss 1	Toss 2	Toss 3	Toss 4	Toss 5	
One way:	$\frac{1}{2}$	$\frac{1}{2}$	$\frac{1}{2}$	$\frac{1}{2}$	$\frac{1}{2}$	$= \frac{1}{32}$
Another way:	$\frac{1}{2}$	$\frac{1}{2}$	$\frac{1}{2}$	$\frac{1}{2}$	$\frac{1}{2}$	$= \frac{1}{32}$
Another way:	$\frac{1}{2}$	$\frac{1}{2}$	$\frac{1}{2}$	$\frac{1}{2}$	$\frac{1}{2}$	$= \frac{1}{32}$
Another way:	$\frac{1}{2}$	$\frac{1}{2}$	$\frac{1}{2}$	$\frac{1}{2}$	$\frac{1}{2}$	$= \frac{1}{32}$
Another way:	$\frac{1}{2}$	$\frac{1}{2}$	$\frac{1}{2}$	$\frac{1}{2}$	$\frac{1}{2}$	$= \frac{1}{32}$
Another way:	$\frac{1}{2}$	$\frac{1}{2}$	$\frac{1}{2}$	$\frac{1}{2}$	$\frac{1}{2}$	$= \frac{1}{32}$
					+	
					$\frac{6}{32}$	$= \frac{3}{16}$

Especially in a complex problem such as this one, keep track of where you are in the solution process—you're not done yet! The fraction $\frac{3}{16}$ represents the probability of no heads or one heads; you want the probability of getting two, three, four, or five heads. Recall that $1 - \text{Prob}_{0 \text{ heads or } 1 \text{ heads}} = \text{Prob}_{2,\,3,\,4, \text{ or } 5 \text{ heads}}$, so $1 - \frac{3}{16} = \frac{16}{16} - \frac{3}{16} = \frac{13}{16}$. Notice that this is quite a high probability—more than $\frac{12}{16}$, or 75%. What does that tell you? That your answer was so high a probability means that you were smart to solve using the indirect, "1 minus" approach rather than the direct approach, which would have entailed writing out all the different ways a fair coin tossed five times could have landed heads up two times, and then three times, and then four times, and then five times.

10. D

Re-express the equation in exponential form. The base is x, the exponent is 3, and the result is 6:

$$\log_x 6 = 3$$
$$x^3 = 6$$
$$x = \sqrt[3]{6} \approx 1.817$$

11. A

First factor the numerator and denominator and cancel common factors:

$$\frac{4x^2 + 8x - 5}{1 - 4x^2} = \frac{(2x-1)(2x+5)}{(1-2x)(1+2x)}$$
$$= \frac{-(1-2x)(2x+5)}{(1-2x)(1+2x)}$$
$$= \frac{-(2x+5)}{1+2x}$$
$$= \frac{-2x-5}{2x+1}$$

Then plug in $x = \frac{1}{2}$:

$$\frac{-2x-5}{2x+1} = \frac{-2\left(\frac{1}{2}\right)-5}{2\left(\frac{1}{2}\right)+1}$$
$$= \frac{-1-5}{1+1}$$
$$= \frac{-6}{2} = -3$$

12. D

Use the formula. The first term a_1 is 2, and the ratio between consecutive terms is

$$r = \frac{a_2}{a_1} = \frac{-\frac{1}{2}}{2} = -\frac{1}{4}$$

Plug $a_1 = 2$ and $r = -\frac{1}{4}$ into the formula:

$$\text{Sum} = \frac{a_1}{1-r}$$
$$= \frac{2}{1-\left(-\frac{1}{4}\right)} = \frac{2}{\frac{5}{4}}$$
$$= \frac{2}{1} \times \frac{4}{5} = \frac{8}{5} = 1\frac{3}{5}$$

Part Three

PRACTICE TESTS

HOW TO TAKE THE PRACTICE TESTS

Before taking a practice test, find a quiet room where you can work uninterrupted for one hour. Make sure you have several No. 2 pencils with erasers.

Use the answer grid provided to record your answers. Guidelines for scoring your test appear on the reverse side of the answer grid. Time yourself. Spend no more than one hour on the 50 questions. Once you start the practice test, don't stop until you've reached the one-hour time limit. You'll find an answer key and complete answer explanations following the test. Be sure to read the explanations for all questions, even those you answered correctly.

Good luck!

HOW TO CALCULATE YOUR SCORE

Step 1: Figure out your raw score. Use the answer key to count the number of questions you answered correctly and the number of questions you answered incorrectly. (Do not count any questions you left blank.) Multiply the number wrong by 0.25 and subtract the result from the number correct. Round the result to the nearest whole number. This is your raw score.

SAT Subject Test: Mathematics Level 2 — Practice Test 1

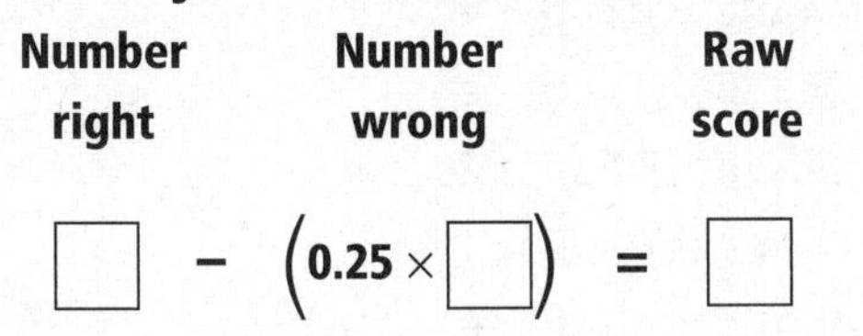

Step 2: Find your scaled score. In the Score Conversion Table below, find your raw score (rounded to the nearest whole number) in one of the columns to the left. The score directly to the right of that number will be your scaled score.

A note on your practice test scores: Don't take these scores too literally. Practice test conditions cannot precisely mirror real test conditions. Your actual SAT Subject Test: Mathematics Level 2 score will almost certainly vary from your practice test scores. However, your scores on the practice tests will give you a rough idea of your range on the actual exam.

Conversion Table

Raw	Scaled	Raw	Scaled	Raw	Scaled	Raw	Scaled
50	800	34	690	18	550	2	370
49	800	33	680	17	540	1	350
48	800	32	570	16	530	0	340
47	800	31	660	15	520	–1	330
46	800	30	650	14	520	–2	310
45	800	29	640	13	510	–3	300
44	790	28	630	12	500	–4	290
43	780	27	620	11	490	–5	280
42	770	26	610	10	480	–6	260
41	760	25	600	9	460	–7	250
40	750	24	590	8	450	–8	240
39	740	23	580	7	440	–9	220
38	730	22	580	6	420	–10	210
37	720	21	570	5	410	–11	200
36	710	20	560	4	400	–12	200
35	700	19	550	3	380		

Answer Grid
Practice Test 1

1.	Ⓐ Ⓑ Ⓒ Ⓓ Ⓔ	26.	Ⓐ Ⓑ Ⓒ Ⓓ Ⓔ
2.	Ⓐ Ⓑ Ⓒ Ⓓ Ⓔ	27.	Ⓐ Ⓑ Ⓒ Ⓓ Ⓔ
3.	Ⓐ Ⓑ Ⓒ Ⓓ Ⓔ	28.	Ⓐ Ⓑ Ⓒ Ⓓ Ⓔ
4.	Ⓐ Ⓑ Ⓒ Ⓓ Ⓔ	29.	Ⓐ Ⓑ Ⓒ Ⓓ Ⓔ
5.	Ⓐ Ⓑ Ⓒ Ⓓ Ⓔ	30.	Ⓐ Ⓑ Ⓒ Ⓓ Ⓔ
6.	Ⓐ Ⓑ Ⓒ Ⓓ Ⓔ	31.	Ⓐ Ⓑ Ⓒ Ⓓ Ⓔ
7.	Ⓐ Ⓑ Ⓒ Ⓓ Ⓔ	32.	Ⓐ Ⓑ Ⓒ Ⓓ Ⓔ
8.	Ⓐ Ⓑ Ⓒ Ⓓ Ⓔ	33.	Ⓐ Ⓑ Ⓒ Ⓓ Ⓔ
9.	Ⓐ Ⓑ Ⓒ Ⓓ Ⓔ	34.	Ⓐ Ⓑ Ⓒ Ⓓ Ⓔ
10.	Ⓐ Ⓑ Ⓒ Ⓓ Ⓔ	35.	Ⓐ Ⓑ Ⓒ Ⓓ Ⓔ
11.	Ⓐ Ⓑ Ⓒ Ⓓ Ⓔ	36.	Ⓐ Ⓑ Ⓒ Ⓓ Ⓔ
12.	Ⓐ Ⓑ Ⓒ Ⓓ Ⓔ	37.	Ⓐ Ⓑ Ⓒ Ⓓ Ⓔ
13.	Ⓐ Ⓑ Ⓒ Ⓓ Ⓔ	38.	Ⓐ Ⓑ Ⓒ Ⓓ Ⓔ
14.	Ⓐ Ⓑ Ⓒ Ⓓ Ⓔ	39.	Ⓐ Ⓑ Ⓒ Ⓓ Ⓔ
15.	Ⓐ Ⓑ Ⓒ Ⓓ Ⓔ	40.	Ⓐ Ⓑ Ⓒ Ⓓ Ⓔ
16.	Ⓐ Ⓑ Ⓒ Ⓓ Ⓔ	41.	Ⓐ Ⓑ Ⓒ Ⓓ Ⓔ
17.	Ⓐ Ⓑ Ⓒ Ⓓ Ⓔ	42.	Ⓐ Ⓑ Ⓒ Ⓓ Ⓔ
18.	Ⓐ Ⓑ Ⓒ Ⓓ Ⓔ	43.	Ⓐ Ⓑ Ⓒ Ⓓ Ⓔ
19.	Ⓐ Ⓑ Ⓒ Ⓓ Ⓔ	44.	Ⓐ Ⓑ Ⓒ Ⓓ Ⓔ
20.	Ⓐ Ⓑ Ⓒ Ⓓ Ⓔ	45.	Ⓐ Ⓑ Ⓒ Ⓓ Ⓔ
21.	Ⓐ Ⓑ Ⓒ Ⓓ Ⓔ	46.	Ⓐ Ⓑ Ⓒ Ⓓ Ⓔ
22.	Ⓐ Ⓑ Ⓒ Ⓓ Ⓔ	47.	Ⓐ Ⓑ Ⓒ Ⓓ Ⓔ
23.	Ⓐ Ⓑ Ⓒ Ⓓ Ⓔ	48.	Ⓐ Ⓑ Ⓒ Ⓓ Ⓔ
24.	Ⓐ Ⓑ Ⓒ Ⓓ Ⓔ	49.	Ⓐ Ⓑ Ⓒ Ⓓ Ⓔ
25.	Ⓐ Ⓑ Ⓒ Ⓓ Ⓔ	50.	Ⓐ Ⓑ Ⓒ Ⓓ Ⓔ

right

wrong

Use the answer key following the test to count up the number of questions you got right and the number you got wrong. (Remember not to count omitted questions as wrong.) "How to Calculate Your Score" on the back of this page will show you how to find your score.

Practice Test 1

50 Questions (1 hour)

Directions: For each question, choose the BEST answer from the choices given. If the precise answer is not among the choices, choose the one that best approximates the answer. Then fill in the corresponding oval on the answer sheet.

Notes:

(1) To answer some of these questions, you will need a calculator. You must use at least a scientific calculator, but programmable and graphing calculators are also allowed.

(2) Make sure your calculator is in the correct mode (degree or radian) for the question being asked.

(3) Figures in this test are drawn as accurately as possible UNLESS it is stated in a specific question that the figure is not drawn to scale. All figures are assumed to lie in a plane unless otherwise specified.

(4) The domain of any function f is assumed to be the set of all real numbers x for which $f(x)$ is a real number, unless otherwise indicated.

Reference Information: Use the following formulas as needed.

Right circular cone: If r = radius and h = height, then Volume = $\frac{1}{3}\pi r^2 h$; and if c = circumference of the base and ℓ = slant height, then Lateral Area = $\frac{1}{2}c\ell$.

Sphere: If r = radius, then Volume = $\frac{4}{3}\pi r^3$ and Surface Area = $4\pi r^2$.

Pyramid: If B = area of the base and h = height, then Volume = $\frac{1}{3}Bh$.

1. If $x^3 = 7^5$, what is the value of x?

(A) 3.2
(B) 11.6
(C) 25.6
(D) 243.0
(E) 26,041.6

2. If $a \Delta b \Delta c = \frac{ab}{c}$, which of the following equals 5?

(A) $4 \Delta 3 \Delta 2$
(B) $5 \Delta 2 \Delta 5$
(C) $6 \Delta 4 \Delta 2$
(D) $8 \Delta 4 \Delta 2$
(E) $10 \Delta 2 \Delta 4$

3. If $f(x) = e^x$ and $g(x) = \frac{x}{2}$, then $g(f(2)) =$

(A) 2.7
(B) 3.7
(C) 4.2
(D) 5.4
(E) 6.1

4. If $\frac{x+2y}{y} = 5$, what is the value of $\frac{y}{x}$?

(A) -3 (B) $-\frac{1}{3}$ (C) $\frac{1}{3}$ (D) 3 (E) 4

DO YOUR FIGURING HERE.

GO ON TO THE NEXT PAGE

5. In Figure 1, if $\cos\theta = 0.75$, $\tan\theta =$

(A) 0.60 (B) 0.67 (C) 0.75 (D) 0.88 (E) 1.33

6. Which of the following is an equation of the line that has a y-intercept of 6 and an x-intercept of -2?

(A) $3x - y = 6$
(B) $3x - y = -6$
(C) $3x + y = 6$
(D) $6x + y = 3$
(E) $6x - y = 3$

7. For all $y \neq 0$, $\frac{1}{y} + \frac{1}{2y} + \frac{1}{3y} =$

(A) $\frac{1}{2y}$

(B) $\frac{1}{6y}$

(C) $\frac{5}{6y}$

(D) $\frac{11}{6y}$

(E) $\frac{1}{6y^3}$

DO YOUR FIGURING HERE.

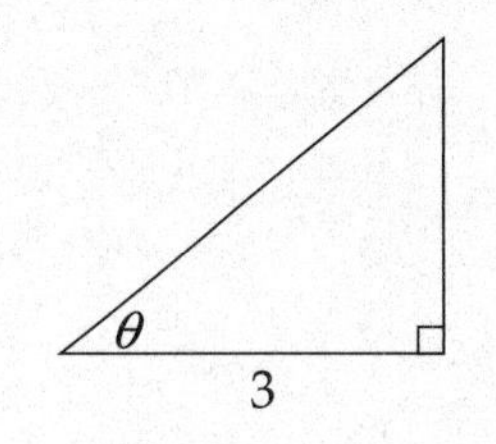

Figure 1

GO ON TO THE NEXT PAGE

8. In a class of 10 boys and 15 girls, the average score on a biology test is 90. If the average score for the girls is x, what is the average score for the boys in terms of x?

(A) $200 - \frac{2}{3}x$

(B) $225 - \frac{3}{2}x$

(C) $250 - 2x$

(D) $250 - 3x$

(E) $275 - 2x$

9. Which of the following graphs is symmetric about the origin?

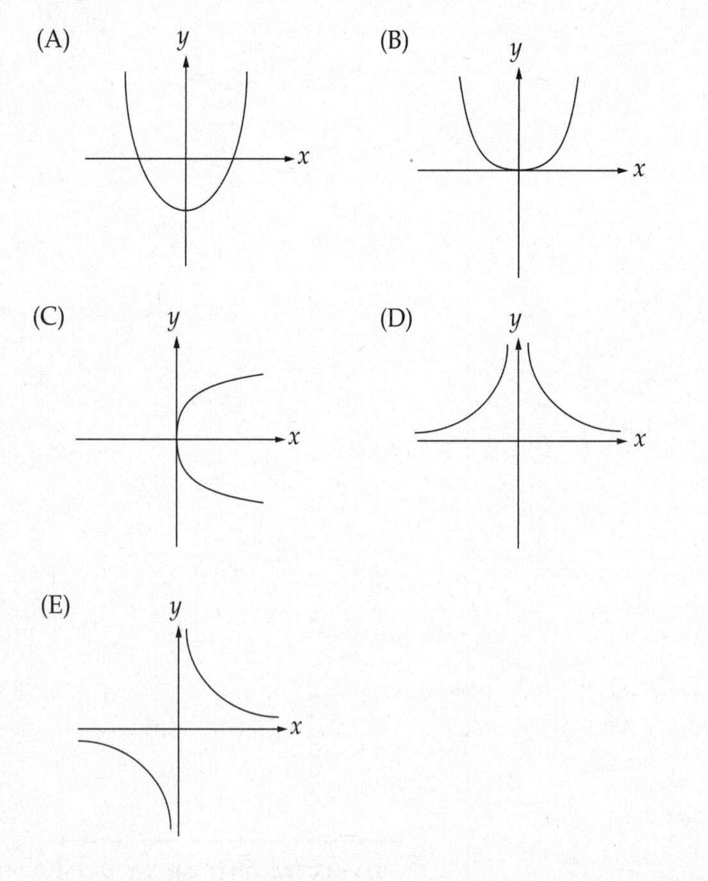

DO YOUR FIGURING HERE.

GO ON TO THE NEXT PAGE

10. George is going on vacation and wishes to take along 2 books to read. If he has 5 different books to choose from, how many different combinations of 2 books can he bring?

(A) 2 (B) 5 (C) 10 (D) 15 (E) 20

11. If $\sqrt{3-x} - x = 3, x =$

(A) -1 or -6
(B) 1 or -6
(C) -1 only
(D) -6 only
(E) There is no solution.

12. The lines with the equations $y = m_1x + 4$ and $y = m_2x + 3$ will intersect to the right of the y-axis if and only if

(A) $m_1 = m_2$
(B) $m_1 < m_2$
(C) $m_1 > m_2$
(D) $m_1 + m_2 = 0$
(E) $m_1 \neq m_2$

13. If the probability that it will rain sometime on Monday is $\frac{1}{3}$ and the independent probability that it will rain sometime on Tuesday is $\frac{1}{2}$, what is the probability that it will rain on both days?

(A) $\frac{1}{6}$ (B) $\frac{1}{5}$ (C) $\frac{1}{3}$ (D) $\frac{2}{5}$ (E) $\frac{5}{6}$

DO YOUR FIGURING HERE.

GO ON TO THE NEXT PAGE

14. If $\sin 2A = \frac{1}{2}$, then $\frac{1}{2\sin A\cos A} =$

(A) 1 (B) $\frac{3}{2}$ (C) 2 (D) 3 (E) 4

15. What values for x would make $\frac{1}{\sqrt{x+1}}$ undefined?

(A) −1 only
(B) 1 only
(C) All real numbers greater than −1
(D) All real numbers less than −1
(E) All real numbers less than or equal to −1

16. If $-2 \le x \le 2$, the maximum value of $f(x) = 1 - x^2$ is

(A) 2 (B) 1 (C) 0 (D) −1 (E) −2

17. If $f(x) = 1 - 4x$, and $f^{-1}(x)$ is the inverse of $f(x)$, then $f(-3)\,f^{-1}(-3) =$

(A) 1 (B) 3 (C) 4 (D) 10 (E) 13

18. Which of the following polynomials, when divided by $3x + 4$, equals $2x^2 + 5x - 3$ with remainder 3?

(A) $6x^3 + 23x^2 - 11x - 12$
(B) $6x^3 + 23x^2 - 11x - 9$
(C) $6x^3 + 23x^2 - 11x - 15$
(D) $6x^3 + 23x^2 + 11x - 12$
(E) $6x^3 + 23x^2 + 11x - 9$

DO YOUR FIGURING HERE.

GO ON TO THE NEXT PAGE

19. Let $\lfloor x \rfloor$ be defined to be the "floor" of x, where $\lfloor x \rfloor$ is the greatest integer that is less than or equal to x, and let $\lceil x \rceil$ be the "ceiling" of x, where $\lceil x \rceil$ is the least integer that is greater than or equal to x. If $f(x) = \lceil x \rceil + \lfloor x \rfloor$ and x is not an integer, then $f(x)$ is also equal to

(A) $2\lceil x \rceil - 2$
(B) $2\lceil x \rceil$
(C) $2\lfloor x \rfloor$
(D) $2\lfloor x \rfloor + 1$
(E) $2\lfloor x \rfloor + 2$

20. If $\log_2 x + \log_2 x = 7$, then $x =$

(A) 1.21 (B) 1.40 (C) 11.31 (D) 18.52 (E) 22.63

21. If $f(x) = \dfrac{\sqrt{x-1}}{x}$, what is the domain of $f(x)$?

(A) All real numbers except for 0
(B) All real numbers greater than or equal to 1
(C) All real numbers less than or equal to 1
(D) All real numbers greater than or equal to –1 but less than or equal to 1
(E) All real numbers less than or equal to –1

22. How many ways can 2 identical red chairs and 4 identical blue chairs be arranged in one row?

(A) 6 (B) 15 (C) 21 (D) 24 (E) 30

DO YOUR FIGURING HERE.

GO ON TO THE NEXT PAGE

23. If $a + b > 0$ and $c + d > 0$, which of the following must be true?

(A) $a + b + c > 0$
(B) $ac + bd > 0$
(C) $a^2 + b^2 > 0$
(D) $d(a + b) > 0$
(E) $a + b > c + d$

24. If $x > 0$, $a = x \cos \theta$, and $b = x \sin \theta$, then $\sqrt{a^2 + b^2} =$

(A) 1
(B) x
(C) $2x$
(D) $x (\cos \theta + \sin \theta)$
(E) $x \cos \theta \sin \theta$

25. If $0° < x < 90°$ and $5 \sin^2 x = 7 \sin x - 2$, what is the value of $\sin x$?

(A) 1.00 (B) 0.71 (C) 0.40 (D) 0.38 (E) 0.35

26. $3 - 2i$ and $3 + 2i$ are roots to which of the following quadratic equations?

(A) $x^2 + 6x + 13 = 0$
(B) $x^2 - 6x + 13 = 0$
(C) $x^2 - 6x - 13 = 0$
(D) $x^2 + 6x + 7 = 0$
(E) $x^2 + 6x - 4 = 0$

27. In Figure 2, if point P is located on the unit circle, then $x + y =$

(A) 0.37 (B) 0.50 (C) 0.78 (D) 0.87 (E) 1.37

DO YOUR FIGURING HERE.

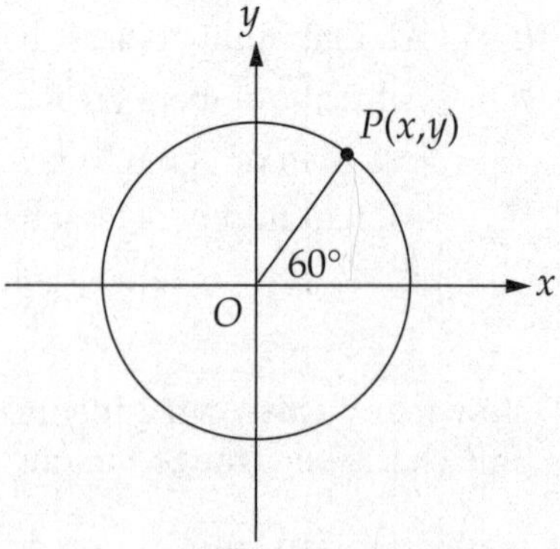

y
P(x,y)
60°
O
x

Figure 2

GO ON TO THE NEXT PAGE

28. If $0 \le t \le 1$, which of the following graphs is the graph of y versus x where x and y are related by the parametric equations $y = t^2$ and $x = \sqrt{t}$?

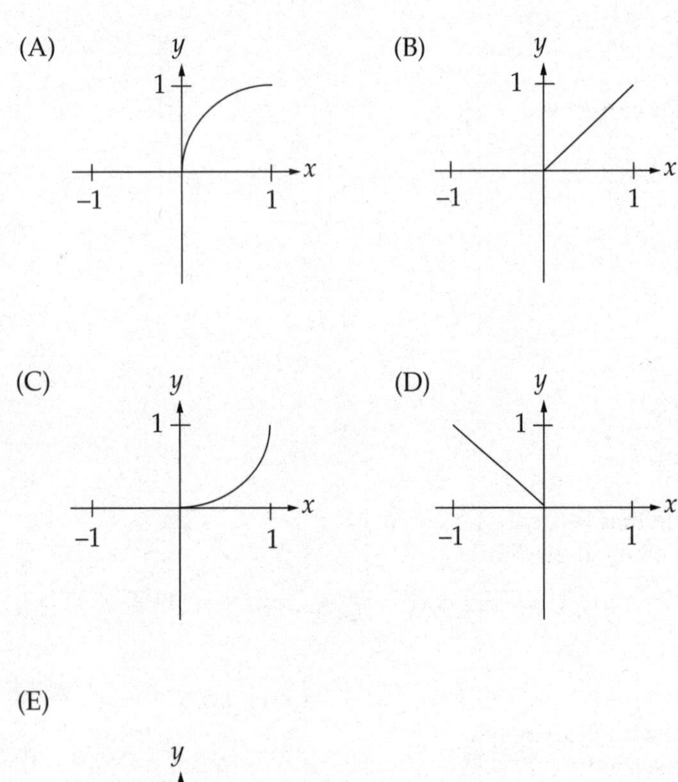

(E)

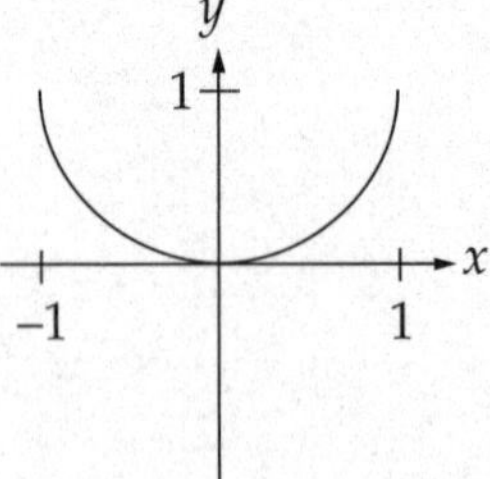

DO YOUR FIGURING HERE.

GO ON TO THE NEXT PAGE

29. In Figure 3, if isosceles right triangle ABC and square $ACDE$ share side AC, what is the degree measure of angle EBC ?

(A) 27 (B) 30 (C) 60 (D) 63 (E) 75

30. In Figure 4, which of the following denotes the correct vector arithmetic?

(A) $\overrightarrow{x} + \overrightarrow{y} = \overrightarrow{z}$

(B) $\overrightarrow{y} + \overrightarrow{z} = \overrightarrow{x}$

(C) $\overrightarrow{x} + \overrightarrow{z} = \overrightarrow{y}$

(D) $\overrightarrow{z} - \overrightarrow{x} = \overrightarrow{y}$

(E) $\overrightarrow{z} - \overrightarrow{y} = \overrightarrow{x}$

31. The horizontal distance, in feet, of a projectile that is fired with an initial velocity v, in feet per second, at an angle θ with the horizontal, is given by

$$H(v, \theta) = \frac{v^2 \sin(2\theta)}{32}$$

If a football is kicked at an angle of 50 degrees with the horizontal and an initial velocity of 30 feet per second, what is the horizontal distance, in feet, from the point where the football is kicked to the point where the football first hits the ground?

(A) 28 (B) 30 (C) 33 (D) 36 (E) 39

32. If a right circular cone has a lateral surface area of 6π and a slant height of 6, what is the radius of the base?

(A) 0.50

(B) 0.75

(C) 1.00

(D) 1.25

(E) 1.50

DO YOUR FIGURING HERE.

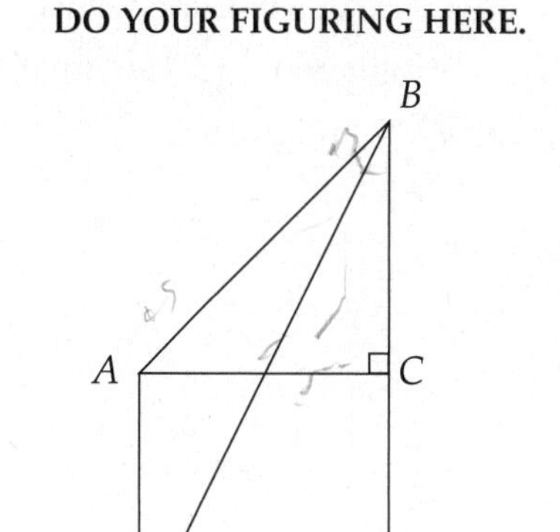

Figure 3

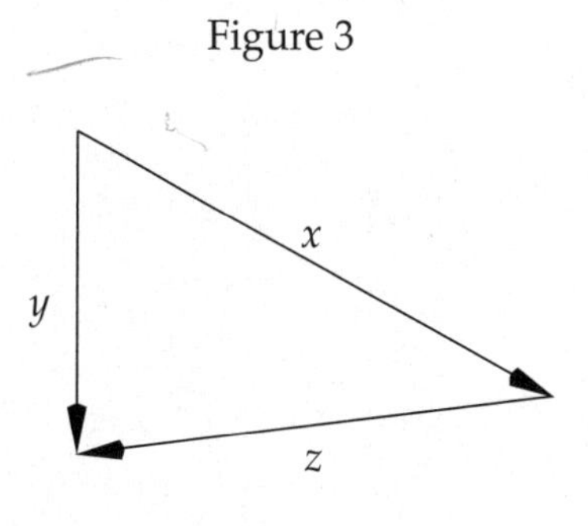

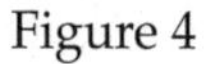

Figure 4

GO ON TO THE NEXT PAGE

33. If two fair dice are tossed, what is the probability that the two numbers that turn up are consecutive integers?

(A) 0.14 (B) 0.17 (C) 0.28 (D) 0.33 (E) 0.50

34. Which of the following is an equation of the ellipse centered at (–2,3) with a minor axis of 4 parallel to the x-axis and a major axis of 6 parallel to the y-axis?

(A) $\frac{(x-2)^2}{4}+\frac{(y-3)^2}{9}=1$

(B) $\frac{(x+2)^2}{4}+\frac{(y-3)^2}{9}=1$

(C) $\frac{(x-2)^2}{4}+\frac{(y+3)^2}{9}=1$

(D) $\frac{(x+2)^2}{4}+\frac{(y+3)^2}{9}=1$

(E) $\frac{(x-2)^2}{9}+\frac{(y+3)^2}{4}=1$

35. If $f(x) \geq 0$ and $g(x) \geq 0$ for all real x, which of the following statements must be true?

I. $f(x) + g(x) \geq 0$

II. $f(x) - g(x) \geq 0$

III. $f(x)g(x) \geq 0$

(A) I only
(B) II only
(C) III only
(D) I and II
(E) I and III

DO YOUR FIGURING HERE.

GO ON TO THE NEXT PAGE

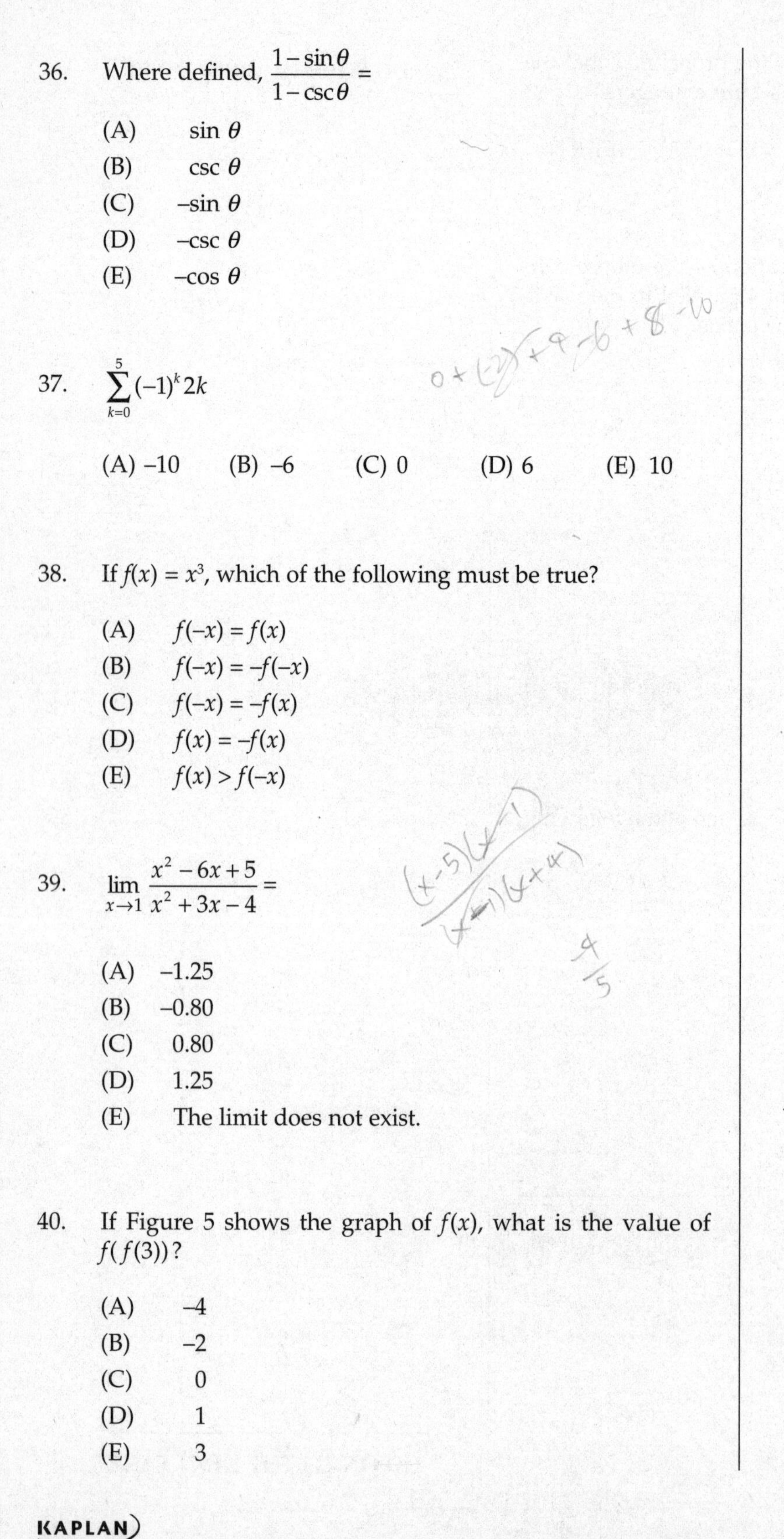

36. Where defined, $\dfrac{1-\sin\theta}{1-\csc\theta} =$

(A) $\sin\theta$
(B) $\csc\theta$
(C) $-\sin\theta$
(D) $-\csc\theta$
(E) $-\cos\theta$

37. $\sum_{k=0}^{5}(-1)^k 2k$

(A) –10 (B) –6 (C) 0 (D) 6 (E) 10

38. If $f(x) = x^3$, which of the following must be true?

(A) $f(-x) = f(x)$
(B) $f(-x) = -f(-x)$
(C) $f(-x) = -f(x)$
(D) $f(x) = -f(x)$
(E) $f(x) > f(-x)$

39. $\lim_{x \to 1} \dfrac{x^2 - 6x + 5}{x^2 + 3x - 4} =$

(A) –1.25
(B) –0.80
(C) 0.80
(D) 1.25
(E) The limit does not exist.

40. If Figure 5 shows the graph of $f(x)$, what is the value of $f(f(3))$?

(A) –4
(B) –2
(C) 0
(D) 1
(E) 3

DO YOUR FIGURING HERE.

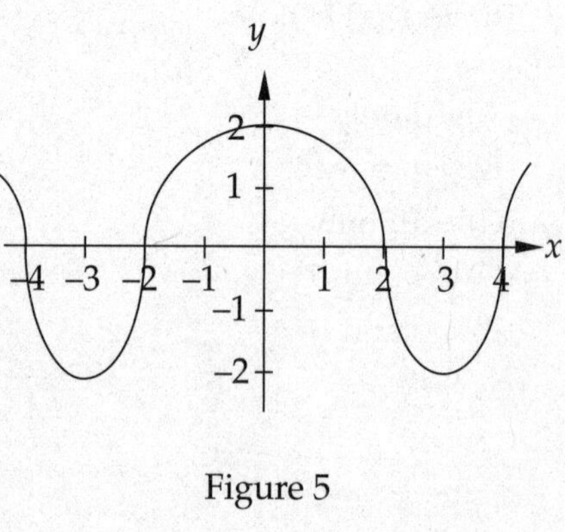

Figure 5

GO ON TO THE NEXT PAGE

41. If all the terms of a geometric series are positive, the first term of the series is 2, and the third term is 8, how many digits are there in the 40th term?

(A) 10
(B) 11
(C) 12
(D) 13
(E) 14

42. In Figure 6, what is the degree measure, to the nearest integer, of angle ABO ?

(A) 50
(B) 48
(C) 45
(D) 43
(E) 40

43. If $\log_2(x - 16) = \log_4(x - 4)$, which of the following could be the value of x ?

(A) 12 (B) 13 (C) 16 (D) 20 (E) 24

44. If a sphere of radius 3 is inscribed in a cube such that it is tangent to all six faces of the cube, the volume contained outside the sphere and inside the cube is

(A) 97 (B) 103 (C) 109 (D) 115 (E) 121

45. If $f(x) = \sin(\arctan x)$, $g(x) = \tan(\arcsin x)$, and $0 \leq x < \frac{\pi}{2}$, then $f\left(g\left(\frac{\pi}{10}\right)\right) =$

(A) 0.314
(B) 0.354
(C) 0.577
(D) 0.707
(E) 0.866

DO YOUR FIGURING HERE.

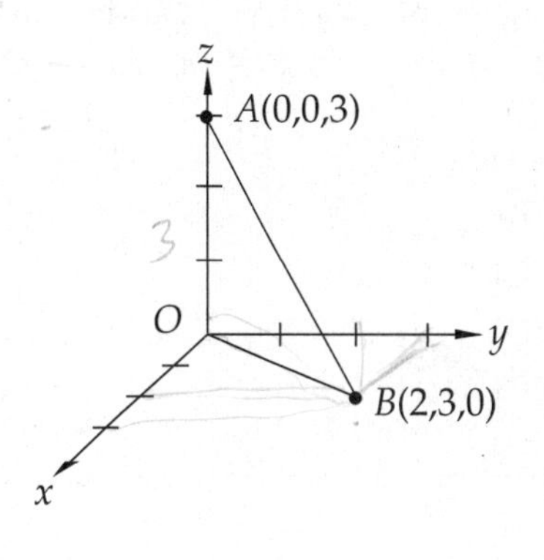

Figure 6

GO ON TO THE NEXT PAGE

DO YOUR FIGURING HERE.

46. If $f(x) = \frac{1}{(x+1)!}$, what is the smallest integer x such that $f(x) < 0.000005$?

(A) 7 (B) 8 (C) 9 (D) 10 (E) 11

47. In Figure 7, point O has coordinates (0,0), point P lies on the graph of $y = 6 - x^2$, and point B has coordinates $(2\sqrt{3},0)$. If $OP = BP$, the area of triangle OPB is

(A) 1.7
(B) 3.0
(C) 3.5
(D) 4.7
(E) 5.2

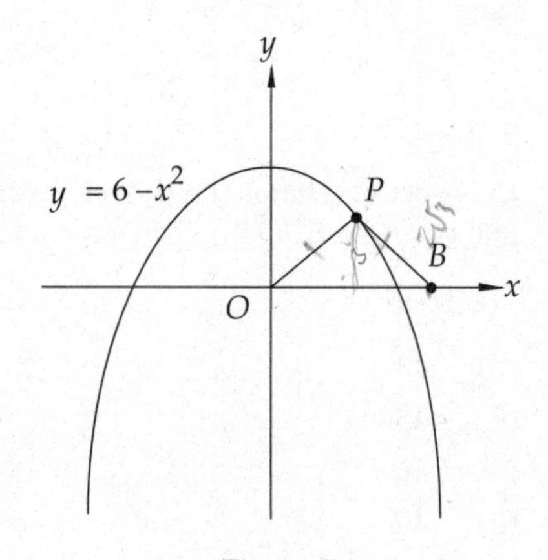

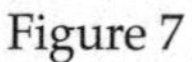
Figure 7

48. If cos $2x$ = sin x, and x is in radians, which of the following is a possible value of x ?

(A) 0.39 (B) 0.52 (C) 1.05 (D) 1.60 (E) 2.09

49. In Figure 8, if a wooden right circular cylinder with radius 2 meters and height 6 meters has a cylindrical hole of diameter 2 meters drilled through the center as shown, what is the entire surface area (including the top and bottom faces), in square meters, of the resulting figure?

(A) 38π
(B) 40π
(C) 42π
(D) 44π
(E) 46π

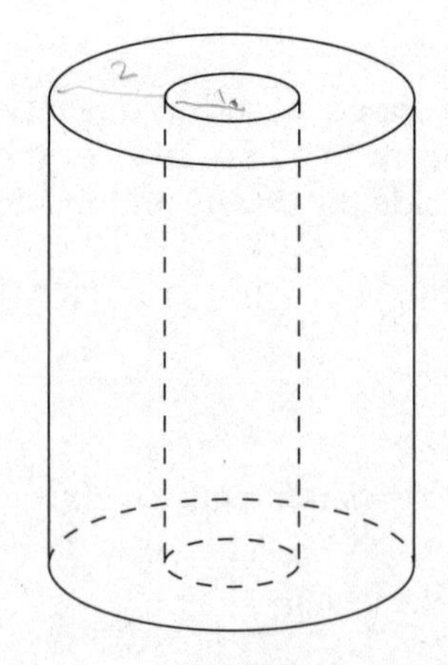

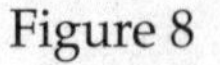
Figure 8

50. What is the greatest possible number of points of intersection between a parabola and a circle?

(A) 2 (B) 3 (C) 4 (D) 6 (E) 8

STOP!

If you finish before time is up, you may check your work.

Turn the page
for answers and explanations
to Practice Test 1.

Answer Key
Practice Test 1

1. C
2. E
3. B
4. C
5. D
6. B
7. D
8. B
9. E
10. C
11. C
12. B
13. A
14. C
15. E
16. B
17. E
18. E
19. D
20. C
21. B
22. B
23. C
24. B
25. C
26. B
27. E
28. C
29. A
30. C
31. A
32. C
33. C
34. B
35. E
36. C
37. B
38. C
39. B
40. C
41. D
42. E
43. D
44. B
45. A
46. B
47. E
48. B
49. C
50. C

ANSWERS AND EXPLANATIONS

1. C

Use your calculator. First find that $7^5 = 16{,}807$. Then find that the cube root of 16,807 is about 25.6.

2. E

Go down the answer choices and try multiplying the first two numbers and dividing by the third until you find the choice that yields 5.

The answer is (E) because $\frac{10 \times 2}{4} = 5$.

3. B

Perform the inside function first:

$$f(x) = e^x$$
$$f(2) = e^2 \approx 7.389$$

Then perform the outside function on the result:

$$g(x) = \frac{x}{2}$$
$$g(7.389) = \frac{7.389}{2} \approx 3.7$$

4. C

Manipulate the equation to get $\frac{y}{x}$ on one side:

$$\frac{x+2y}{y} = 5$$
$$x + 2y = 5y$$
$$x = 3y$$
$$1 = \frac{3y}{x}$$
$$\frac{y}{x} = \frac{1}{3}$$

5. D

First use your calculator to solve for θ:

$$\cos\theta = 0.75$$
$$\theta = \arccos(0.75) \approx 41.4°$$

Then find the tangent of the result:

$$\tan(41.4°) \approx 0.88$$

6. B

A y-intercept of 6 means that one point on the line is (0,6). Plug $x = 0$ and $y = 6$ into the answer choices, and you'll find that only (B) and (C) work. Now test those two choices with the x-intercept (–2,0). Of (B) and (C), only (B) works this time.

7. D

To add fractions, you need a common denominator. Here the LCD is $6y$:

$$\frac{1}{y} + \frac{1}{2y} + \frac{1}{3y} = \frac{6}{6y} + \frac{3}{6y} + \frac{2}{6y} = \frac{11}{6y}$$

8. B

Let y represent the average score for the boys. Fifteen girls average x, so their 15 scores add up to $15x$. Ten boys average y, so their 10 scores add up to $10y$. The 25 students average 90, so their scores add up to $25 \times 90 = 2{,}250$. Now you can set up an equation and solve for y. The girls' total and the boys' total add up to the grand total of 2,250:

$$15x + 10y = 2{,}250$$
$$10y = 2{,}250 - 15x$$
$$y = \frac{2{,}250 - 15x}{10} = 225 - \frac{3}{2}x$$

9. E

To be symmetric about the origin means that, for any point A on the graph, there is another point B on the graph such that the origin is the midpoint of AB. So you should be able to start at any point on the graph, draw a straight line segment to the origin, continue straight the same distance beyond the origin, and be at another point on the graph. Thus, for example, you can see that (A) does not work:

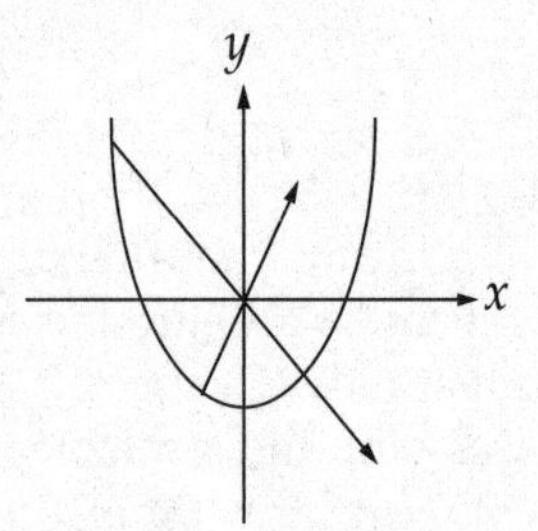

But (E) does work:

10. C

George has 5 choices for the first book and then 4 choices for the second. That's 20 permutations for taking 2 books out of 5. But this is a combinations question. Order doesn't matter. Whether he reads A or B first, it's the same combination of 2 books, so you have to divide the 20 permutations by 2. If you like formulas, here's how to do this one:

$$_nC_r = \frac{n!}{r!(n-r)!}$$

$$_5C_2 = \frac{5!}{2!3!} = 10$$

11. C

One of the steps in solving this equation is to square both sides. That step can result in "extraneous" solutions. Look:

$$\sqrt{3-x} - x = 3$$

$$\sqrt{3-x} = 3 + x$$

$$(\sqrt{3-x})^2 = (3+x)^2$$

$$3 - x = 9 + 6x + x^2$$

$$x^2 + 7x + 6 = 0$$

$$(x+1)(x+6) = 0$$

$$x = -1 \text{ or } -6$$

But if you plug those solutions back into the original equation, you'll find that $x = -6$ doesn't work. The only solution is $x = -1$.

Alternatively, you could Backsolve using the numbers provided in the answer choices.

12. B

You can think this one through conceptually: The first equation intercepts the y-axis at (0, 4), and the second equation intercepts the y-axis at (0, 3). That they intersect to the right of the y-axis means that the x-coordinate of the point of intersection is positive:

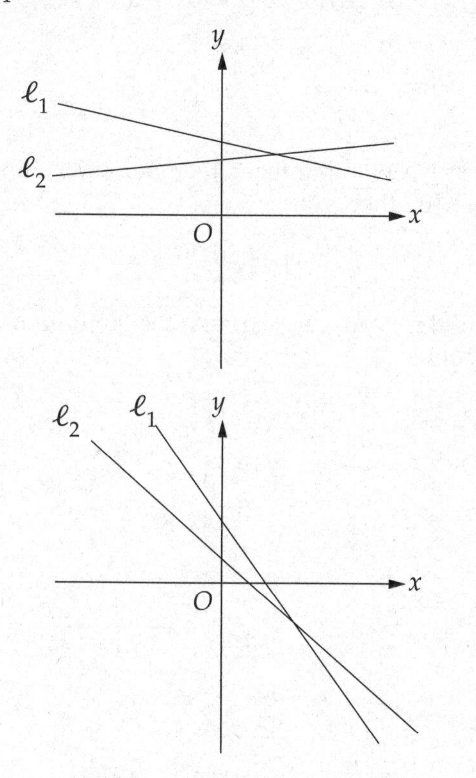

In both of the cases shown, ℓ_2 has the greater slope, so $m_2 > m_1$.

Alternatively, you could do this one algebraically. The point of intersection is the point at which $m_1x + 4 = m_2x + 3$:

$$m_1x+4=m_2x+3$$
$$m_2x-m_1x=4-3$$
$$x(m_2-m_1)=1$$
$$m_2-m_1=\frac{1}{x}$$

Since x is positive, so is $\frac{1}{x}$:

$$m_2-m_1=\frac{1}{x}$$
$$m_2-m_1>0$$
$$m_2>m_1$$

13. A

The probability of independent events occurring is the product of the separate probabilities: $\frac{1}{3}\times\frac{1}{2}=\frac{1}{6}$.

14. C

You can use your calculator. Arcsin $\frac{1}{2}$ = 30°, so $2A$ = 30° and A = 15°. Now use your calculator to find that $\frac{1}{2\sin 15^\circ \cos 15^\circ}=2$. But the solution's even quicker if you remember the double-angle sine formula:

$$\sin 2A = 2\sin A \cos A$$

Therefore,

$$\frac{1}{2\sin A\cos A}=\frac{1}{\sin 2A}=\frac{1}{0.5}=2$$

15. E

The expression will be undefined when the denominator is zero or when the expression under the radical is negative. The denominator is zero when

$$\sqrt{x+1}=0$$
$$x+1=0$$
$$x=-1$$

The expression under the radical is negative when

$$x+1<0$$
$$x<-1$$

So the expression is undefined for all $x \leq -1$.

16. B

You can think this one through conceptually. The expression $1 - x^2$ will be at its maximum when the x^2 that's subtracted from the 1 is as small as it can be. The square of a real number can't be any smaller than 0—and $x = 0$ is within the specified domain—so

$$\text{maximum} = 1 - 0^2 = 1$$

17. E

First find the inverse of $f(x)$:

$$f(x)=1-4x$$
$$y=1-4x$$
$$y+4x=1$$
$$4x=1-y$$
$$x=\frac{1-y}{4}$$
$$f^{-1}(x)=\frac{1-x}{4}$$

Now find $f(-3)$ and $f^{-1}(-3)$:

$$f(x)=1-4x$$
$$f(-3)=1-4(-3)=1-(-12)=13$$

$$f^{-1}(x)=\frac{1-x}{4}$$
$$f^{-1}(-3)=\frac{1-(-3)}{4}=\frac{4}{4}=1$$

And now, to get $f(-3)f^{-1}(-3)$, multiply:

$$f(-3)\,f^{-1}(-3)= 13(1) = 13$$

18. E

To find the original polynomial, multiply $3x + 4$ by $2x^2 + 5x - 3$ and then add the remainder 3 to the result:

$$(3x+4)(2x^2+5x-3)+3$$
$$=6x^3+15x^2-9x+8x^2+20x-12+3$$
$$=6x^3+23x^2+11x-9$$

19. D

When x is not an integer, the floor and ceiling are 1 apart. In other words,

$$\lceil x \rceil = \lfloor x \rfloor + 1$$

With this you can express the definition of the function:

$$f(x)=\lfloor x \rfloor + \lceil x \rceil$$
$$=\lfloor x \rfloor + (\lfloor x \rfloor + 1)$$
$$=2\lfloor x \rfloor + 1$$

20. C

$$\log_2 x + \log_2 x = 7$$
$$2\log_2 x = 7$$
$$\log_2 x = 3.5$$
$$x = 2^{3.5} \approx 11.31$$

21. B

To be in the domain of this function, x must not be anything that makes the expression under the radical negative or that makes the denominator zero. The expression under the radical is $x - 1$, and it must be nonnegative:

$$x-1\geq 0$$
$$x\geq 1$$

The denominator's simply x, which then cannot be zero. It's already been established, however, that x must be greater than or equal to 1, so that's the domain.

22. B

There is a formula that applies to this situation. The number of distinct permutations of n things, a of which are indistinguishable, b of which are indistinguishable, etc., is

$$\frac{n!}{a!b!\dots}$$

Here there are 6 chairs, 2 of which are indistinguishable and 4 of which are indistinguishable, so the number of permutations is

$$\frac{6!}{2!4!} = \frac{6\cdot5\cdot4\cdot3\cdot2\cdot1}{2\cdot1\cdot4\cdot3\cdot2\cdot1} = \frac{6\cdot5}{2} = 15$$

23. C

The best way to go about this one is to check out each answer choice, trying to think of a case where that choice is not true. The correct answer is the one that has no counterexample. That $a + b > 0$ and $c + d > 0$ would imply, for example, that the total sum $a + b + c + d$ would also be positive, but that's not the same as saying (A), $a + b + c > 0$. If $a = 3, b = -2, c = -4$, and $d = 5$, then (A) is not true. Nor is (B). (C) is true for this set of numbers and for any possible set of numbers because $a^2 + b^2$ will be greater than zero as long as a and b are not both zero.

24. B

$$\begin{aligned}\sqrt{a^2+b^2} &= \sqrt{(x\cos\theta)^2+(x\sin\theta)^2}\\ &= \sqrt{x^2(\cos^2\theta+\sin^2)}\\ &= \sqrt{x^2}\\ &= |x|\end{aligned}$$

It's given that $x > 0$, so $|x| = x$.

25. C

What you have here is a quadratic equation in which the unknown is sin x. To make things simpler, replace sin x with y and solve for y:

$$\begin{aligned}5\sin^2 x &= 7\sin x - 2\\ 5y^2 &= 7y-2\\ 5y^2-7y+2 &= 0\\ y &= \frac{7\pm\sqrt{49-40}}{10}\\ &= \frac{7\pm3}{10}\\ &= \frac{4}{10} \text{ or } \frac{10}{10}\\ &= 0.40 \text{ or } 1\end{aligned}$$

It's given that x is a positive acute angle, so $0 < \sin x < 1$, and only 0.40 fits.

26. B

If the solutions to $ax^2 + bx + c = 0$ are $3 \pm 2i$, then

$$\frac{-b}{2a} = 3 \text{ and } \frac{\sqrt{b^2-4ac}}{2a} = 2i$$

In all the answer choices, $a = 1$, so you can say more simply

$$\frac{-b}{2} = 3 \qquad b = -6$$

$$\begin{aligned}\frac{\sqrt{b^2-4c}}{2} &= 2i\\ \sqrt{b^2-4c} &= 4i\\ b^2-4c &= -16\\ 36-4c &= -16\\ -4c &= -52\\ c &= 13\end{aligned}$$

So the answer is the equation with $a = 1$, $b = -6$, and $c = 13$: $x^2 - 6x + 13 = 0$.

Another way to approach this problem is to generate the expression from the roots. If the solutions are $3\pm 2i$, then the factored expression is $[x-(3+2i)][x-(3-2i)]$.

Multiplying this out gives

$$\begin{aligned}&x^2 - x(3-2i)-(3+2i)x+(3+2i)(3-2i)\\ &= x^2-3x+2ix-3x-2ix+9-6i+6i-4i^2\end{aligned}$$

Combining like terms gives you

$$x^2-6x+9-4i^2 = x^2-6x+13$$

27. E

Draw in a right triangle:

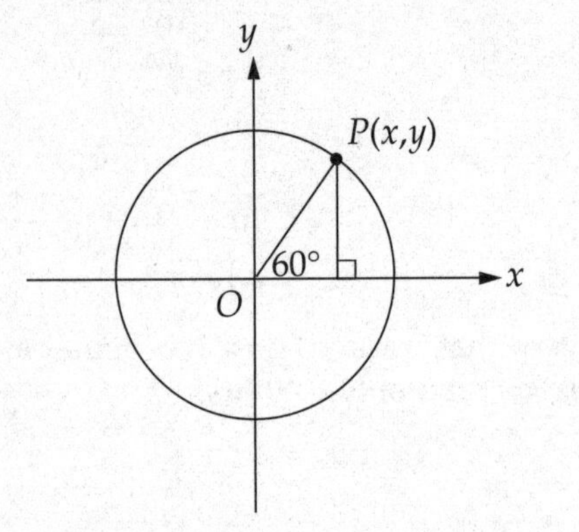

That's a 30-60-90 triangle. The hypotenuse is the radius, so it's 1. That means that the short leg is $\frac{1}{2}$ and the long leg is $\frac{\sqrt{3}}{2}$:

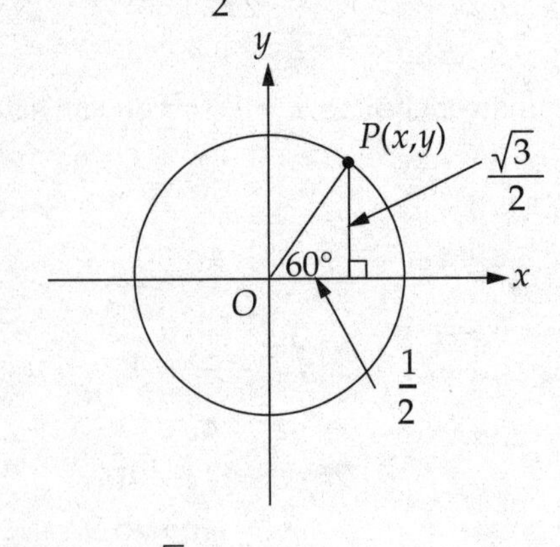

So, $x + y = \frac{1}{2} + \frac{\sqrt{3}}{2} \approx 1.37$.

28. C

Combine the equations so as to lose t and get y in terms of x:

$$x = \sqrt{t}$$
$$x^2 = t$$
$$y = t^2$$
$$y = \left(x^2\right)^2 = x^4$$

So you might be tempted by (E), which looks like the graph of $y = x^4$. But the stem says that $0 \le t \le 1$, so the only possible values of $x = \sqrt{t}$ are $0 \le x \le 1$, so the correct graph is (C).

29. A

Mark up the figure. Call the sides of the square and the legs of the triangle each 1. What you're looking for is the measure of the angle marked $x°$ in right triangle BDE:

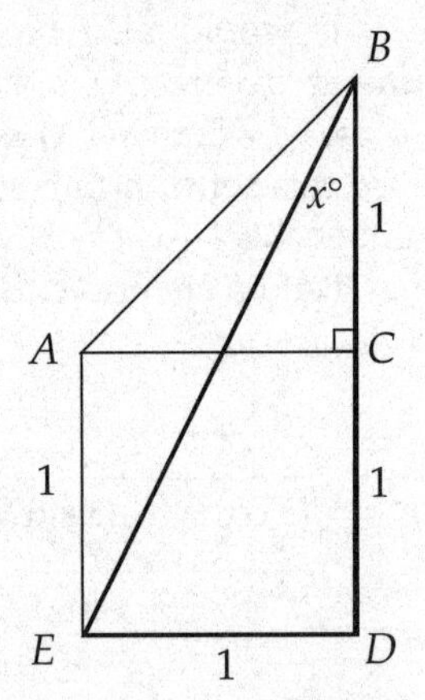

The leg opposite x is 1, and the leg adjacent to x is 2, so

$$\tan x = \frac{\text{opposite}}{\text{adjacent}} = \frac{1}{2}$$
$$x = \arctan\frac{1}{2} \approx 27$$

30. C

The figure shows the head of $\vec{x}$ touching the tail of $\vec{z}$, so those are the two vectors being added. The result is $\vec{y}$ because it then connects the tail of $\vec{x}$ to the head of $\vec{z}$. Therefore,

$$\vec{x} + \vec{z} = \vec{y}$$

31. A

Just plug $\theta = 50$ and $v = 30$ into the formula and crank out the answer:

$$\begin{aligned} H &= \frac{v^2 \sin(2\theta)}{32} \\ &= \frac{(30^2)\sin(2 \times 50^\circ)}{32} \\ &= \frac{900 \sin 100^\circ}{32} \\ &\approx 28 \end{aligned}$$

32. C

A formula for the lateral area of a cone is given in the directions:

$$\text{Lateral area} = \frac{1}{2} c\ell$$

Here the lateral area is 6π and $\ell = 6$, so you can solve for c:

$$\begin{aligned} 6\pi &= \frac{1}{2}c(6) \\ \pi &= \frac{1}{2}c \\ c &= 2\pi \end{aligned}$$

Now you can use the base circumference $c = 2\pi$ to find the base radius:

$$\begin{aligned} \text{Circumference} &= 2\pi r \\ 2\pi &= 2\pi r \\ r &= 1 \end{aligned}$$

33. C

The total number of possible outcomes is $6 \times 6 = 36$. Of those outcomes, the following are consecutive integers:

1 and 2

2 and 1

2 and 3

3 and 2

3 and 4

4 and 3

4 and 5

5 and 4

5 and 6

6 and 5

That's 10 favorable outcomes:

$$\begin{aligned} \text{Probability} &= \frac{\text{Favorable outcomes}}{\text{Total possible outcomes}} \\ &= \frac{10}{36} \approx 0.28 \end{aligned}$$

34. B

The equation of an ellipse centered at the point (p, q) and with axes $2a$ and $2b$ is

$$\frac{(x-p)^2}{a^2} + \frac{(y-q)^2}{b^2} = 1$$

Here $p = -2$, $q = 3$, $a = 2$, and $b = 3$, so the equation is

$$\frac{(x+2)^2}{2^2} + \frac{(y-3)^2}{3^2} = 1$$

which is the same as choice (B).

35. E

Don't let the functions symbolism confuse you. Just think of these as two quantities—$f(x)$ and $g(x)$—that are both nonnegative. Statement I says their sum is nonnegative. That's true—add any two nonnegatives and you'll get a nonnegative sum. Statement II says the difference $f(x) - g(x)$ is nonnegative. Well, that's true only if $f(x) \geq g(x)$. But there's no reason that $g(x)$ couldn't be greater than $f(x)$. Statement III says their product is nonnegative. That's true—multiply any two nonnegatives and you'll get a nonnegative product. Statements I and III are true.

36. C

Express cosecant as 1 over sine:

$$\frac{1-\sin\theta}{1-\csc\theta} = \frac{1-\sin\theta}{1-\frac{1}{\sin\theta}}$$

$$= \frac{1-\sin\theta}{\frac{\sin\theta}{\sin\theta}-\frac{1}{\sin\theta}}$$

$$= \frac{1-\sin\theta}{\frac{\sin\theta-1}{\sin\theta}}$$

$$= \frac{(1-\sin\theta)(\sin\theta)}{\sin\theta-1}$$

$$= \frac{-(\sin\theta-1)(\sin\theta)}{\sin\theta-1}$$

$$= -\sin\theta$$

37. B

Just plug in the six possible values for k, compute the results, and add them up:

$$\begin{aligned}
k=0 &\Rightarrow (-1)^0 2(0) = & 0\\
k=1 &\Rightarrow (-1)^1 2(1) = & -2\\
k=2 &\Rightarrow (-1)^2 2(2) = & 4\\
k=3 &\Rightarrow (-1)^3 2(3) = & -6\\
k=4 &\Rightarrow (-1)^4 2(4) = & 8\\
k=5 &\Rightarrow (-1)^5 2(5) = & \underline{-10}\\
& & 12-18=-6
\end{aligned}$$

38. C

The function is cubing. Think about each answer choice. (A): The cube of minus x equals the cube of x ? No. (B): The cube of minus x equals the opposite of the cube of minus x ? No. (C): The cube of minus x equals the opposite of the cube of x ? That sounds plausible. In fact, (C) is true. It doesn't matter whether you take the opposite of a number first and then cube it, or cube the number first and then take its opposite—you'll get the same result both ways.

39. B

The first step in finding a limit is generally to factor:

$$\frac{x^2-6x+5}{x^2+3x-4}=\frac{(x-5)(x-1)}{(x+4)(x-1)}$$

If x ever actually gets to 1, the expression becomes undefined—zero over zero. But if you cancel out the $(x-1)$ from the top and bottom:

$$\frac{(x-5)(x-1)}{(x+4)(x-1)}=\frac{x-5}{x+4}$$

you can plug in $x = 1$ and find the limit:

$$\lim_{x\to 1}\frac{x-5}{x+4}=\frac{1-5}{1+4}$$
$$=\frac{-4}{5}=-0.80$$

40. C

Don't try to figure out an equation to fit this weird graph. Just read the values right off the graph. First, find $f(3)$. Go to +3 on the x-axis and see what y is there. It's –2. Now find $f(-2)$. Go to –2 on the x-axis and see what y is there. It's 0:

$$f(3)=-2$$
$$f(f(3))=f(-2)=0$$

41. D

The first term is 2^1 and the third term is 2^3, so the 40th term is 2^{40}. Use your calculator and you'll get an answer in scientific notation something like this:

1.0995 E12

That is,

$$1.0995\times 10^{12}$$

That's 1 followed by 12 digits, for a total of 13 digits.

42. E

The length of OA is 3, and the length of OB is $\sqrt{2^2+3^2}=\sqrt{13}$. OA over OB is the tangent of the angle you're looking for:

$$\tan x=\frac{OA}{OB}=\frac{3}{\sqrt{13}}$$
$$x=\arctan\left(\frac{3}{\sqrt{13}}\right)\approx 40^\circ$$

43. D

Put everything in terms of $\log_2$:

$$\log_2(x-16) = \log_4(x-4)$$
$$\log_2(x-16) = \frac{\log_2(x-4)}{\log_2 4}$$
$$\log_2(x-16) = \frac{\log_2(x-4)}{2}$$
$$2\log_2(x-16) = \log_2(x-4)$$
$$\log_2(x-16)^2 = \log_2(x-4)$$
$$(x-16)^2 = x-4$$
$$x^2 - 32x + 256 = x - 4$$
$$x^2 - 33x + 260 = 0$$
$$(x-13)(x-20) = 0$$
$$x = 13 \text{ or } 20$$

Of those two apparent solutions, one is impossible. The log of a negative number is undefined, so $x = 13$ is an extraneous solution: You can't take the log of (13 – 16). The only solution is 20.

44. B

The cube is $6 \times 6 \times 6$, so its volume is 216. The sphere has radius 3, so

$$\text{Volume of sphere} = \frac{4}{3}\pi r^3$$
$$= \frac{4}{3}\pi(3^3)$$
$$= 36\pi$$

The difference is $216 - 36\pi \approx 103$.

45. A

You could use your calculator and do this one step by step. Set your calculator to radian mode. First perform the inside function:

$$g\left(\frac{\pi}{10}\right) = \tan\left(\arcsin\left(\frac{\pi}{10}\right)\right)$$
$$\approx \tan(0.3196)$$
$$\approx 0.3309$$

Then perform the outside function on the result:

$$f(0.3309) = \sin(\arctan(0.3309))$$
$$\approx \sin(0.3196)$$
$$\approx 0.314$$

Far quicker and simpler would be to realize that if you take the sin of the arctan of the tan of the arcsin, you'll end up back where you started.

The answer to this question is just the decimal approximation of the fraction $\frac{\pi}{10}$.

46. B

First convert 0.000005 into a fraction:

$$0.000005 = \frac{5}{1{,}000{,}000} = \frac{1}{200{,}000}$$

You're looking for the smallest integer value of x that will make $\frac{1}{(x+1)!}$ less than $\frac{1}{200{,}000}$.

In other words, you're looking for the smallest integer x that will make $(x + 1)!$ greater than 200,000. Use your calculator and try a few possibilities. $1 \times 2 \times 3 \times 4 \times 5 \times 6 \times 7 \times 8 = 40{,}320$. Not big enough. But multiply that by 9 and you're up to 362,880. So $9! > 200{,}000$, $x + 1 = 9$, and therefore $x = 8$.

47. E

To find the area of the triangle, you want the base and the height.

The base is the length OB, which is simply the x-coordinate of point B: $2\sqrt{3}$. The height is the y-coordinate of point P, which is equal to $6 - x^2$. The triangle is isosceles, so the altitude from P to base OB divides the base in half and the x-coordinate for point P is $\sqrt{3}$. Plug $x = \sqrt{3}$ into the equation $y = 6 - x^2$ to find the height:

$$\text{height} = 6 - \left(\sqrt{3}\right)^2 = 6 - 3 = 3$$

So if the base is $2\sqrt{3}$ and the height is 3:

$$\begin{aligned}\text{Area} &= \frac{1}{2}\text{(base)(height)}\\ &= \frac{1}{2}\left(2\sqrt{3}\right)(3) = 3\sqrt{3} \approx 5.2\end{aligned}$$

48. B

Use the relationship $\cos 2x = 1 - 2\sin^2 x$ to get everything in terms of sine. And be sure your calculator is in radian mode.

$$\begin{aligned}\cos 2x &= \sin x\\ 1 - 2\sin^2 x &= \sin x\\ 2\sin^2 x + \sin x - 1 &= 0\\ (2\sin x - 1)(\sin x + 1) &= 0\\ \sin x &= \frac{1}{2} \text{ or } -1\\ x &= \arcsin\left(\frac{1}{2}\right) \text{ or } \arcsin(-1)\\ &\approx 0.52 \text{ or } -1.57\end{aligned}$$

Of those solutions, only 0.52 is listed in the answer choices.

49. C

The entire surface area you're looking for consists of the lateral areas of the outside cylinder and the inside cylinder, plus the areas of the larger top and bottom circles, minus the areas of the smaller top and bottom circles. The lateral area of the outside cylinder is $2\pi rh = 2\pi(2)(6) = 24\pi$. The lateral area of the inside cylinder is $2\pi rh = 2\pi(1)(6) = 12\pi$. The areas of the larger top and bottom circles are each $\pi r^2 = \pi(2^2) = 4\pi$. And the areas of the smaller top and bottom circles are each $\pi r^2 = \pi(1^2) = \pi$. The total surface area, then, is

$$24\pi + 12\pi + 2(4\pi) - 2(\pi) = 42\pi$$

50. C

Visualize the situation and/or make a few sketches. Try to imagine as many points of intersection as possible. Here's a way to get four:

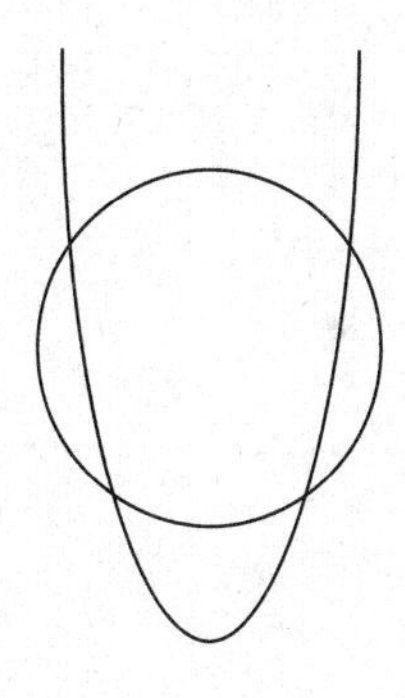

There's no way to get more.

HOW TO CALCULATE YOUR SCORE

Step 1: Figure out your raw score. Use the answer key to count the number of questions you answered correctly and the number of questions you answered incorrectly. (Do not count any questions you left blank.) Multiply the number wrong by 0.25 and subtract the result from the number correct. Round the result to the nearest whole number. This is your raw score.

SAT Subject Test: Mathematics Level 2 — Practice Test 2

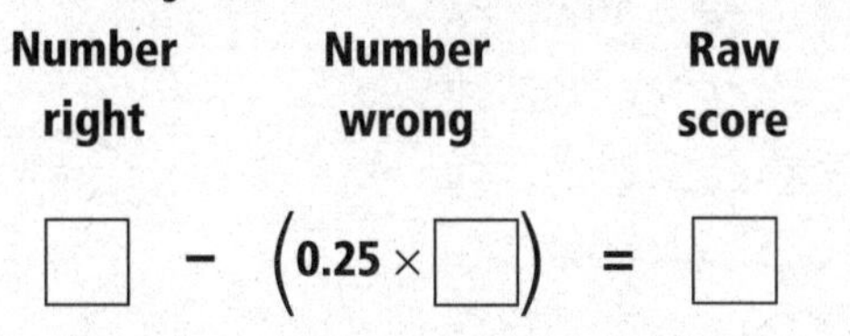

Step 2: Find your scaled score. In the Score Conversion Table below, find your raw score (rounded to the nearest whole number) in one of the columns to the left. The score directly to the right of that number will be your scaled score.

A note on your practice test scores: Don't take these scores too literally. Practice test conditions cannot precisely mirror real test conditions. Your actual SAT Subject Test: Mathematics Level 2 score will almost certainly vary from your practice test scores. However, your scores on the practice tests will give you a rough idea of your range on the actual exam.

Conversion Table

Raw	Scaled	Raw	Scaled	Raw	Scaled	Raw	Scaled
50	800	34	690	18	550	2	370
49	800	33	680	17	540	1	350
48	800	32	570	16	530	0	340
47	800	31	660	15	520	–1	330
46	800	30	650	14	520	–2	310
45	800	29	640	13	510	–3	300
44	790	28	630	12	500	–4	290
43	780	27	620	11	490	–5	280
42	770	26	610	10	480	–6	260
41	760	25	600	9	460	–7	250
40	750	24	590	8	450	–8	240
39	740	23	580	7	440	–9	220
38	730	22	580	6	420	–10	210
37	720	21	570	5	410	–11	200
36	710	20	560	4	400	–12	200
35	700	19	550	3	380		

Answer Grid
Practice Test 2

1.	Ⓐ Ⓑ Ⓒ Ⓓ Ⓔ	26.	Ⓐ Ⓑ Ⓒ Ⓓ Ⓔ
2.	Ⓐ Ⓑ Ⓒ Ⓓ Ⓔ	27.	Ⓐ Ⓑ Ⓒ Ⓓ Ⓔ
3.	Ⓐ Ⓑ Ⓒ Ⓓ Ⓔ	28.	Ⓐ Ⓑ Ⓒ Ⓓ Ⓔ
4.	Ⓐ Ⓑ Ⓒ Ⓓ Ⓔ	29.	Ⓐ Ⓑ Ⓒ Ⓓ Ⓔ
5.	Ⓐ Ⓑ Ⓒ Ⓓ Ⓔ	30.	Ⓐ Ⓑ Ⓒ Ⓓ Ⓔ
6.	Ⓐ Ⓑ Ⓒ Ⓓ Ⓔ	31.	Ⓐ Ⓑ Ⓒ Ⓓ Ⓔ
7.	Ⓐ Ⓑ Ⓒ Ⓓ Ⓔ	32.	Ⓐ Ⓑ Ⓒ Ⓓ Ⓔ
8.	Ⓐ Ⓑ Ⓒ Ⓓ Ⓔ	33.	Ⓐ Ⓑ Ⓒ Ⓓ Ⓔ
9.	Ⓐ Ⓑ Ⓒ Ⓓ Ⓔ	34.	Ⓐ Ⓑ Ⓒ Ⓓ Ⓔ
10.	Ⓐ Ⓑ Ⓒ Ⓓ Ⓔ	35.	Ⓐ Ⓑ Ⓒ Ⓓ Ⓔ
11.	Ⓐ Ⓑ Ⓒ Ⓓ Ⓔ	36.	Ⓐ Ⓑ Ⓒ Ⓓ Ⓔ
12.	Ⓐ Ⓑ Ⓒ Ⓓ Ⓔ	37.	Ⓐ Ⓑ Ⓒ Ⓓ Ⓔ
13.	Ⓐ Ⓑ Ⓒ Ⓓ Ⓔ	38.	Ⓐ Ⓑ Ⓒ Ⓓ Ⓔ
14.	Ⓐ Ⓑ Ⓒ Ⓓ Ⓔ	39.	Ⓐ Ⓑ Ⓒ Ⓓ Ⓔ
15.	Ⓐ Ⓑ Ⓒ Ⓓ Ⓔ	40.	Ⓐ Ⓑ Ⓒ Ⓓ Ⓔ
16.	Ⓐ Ⓑ Ⓒ Ⓓ Ⓔ	41.	Ⓐ Ⓑ Ⓒ Ⓓ Ⓔ
17.	Ⓐ Ⓑ Ⓒ Ⓓ Ⓔ	42.	Ⓐ Ⓑ Ⓒ Ⓓ Ⓔ
18.	Ⓐ Ⓑ Ⓒ Ⓓ Ⓔ	43.	Ⓐ Ⓑ Ⓒ Ⓓ Ⓔ
19.	Ⓐ Ⓑ Ⓒ Ⓓ Ⓔ	44.	Ⓐ Ⓑ Ⓒ Ⓓ Ⓔ
20.	Ⓐ Ⓑ Ⓒ Ⓓ Ⓔ	45.	Ⓐ Ⓑ Ⓒ Ⓓ Ⓔ
21.	Ⓐ Ⓑ Ⓒ Ⓓ Ⓔ	46.	Ⓐ Ⓑ Ⓒ Ⓓ Ⓔ
22.	Ⓐ Ⓑ Ⓒ Ⓓ Ⓔ	47.	Ⓐ Ⓑ Ⓒ Ⓓ Ⓔ
23.	Ⓐ Ⓑ Ⓒ Ⓓ Ⓔ	48.	Ⓐ Ⓑ Ⓒ Ⓓ Ⓔ
24.	Ⓐ Ⓑ Ⓒ Ⓓ Ⓔ	49.	Ⓐ Ⓑ Ⓒ Ⓓ Ⓔ
25.	Ⓐ Ⓑ Ⓒ Ⓓ Ⓔ	50.	Ⓐ Ⓑ Ⓒ Ⓓ Ⓔ

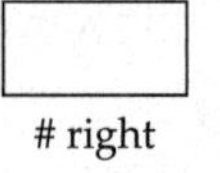

right

wrong

Use the answer key following the test to count up the number of questions you got right and the number you got wrong. (Remember not to count omitted questions as wrong.) "How to Calculate Your Score" on the back of this page will show you how to find your score.

Practice Test 2

50 Questions (1 hour)

Directions: For each question, choose the BEST answer from the choices given. If the precise answer is not among the choices, choose the one that best approximates the answer. Then fill in the corresponding oval on the answer sheet.

Notes:

(1) To answer some of these questions, you will need a calculator. You must use at least a scientific calculator, but programmable and graphing calculators are also allowed.

(2) Make sure your calculator is in the correct mode (degree or radian) for the question being asked.

(3) Figures in this test are drawn as accurately as possible UNLESS it is stated in a specific question that the figure is not drawn to scale. All figures are assumed to lie in a plane unless otherwise specified.

(4) The domain of any function f is assumed to be the set of all real numbers x for which $f(x)$ is a real number, unless otherwise indicated.

Reference Information: Use the following formulas as needed.

Right circular cone: If r = radius and h = height, then Volume $= \frac{1}{3}\pi r^2 h$; and if c = circumference of the base and ℓ = slant height, then Lateral Area $= \frac{1}{2}c\ell$.

Sphere: If r = radius, then Volume $= \frac{4}{3}\pi r^3$ and Surface Area $= 4\pi r^2$.

Pyramid: If B = area of the base and h = height, then Volume $= \frac{1}{3}Bh$.

DO YOUR FIGURING HERE.

1. If $\frac{x+y}{0.01}=7$, then $\frac{1}{2x+2y}=$

(A) 0.14 (B) 0.28 (C) 3.50 (D) 7.00 (E) 7.14

2. $\frac{(100^{12})(10^4)}{10^2}=$

(A) 10^8 (B) 10^{14} (C) 10^{24} (D) 10^{26} (E) 10^{48}

3. If $\frac{x^2}{4}=\frac{6}{x}$, then $x=$

(A) 2.59 (B) 2.88 (C) 3.03 (D) 3.89 (E) 8.00

4. Which of the following is an equation of a line that will have points in all the quadrants except the first?

(A) $y = 2x$
(B) $y = 2x + 3$
(C) $y = 2x - 3$
(D) $y = -2x + 3$
(E) $y = -2x - 3$

5. If $b = 3 - a$ and $b \neq a$, then $\frac{a^2-b^2}{b-a}=$

(A) 3 (B) 1 (C) 0 (D) –1 (E) –3

6. If $f(x) = e^x + 2x$, then $f(\ln 2) =$

(A) 1.20
(B) 2.69
(C) 2.77
(D) 3.39
(E) 4.00

GO ON TO THE NEXT PAGE

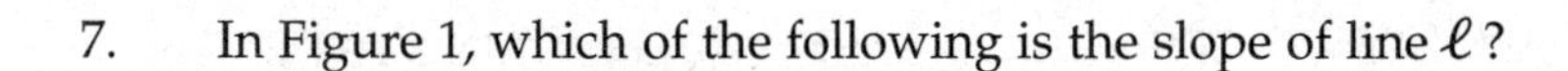

7. In Figure 1, which of the following is the slope of line ℓ?

(A) –3

(B) –2

(C) $-\frac{1}{2}$

(D) $\frac{1}{2}$

(E) 2

DO YOUR FIGURING HERE.

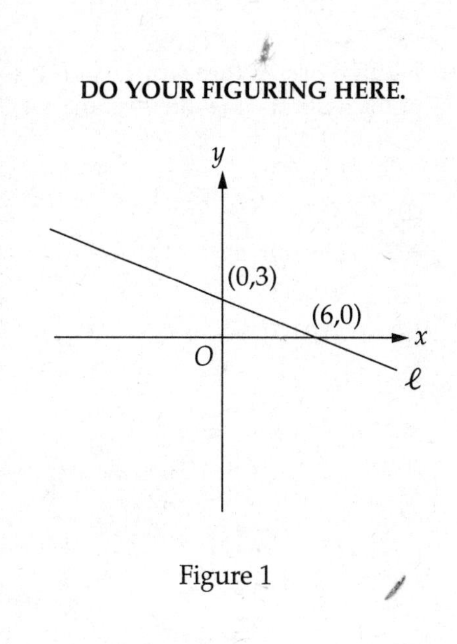

Figure 1

8. Which of the following is the complete solution set to the inequality $|x|+|x-3|>3$?

(A) $\{x : x > 3 \text{ or } x < 0\}$
(B) $\{x : -3 < x < 3\}$
(C) $\{x : -3 > x\}$
(D) $\{x : -3 < x\}$
(E) $\{x :$ The set of all real numbers$\}$

9. Which of the following is the solution set for $(3x - 6)(2 + x) < 0$?

(A) $\{x: x < 2\}$
(B) $\{x: x > 2\}$
(C) $\{x: x > -2\}$
(D) $\{x: x < -2 \text{ or } x > 2\}$
(E) $\{x: -2 < x < 2\}$

GO ON TO THE NEXT PAGE

10. If a line passes through the points (5,3) and (8,–1), at what point will this line intersect the y-axis?

(A) (0,8.33)
(B) (0,8.67)
(C) (0,9.00)
(D) (0,9.33)
(E) (0,9.67)

DO YOUR FIGURING HERE.

11. If $f(x) = 2x + 1$, and $f(x + 2) + f(x) = x$, the value of x is

(A) –2 (B) –1 (C) $-\frac{1}{2}$ (D) $\frac{1}{2}$ (E) 1

12. Set S is the set of all points (x,y) in the coordinate plane such that x and y are both integers with absolute value less than 4. If one of these points is chosen at random, what is the probability that this point will be 2 units or less from the origin?

(A) 0.189
(B) 0.227
(C) 0.265
(D) 0.314
(E) 0.356

GO ON TO THE NEXT PAGE

13. In Figure 2, what is the length of AC ?

(A) 2.94
(B) 3.49
(C) 3.81
(D) 4.05
(E) 4.26

14. If $a = \sqrt[3]{t}$ and $b = t^2$, then $\frac{b}{a^5} =$

(A) $t^{-\frac{1}{3}}$

(B) $t^{\frac{1}{3}}$

(C) $t^{\frac{5}{6}}$

(D) $t^{\frac{6}{5}}$

(E) $t^{\frac{10}{3}}$

15. If A, B, C, D, E, and F are 6 distinct points on the circumference of a circle, how many different chords can be drawn using any 2 of the 6 points?

(A) 6 (B) 12 (C) 15 (D) 30 (E) 36

DO YOUR FIGURING HERE.

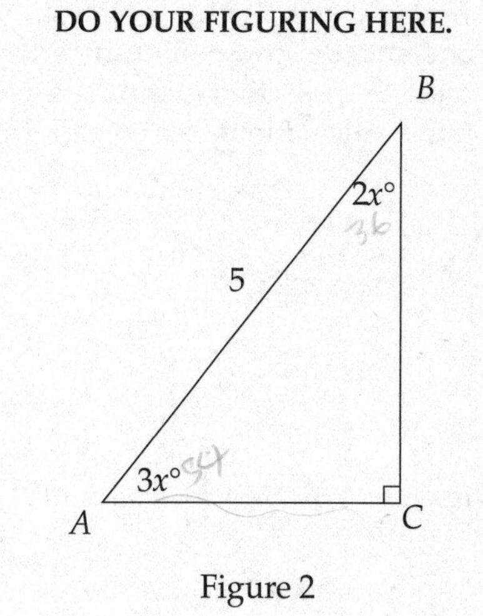

Figure 2

GO ON TO THE NEXT PAGE

16. A new computer can perform x calculations in y seconds and an older computer can perform r calculations in s minutes. If these two computers work simultaneously, how many calculations can be performed in t minutes?

(A) $t\left(\frac{x}{60y}+\frac{r}{s}\right)$

(B) $t\left(\frac{60x}{y}+\frac{r}{s}\right)$

(C) $t\left(\frac{x}{y}+\frac{r}{s}\right)$

(D) $t\left(\frac{x}{y}+\frac{60r}{s}\right)$

(E) $60t\left(\frac{x}{y}+\frac{r}{s}\right)$

DO YOUR FIGURING HERE.

17. Which of the following could be the equation of the parabola in Figure 3?

(A) $y = (x - 2)(x - 3)$
(B) $y = (x + 2)(x + 3)$
(C) $y = (x + 2)(x - 3)$
(D) $y = (x - 2)(x + 3)$
(E) $x = (y + 2)(y - 3)$

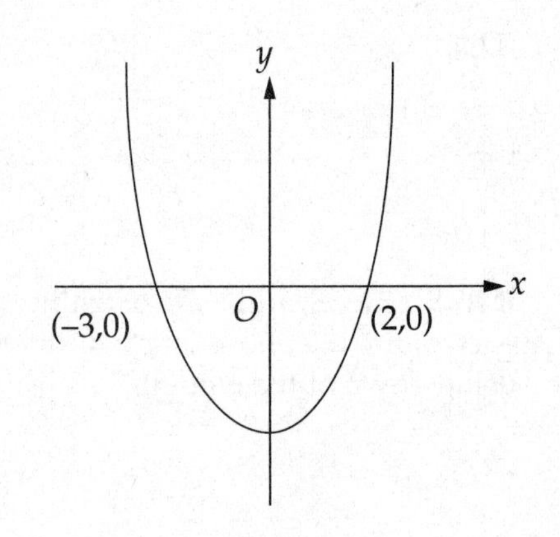

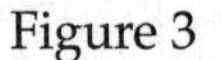
Figure 3

18. If $a + b = 15$, $b + c = 10$, and $a + c = 13$, which of the following is true?

(A) $a < b < c$
(B) $b < a < c$
(C) $c < b < a$
(D) $a < c < b$
(E) $b < c < a$

GO ON TO THE NEXT PAGE

19. In Figure 4, $\frac{1}{\sin\theta}+\frac{1}{\cos\theta}=$

(A) 0.75
(B) 1.20
(C) 1.43
(D) 2.74
(E) 2.92

20. Amanda goes to the toy store to buy 1 ball—either a football, basketball, or soccer ball—and 3 different board games. If the toy store is stocked with all types of balls but only 6 different types of board games, how many different selections of 4 items can Amanda make consisting of 1 type of ball and 3 different board games?

(A) 18 (B) 20 (C) 54 (D) 60 (E) 162

21. If point $P(3,2)$ is rotated 90 degrees counterclockwise with respect to the origin, what will be its new coordinates?

(A) (–2,3)
(B) (–2,–3)
(C) (–3,3)
(D) (–3,2)
(E) (–3,–2)

DO YOUR FIGURING HERE.

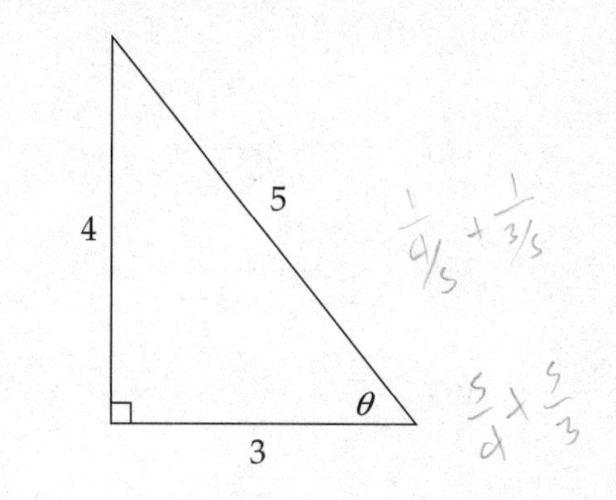

Figure 4

GO ON TO THE NEXT PAGE

22. If $0 < x < \frac{\pi}{2}$ and $\tan x = \frac{a}{2}$, then $\cos x =$

(A) $\frac{2}{\sqrt{a^2 - 4}}$

(B) $\frac{a}{\sqrt{a^2 - 4}}$

(C) $\frac{2}{a + 2}$

(D) $\frac{2}{\sqrt{a^2 + 4}}$

(E) $\frac{a}{\sqrt{a^2 + 4}}$

23. For what value of x will $f(x) = (1 - 2x)^2$ have the minimum value?

(A) -1 (B) $-\frac{1}{2}$ (C) 0 (D) $\frac{1}{2}$ (E) 1

24. If a certain line intersects the origin and is perpendicular to the line with the equation $y = 2x + 5$ at point P, what is the distance from the origin to point P ?

(A) 2.24
(B) 2.45
(C) 2.67
(D) 3.25
(E) 3.89

DO YOUR FIGURING HERE.

GO ON TO THE NEXT PAGE

25. If the volume of a cube is equal to the volume of a sphere, what is the ratio of the edge of the cube to the radius of the sphere?

(A) 1.61
(B) 2.05
(C) 2.33
(D) 2.45
(E) 2.65

26. If $[x]$ represents the greatest integer less than or equal to x, what is the solution to the equation $1 - 2[x] = -3$?

(A) $x = 2$
(B) $2 \le x < 3$
(C) $2 < x \le 3$
(D) $2 < x < 3$
(E) There is no solution.

27. Which of the following lists all and only the vertical asymptotes of the graph $y = \frac{x}{x^2 - 4}$?

(A) $x = 2$ only
(B) $y = 2$ only
(C) $x = 2$ and $x = -2$
(D) $y = 2$ and $y = -2$
(E) $x = 2$, $x = -2$, and $x = 0$

28. If $\cos x \sin x = 0.22$, then $(\cos x - \sin x)^2 =$

(A) 0
(B) 0.11
(C) 0.44
(D) 0.56
(E) 1.00

DO YOUR FIGURING HERE.

GO ON TO THE NEXT PAGE

29. If water is poured at a rate of 12 cubic meters per second into a half-empty rectangular tank with length 5 meters, width 3 meters, and height 25 meters, then how high, in meters, will the water level be after 9 seconds?

(A) 6.0
(B) 7.2
(C) 18.5
(D) 19.7
(E) The tank will be full and overflowing.

30. A circle centered at (3,2) with radius 5 intersects the x-axis at which of the following x-coordinates?

(A) 2.39
(B) 4.58
(C) 7.58
(D) 8.00
(E) 8.39

31. If $0 \leq x \leq \pi$, where is $\frac{\tan x}{\sin x}$ defined?

(A) $0 \leq x \leq \pi$

(B) $0 < x < \pi$

(C) $0 < x < \frac{\pi}{2}$

(D) $\frac{\pi}{2} \leq x \leq \pi$

(E) $0 < x < \frac{\pi}{2}$ and $\frac{\pi}{2} < x < \pi$

DO YOUR FIGURING HERE.

GO ON TO THE NEXT PAGE

32. A rectangular box with an open top is constructed from cardboard to have a square base of area x^2 and height h. If the volume of this box is 50 cubic units, how many square units of cardboard, in terms of x, are needed to build this box?

(A) $5x^2$

(B) $6x^2$

(C) $\frac{200}{x}+x^2$

(D) $\frac{200}{x}+2x^2$

(E) $\frac{250}{x}+2x^2$

33. $\frac{(n+2)!-(n+1)!}{n!}=$

(A) $(n + 2)!$
(B) $(n + 1)!$
(C) $(n + 2)^2$
(D) $(n + 1)^2$
(E) n

DO YOUR FIGURING HERE.

GO ON TO THE NEXT PAGE

34. Bob wishes to borrow some money. He needs to defer to the following formula, where M is the monthly payment, r is the monthly decimal interest rate, P is the amount borrowed, and t is the number of months it will take to repay the loan:

$$M = \frac{rP}{1 - \left(\frac{1}{1+r}\right)^t}$$

If Bob secures a loan of $4,000 that he will pay back in 36 months with a monthly interest rate of 0.01, what is his monthly payment?

(A) $111.11
(B) $119.32
(C) $132.86
(D) $147.16
(E) $175.89

35. A particle is moving along the line $5y = -6x + 30$ at a rate of 2 units per second. If the particle starts at the y-intercept and moves to the right along this line, how many seconds will it take for the particle to reach the x-axis?

(A) 2.50
(B) 3.25
(C) 3.76
(D) 3.91
(E) 7.81

DO YOUR FIGURING HERE.

GO ON TO THE NEXT PAGE

36. In Figure 5, if the area of triangle ABC is 15, what is the length of AC ?

(A) 2.1
(B) 4.1
(C) 6.2
(D) 8.2
(E) 9.6

37. Which of the following functions has a range of $-1 < y < 1$?

(A) $y = \sin x$

(B) $y = \cos x$

(C) $y = \dfrac{x}{1+x}$

(D) $y = \dfrac{x^2}{1+x^2}$

(E) $y = \dfrac{x}{\sqrt{1+x^2}}$

38. What is the sum of the infinite series $1 - \dfrac{1}{3} + \dfrac{1}{9} - \dfrac{1}{27} + \cdots$?

(A) $\dfrac{2}{3}$ (B) $\dfrac{3}{4}$ (C) 1 (D) $\dfrac{4}{3}$ (E) $\dfrac{3}{2}$

DO YOUR FIGURING HERE.

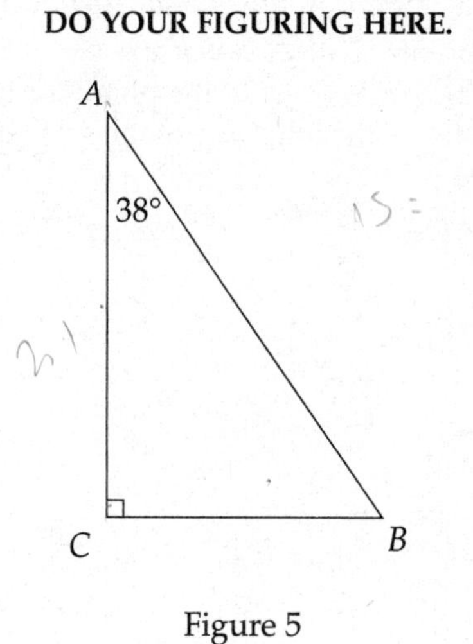

Figure 5

GO ON TO THE NEXT PAGE

39. In Figure 6, the shaded region represents the set C of all points (x, y) such that $x^2 + y^2 \leq 1$. The transformation T maps the point (x, y) to the point $(2x, 4y)$. Which of the following shows the mapping of the set C by the transformation T?

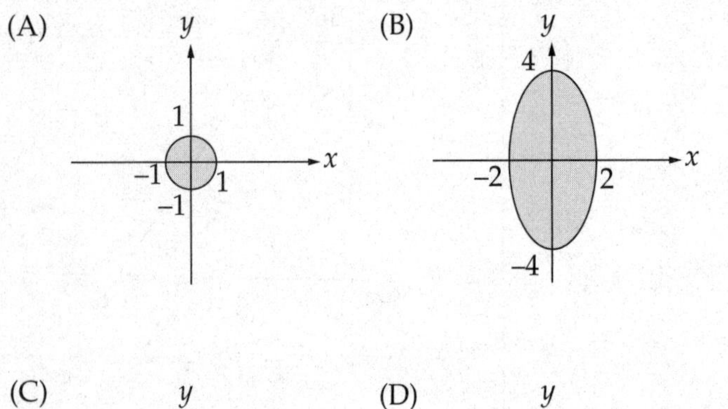

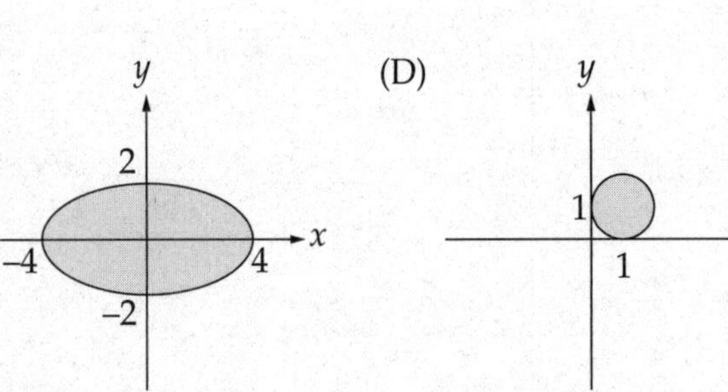

(E)

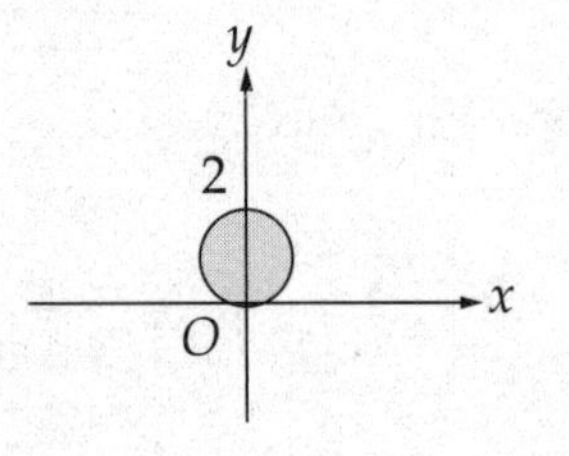

DO YOUR FIGURING HERE.

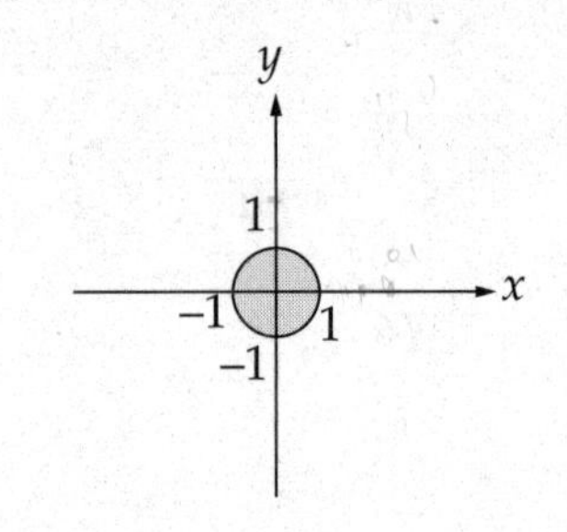

Figure 6

GO ON TO THE NEXT PAGE

DO YOUR FIGURING HERE.

40. $\lim_{n\to\infty} \frac{1-2n^2}{5n^2-n+100} =$

(A) -1

(B) $-\frac{2}{5}$

(C) $\frac{2}{5}$

(D) 1

(E) No limit exists.

41. If $\log_2(x^2 - 3) = 5$, which of the following could be the value of x ?

(A) 3.61
(B) 4.70
(C) 5.29
(D) 5.75
(E) 5.92

42. If 2 is a zero of the function $f(x) = 6x^3 - 11x^2 - 3x + 2$, what are the other zeroes?

(A) $-\frac{1}{3}$ and $-\frac{1}{2}$

(B) $-\frac{1}{3}$ and $\frac{1}{2}$

(C) $\frac{1}{3}$ and $-\frac{1}{2}$

(D) $\frac{1}{3}$ and $\frac{1}{2}$

(E) 2 and 3

GO ON TO THE NEXT PAGE

43. In Figure 7, a circle of radius 1 is placed on an incline where point P, a point on the circle, has the coordinates (–5,–5). The circle is rolled up the incline, and once the circle hits the origin, the circle is then rolled horizontally along the x-axis to the right. What is the x-coordinate of the point where P touches the incline or the x-axis for the fifth time (not including the starting point)?

(A) 8.64
(B) 17.27
(C) 24.34
(D) 27.49
(E) 30.63

DO YOUR FIGURING HERE.

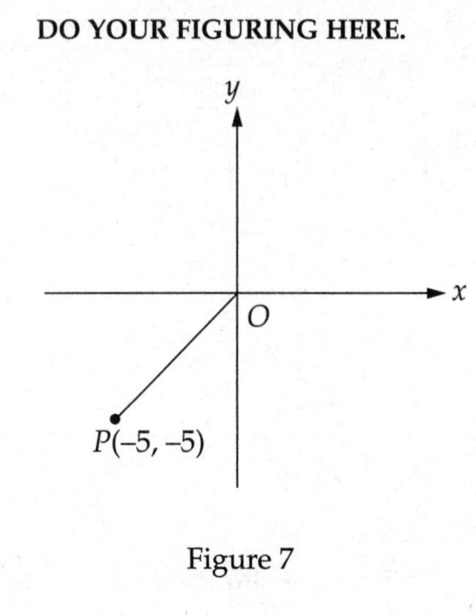

Figure 7

44. If $0 \le x \le 2\pi$ and $\sin x < 0$, which of the following must be true?

I. $\cos x < 0$

II. $\csc x < 0$

III. $|\sin x + \cos x| > 0$

(A) I only
(B) II only
(C) III only
(D) I and II
(E) II and III

45. If $i^2 = -1$, which of the following is a square root of $8 - 6i$?

(A) $3 - i$
(B) $3 + i$
(C) $3 - 4i$
(D) $4 - 3i$
(E) $4 + 3i$

GO ON TO THE NEXT PAGE

46. Figure 8 shows rectangle $ABCD$. Points A and D are on the parabola $y = 2x^2 - 8$, and points B and C are on the parabola $y = 9 - x^2$. If point B has coordinates (–1.50, 6.75), what is the area of rectangle $ABCD$?

(A) 12.50
(B) 17.50
(C) 22.75
(D) 26.50
(E) 30.75

47. If $x \geq 0$ and arcsin x = arccos $(2x)$, then x =

(A) 0.866
(B) 0.707
(C) 0.500
(D) 0.447
(E) 0.245

48. If $f(x) = \frac{1}{2}x - 4$ and $f(g(x)) = g(f(x))$, which of the following can be $g(x)$?

I. $2x - \frac{1}{4}$

II. $2x + 8$

III. $\frac{1}{2}x - 4$

(A) I only
(B) II only
(C) III only
(D) II and III only
(E) I, II, and III

DO YOUR FIGURING HERE.

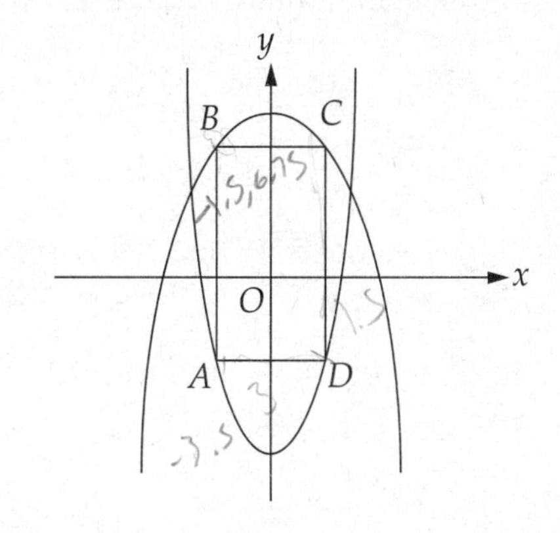

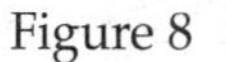

Figure 8

GO ON TO THE NEXT PAGE

49. If a right circular cylinder of height 10 is inscribed in a sphere of radius 6, what is the volume of the cylinder?

(A) 104 (B) 346 (C) 545 (D) 785 (E) 1,131

50. If the diagonals *AC* and *BD* intersect at point *P* in the cube in Figure 9, what is the degree measure of angle *APB* ?

(A) 60
(B) 65
(C) 71
(D) 83
(E) 90

DO YOUR FIGURING HERE.

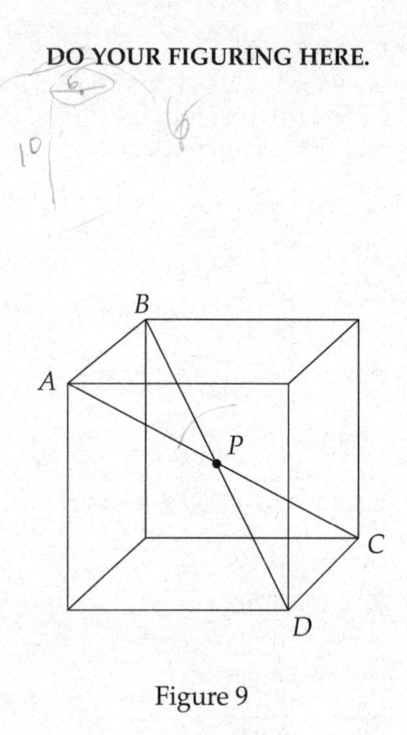

Figure 9

STOP!

If you finish before time is up, you may check your work.

Turn the page for answers and explanations to Practice Test 2.

Answer Key
Practice Test 2

1. E
2. D
3. B
4. E
5. E
6. D
7. C
8. A
9. E
10. E
11. A
12. C
13. A
14. B
15. C
16. B
17. D
18. C
19. E
20. D
21. A
22. D
23. D
24. A
25. A
26. B
27. C
28. D
29. D
30. C
31. E
32. C
33. D
34. C
35. D
36. C
37. E
38. B
39. B
40. B
41. E
42. C
43. C
44. B
45. A
46. E
47. D
48. D
49. B
50. C

ANSWERS AND EXPLANATIONS

1. E

If $\frac{x+y}{0.01} = 7$, then $x + y = 7(0.01) = 0.07$. Therefore,

$$\frac{1}{2x+2y} = \frac{1}{2(x+y)}$$
$$= \frac{1}{2(0.07)} = \frac{1}{0.14} \approx 7.14$$

2. D

Put the whole thing in terms of a power of 10. $(100)^{12} = (10^2)^{12} = 10^{24}$. Therefore,

$$\frac{(100)^{12}(10)^4}{(10)^2} = \frac{(10^{24})(10^4)}{10^2}$$
$$= \frac{10^{28}}{10^2} = 10^{28-2} = 10^{26}$$

3. B

Cross multiply:

$$\frac{x^2}{4} = \frac{6}{x}$$
$$x^2 \times x = 4 \times 6$$
$$x^3 = 24$$

Now use your calculator to find the cube root of 24:

$$x = \sqrt[3]{24} \approx 2.88$$

4. E

The answer choices are all linear equations in convenient $y = mx + b$ form. A line that has points in all quadrants but the first is a line that crosses the y-axis below the origin and heads downhill from there—in other words, a line with both a negative y-intercept and a negative slope. (C) and (E) have negative y-intercepts. (D) and (E) have negative slopes. Only (E) has both.

5. E

Factor the numerator and look for something you can cancel with the denominator:

$$\frac{a^2-b^2}{b-a} = \frac{(a-b)(a+b)}{b-a}$$
$$= \frac{-(b-a)(a+b)}{b-a}$$
$$= -(a+b)$$

It's given that $b = 3 - a$, which is just another way of saying $a + b = 3$, so

$$-(a+b) = -3$$

6. D

To find the value of $f(x)$ for a particular value of x, plug it into the definition:

$$f(x) = e^x + 2x$$
$$f(\ln 2) = e^{\ln 2} + 2\ln 2$$

You'll need your calculator to evaluate part of this expression, but you should realize, without a calculator, that $e^{\ln 2}$ is 2. Use your calculator to find that $2 \ln 2 \approx 1.39$ and, therefore,

$$f(\ln 2) = e^{\ln 2} + 2\ln 2$$
$$\approx 2 + 1.39 = 3.39$$

7. C

You can use the two given points to figure out the slope:

$$\text{Slope} = \frac{y_2 - y_1}{x_2 - x_1}$$
$$= \frac{3-0}{0-6} = \frac{3}{-6} = -\frac{1}{2}$$

8. A

Think about the three different cases.

Case 1, when $x \geq 3$:

$$|x|+|x-3|>3$$
$$x+x-3>3$$
$$2x-3>3$$
$$2x>6$$
$$x>3$$

So all numbers greater than 3 satisfy the inequality, but $x = 3$ itself does not.

Case 2, when $0 \leq x < 3$:

$$|x|+|x-3|>3$$
$$x+(-x)+3>3$$
$$3>3$$

So nothing between 0 and 3 satisfies the inequality.

Case 3, when $x < 0$:

$$|x|+|x-3|>3$$
$$(-x)+(-x)+3>3$$
$$-2x+3>3$$
$$-2x>0$$
$$x<0$$

So all negative numbers satisfy the inequality, and the complete solution set is $\{x : x > 3 \text{ or } x < 0\}$.

9. E

If the product of $(3x - 6)$ and $(2 + x)$ is negative, then one of the two factors is negative and the other is positive. There are two cases:

Case 1, when $3x - 6 < 0$ and $2 + x > 0$:

$$3x-6<0$$
$$3x<0$$
$$x<2$$

and:

$$2+x>0$$
$$x>-2$$

So, all x such that $-2 < x < 2$ satisfy the inequality.

Case 2, when $3x - 6 > 0$ and $2 + x < 0$:

$$3x-6>0$$
$$3x>6$$
$$x>2$$

and:

$$2+x<0$$
$$x<-2$$

There are no numbers that are both less than −2 and greater than 2, so the numbers that work in Case 1 are the complete solution set.

10. E

You can use the two given points to find the slope:

$$\text{Slope} = \frac{y_2 - y_1}{x_2 - x_1} = \frac{-1-3}{8-5} = \frac{-4}{3} = -\frac{4}{3}$$

Next plug the point $(0, y)$ and either one of the given points into the same formula:

$$\text{Slope} = \frac{y_2 - y_1}{x_2 - x_1}$$
$$-\frac{4}{3} = \frac{3-y}{5-0}$$
$$-\frac{4}{3} = \frac{3-y}{5}$$
$$-3(3-y) = (4)(5)$$
$$3 - y = \frac{20}{-3}$$
$$-y = \frac{20}{-3} - 3$$
$$y = \frac{20}{3} + 3 \approx 9.67$$

11. A

Plug both x and $x + 2$ into the definition:

$$f(x) = 2x + 1$$
$$f(x+2) = 2(x+2) + 1$$

Then set the sum equal to x and solve:

$$f(x+2) + f(x) = x$$
$$2(x+2) + 1 + 2x + 1 = x$$
$$2x + 4 + 1 + 2x + 1 - x = 0$$
$$3x + 6 = 0$$
$$3x = -6$$
$$x = -2$$

12. C

Probability is favorable outcomes over total outcomes. Here the total number of possible outcomes is the number of points with both coordinates of integer absolute values less than 4. That's all the points with x-coordinates of –3, –2, –1, 0, 1, 2, or 3 and y-coordinates of –3, –2, –1, 0, 1, 2, or 3. Seven possibilities for x and seven possibilities for y makes $7 \times 7 = 49$ possibilities for (x,y):

Total outcomes = 49

Next figure out how many of those points are 2 units or less from the origin. In other words, you're looking for points (x,y) such that $x^2 + y^2 \leq 4$. There are 9 points on the axes that work:

(0,–2), (0,–1), (0,0), (0,1), (0,2),
(–2,0), (–1,0), (1,0), (2,0)

Additionally there are these four points that fit:

(–1,–1), (–1,1), (1,–1), (1,1)

That's a total of 9 + 4 = 13: Favorable outcomes = 13.

So the probability is $13 \div 49 \approx 0.265$.

13. A

It's a right triangle, so the two acute angles add up to 90 degrees.

Since their degree measures are in a 2-to-3 ratio, they must be $\frac{2}{5}$ of 90 and $\frac{3}{5}$ of 90, or 36 and 54:

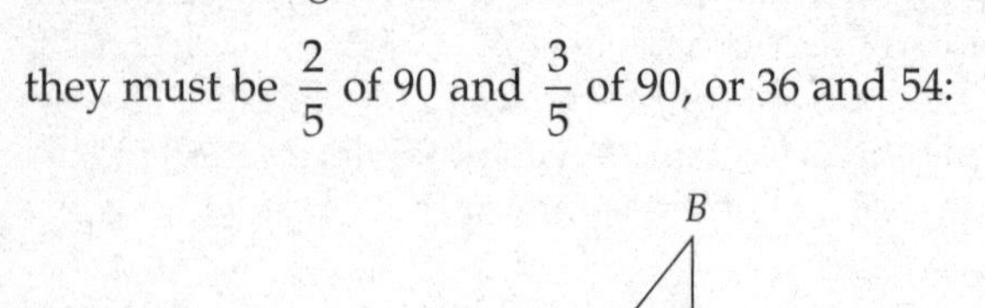

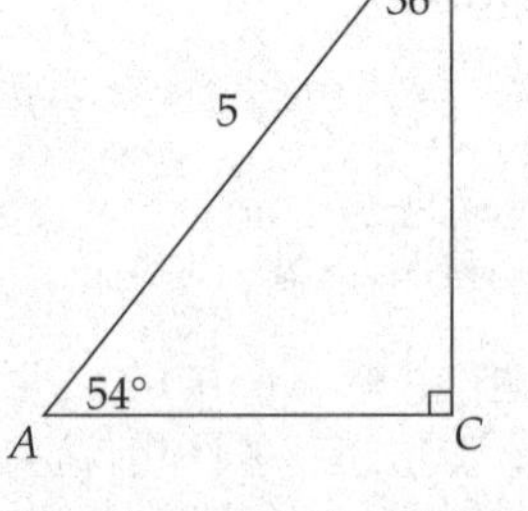

The side AC you're looking for is opposite the 36° angle. Therefore,

$$\frac{AC}{5} = \sin 36^\circ$$

Use your calculator to find that $\sin 36^\circ \approx 0.588$. Therefore,

$$\frac{AC}{5} \approx 0.588$$

$$AC \approx 5(0.588) = 2.94$$

14. B

Plug $a = \sqrt[3]{t} = t^{\frac{1}{3}}$ and $b = t^2$ into the expression:

$$\frac{b}{a^5} = \frac{t^2}{\left(t^{\frac{1}{3}}\right)^5} = \frac{t^2}{t^{\frac{5}{3}}} = t^{2-\frac{5}{3}} = t^{\frac{1}{3}}$$

15. C

Each point is connected to five other points to make 5 chords per point. But this counts every chord twice—*AB* is indistinguishable from *BA*—so after you multiply 6 by 5, you have to divide by 2, yielding 15.

16. B

Take the computers one at a time. Put everything in terms of seconds: t minutes is $60t$ seconds and s minutes is $60s$ seconds.

The new computer performs $\frac{x}{y}$ calculations per second, which is $\frac{60tx}{y}$ calculations in t minutes. The old computer performs $\frac{r}{60s}$ calculations per second, which is $\frac{60tr}{60s} = \frac{tr}{s}$ calculations in t minutes.

The number of calculations they perform together is

$$\frac{60tx}{y} + \frac{tr}{s} = t\left(\frac{60x}{y} + \frac{r}{s}\right)$$

17. D

Try the two given points in the choices. The point (–3,0) satisfies (B) and (D). The point (2,0) satisfies (A) and (D). The only choice that works both times is (D).

18. C

You want to rank a, b, and c. Actually, because you're given three equations, you can solve for the three unknowns. Subtract the equation $b + c = 10$ from the equation $a + b = 15$, and you get $a - c = 5$. Now add that to the equation $a + c = 13$, and you get $2a = 18$, or $a = 9$. Now you can plug that value of a back into the appropriate equations to find that $b = 6$ and $c = 4$. So the correct ranking is $c < b < a$.

19. E

Sine is opposite over hypotenuse and cosine is adjacent over hypotenuse, so $\sin\theta = \frac{4}{5}$ and $\cos\theta = \frac{3}{5}$.

Therefore,

$$\frac{1}{\sin\theta} + \frac{1}{\cos\theta} = \frac{5}{4} + \frac{5}{3} \approx 1.25 + 1.67 = 2.92$$

20. D

Amanda has 3 choices for the ball. As for the board games, she wants to choose 3 out of 6, so the number of board game combinations is

$$_6C_3 = \frac{6!}{(6-3)!3!} = \frac{6\cdot5\cdot4\cdot3\cdot2\cdot1}{3\cdot2\cdot1\cdot3\cdot2\cdot1} = 20$$

For each of 20 board game combinations there are three ball choices, so the total number of options Amanda faces is $3 \times 20 = 60$.

21. A

Sketch a diagram:

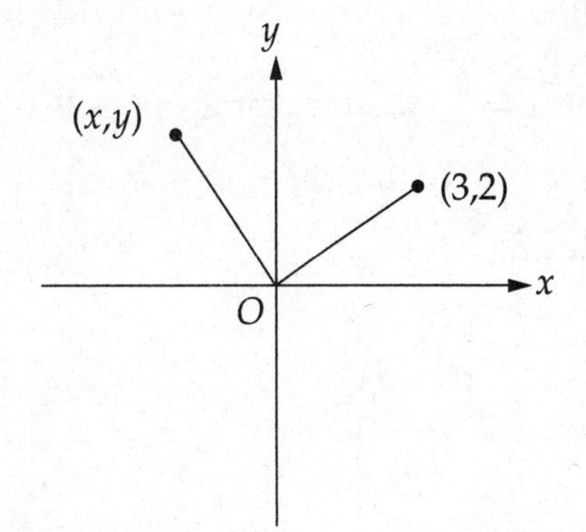

Add a couple of perpendiculars, and you'll make a couple of congruent triangles:

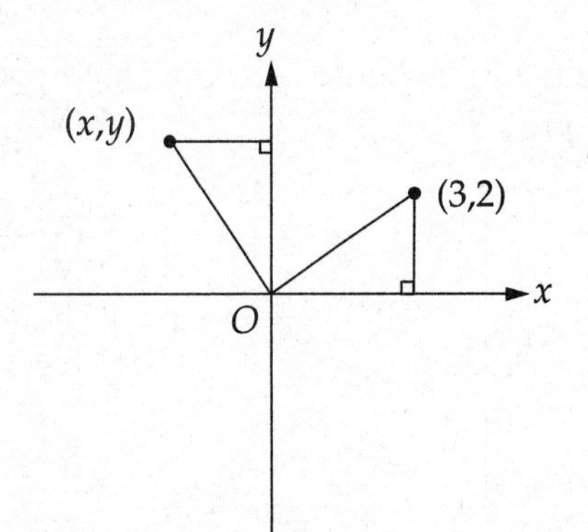

To get to the original point $P(3,2)$, you go right 3 and up 2. To get to the new point (x,y), you go left 2 and up 3, so its coordinates are $(-2,3)$.

22. D

Since x is an acute angle, use a right triangle to solve this one.

Since $\tan x = \frac{a}{2}$, and tangent is opposite over adjacent, you could label the opposite side a and the adjacent side 2:

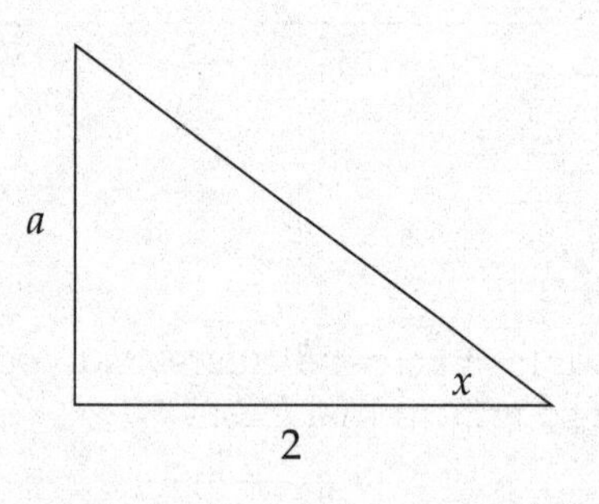

Then, by the Pythagorean theorem, the hypotenuse is $\sqrt{a^2+4}$:

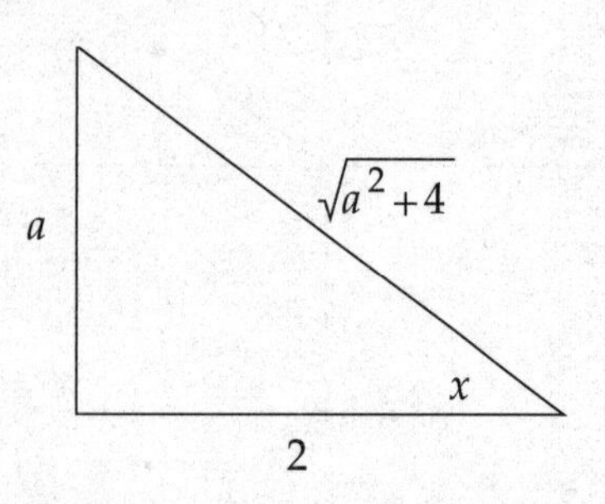

Cosine is adjacent over hypotenuse, so $\cos x$ is $\frac{2}{\sqrt{a^2+4}}$.

23. D

The expression $(1 - 2x)^2$ is something squared. A real number squared cannot be smaller than 0. This function will have its minimum value when the part inside the parentheses is zero:

$$\begin{aligned} 1-2x &= 0 \\ -2x &= -1 \\ x &= \frac{1}{2} \end{aligned}$$

24. A

If the two lines are perpendicular, then their slopes are negative reciprocals.

The slope of $y = 2x + 5$ is 2, so the slope of the other line is $-\frac{1}{2}$. That line goes through the origin, so its y-intercept is 0, and therefore its equation is $y = -\frac{1}{2}x$. Point P—the intersection of these lines—is the point that satisfies both equations:

$$y = 2x+5 \text{ and } y = -\frac{1}{2}x$$

$$\begin{aligned} 2x+5 &= -\frac{1}{2}x \\ 4x+10 &= -x \\ 5x &= -10 \\ x &= -2 \\ y &= -\frac{1}{2}x = -\frac{1}{2}(-2) = 1 \end{aligned}$$

So point P is (–2,1), and the distance from the origin to point P is

$$OP = \sqrt{(-2)^2+1^2} = \sqrt{5} \approx 2.24$$

25. A

Set the formulas equal and solve for $\frac{e}{r}$:

$$\begin{aligned} e^3 &= \frac{4}{3}\pi r^3 \\ \frac{e^3}{r^3} &= \frac{4}{3}\pi \\ \frac{e}{r} &= \sqrt[3]{\frac{4}{3}\pi} \approx 1.61 \end{aligned}$$

26. B

First solve for $[x]$:

$$1 - 2[x] = -3$$
$$-2[x] = -4$$
$$[x] = 2$$

$[x]$ will equal 2 whenever $2 \le x < 3$.

27. C

The easiest way to do this one is to use your graphing calculator.

When you graph $y = \frac{x}{x^2 - 4}$, you get this:

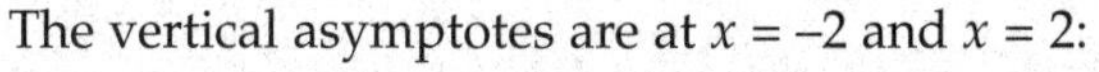

The vertical asymptotes are at $x = -2$ and $x = 2$:

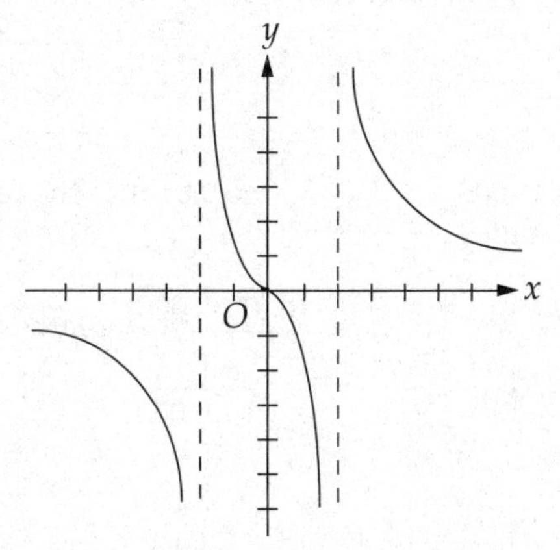

28. D

Expand the expression you're solving for and see what happens:

$$(\cos x - \sin x)^2 = \cos^2 x - 2\cos x \sin x + \sin^2 x$$
$$= (\cos^2 x + \sin^2 x) - 2\cos x \sin x$$
$$= 1 - 2(0.22) = 1 - 0.44 = 0.56$$

29. D

Sketch (or visualize) the situation. Before more water is added, the water level is half of 25, or 12.5 meters:

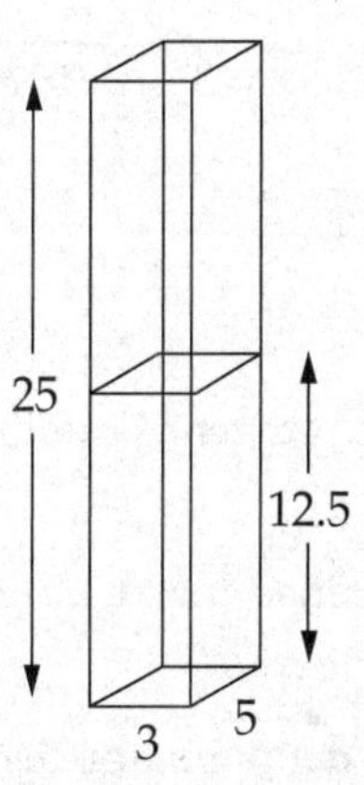

As water is poured in, the rectangular solid that represents the water gets taller, while the dimensions of the base stay the same. As water is poured in for 9 seconds at the rate of 12 cubic meters per second, the amount added is $9 \times 12 = 108$ cubic meters. With a base of 3 meters by 5 meters and the 108 cubic meters of added volume, you can figure out what the added depth is:

$$\text{Added volume} = \text{length} \times \text{width} \times \text{added depth}$$
$$108 = 5 \times 3 \times \text{added depth}$$
$$\text{added depth} = \frac{108}{15} = 7.2$$

That's on top of the pre-existing 12.5 meters. The new water level is $12.5 + 7.2 = 19.7$ meters.

30. C

The equation for a circle centered at (3,2) and with radius $r = 5$ is

$$(x-3)^2 + (y-2)^2 = 25$$

To find where the graph of this equation intersects the x-axis, set $y = 0$ and solve for x:

$$(x-3)^2 + (0-2)^2 = 25$$

$$(x-3)^2 + 4 = 25$$

$$(x-3)^2 = 21$$

$$x - 3 = \pm\sqrt{21}$$

$$x = 3 \pm \sqrt{21} \approx -1.58 \text{ or } 7.58$$

31. E

Between 0 and π, tangent is defined for anything but $\frac{\pi}{2}$. Sine is defined for all values of x, but since the sine is in the denominator here, the expression $\frac{\tan x}{\sin x}$ will be undefined when $\sin x = 0$, which is when $x = 0$ or π. So the expression $\frac{\tan x}{\sin x}$ is defined when $0 < x < \frac{\pi}{2}$ or $\frac{\pi}{2} < x < \pi$.

32. C

Use the given volume = 50 to get h in terms of x:

$$\text{Volume} = \text{length} \times \text{width} \times \text{height}$$

$$50 = x \cdot x \cdot h$$

$$h = \frac{50}{x^2}$$

The area you're looking for is equal to four lateral faces and one bottom face.

Each lateral face has area $xh = x\frac{50}{x^2} = \frac{50}{x}$, and the base has area x^2, so the total area you're looking for is

$$4\left(\frac{50}{x}\right) + x^2 = \frac{200}{x} + x^2$$

33. D

Expand, cancel, and simplify:

$$\frac{(n+2)! - (n+1)!}{n!} = \frac{(n+2)!}{n!} - \frac{(n+1)!}{n!}$$

$$= (n+2)(n+1) - (n+1)$$

$$= n^2 + 3n + 2 - n - 1$$

$$= n^2 + 2n + 1$$

That's the same as (D), $(n+1)^2$.

34. C

Plug $P = 4{,}000$, $t = 36$, and $r = 0.01$ into the formula and calculate:

$$M = \frac{rP}{1 - \left(\frac{1}{1+r}\right)^t} = \frac{(0.01)(4{,}000)}{1 - \left(\frac{1}{1+0.01}\right)^{36}} \approx 132.86$$

35. D

First find the distance from the starting point to the endpoint. The starting point is the y-intercept, which is easy to get if you put the given equation in slope-intercept form:

$$5y = 6x + 30$$
$$y = -\frac{6}{5}x + 6$$

The y-intercept is the point (0,6). The x-intercept is the point at which $y = 0$:

$$5(0) = -6x + 30$$
$$6x = 30$$
$$x = 5$$

So, the x-intercept is the point (5,0). The distance between those points is

$$\text{Distance} = \sqrt{(y_2 - y_1)^2 + (x_2 - x_1)^2}$$
$$= \sqrt{(0-6)^2 + (5-0)^2}$$
$$= \sqrt{36 + 25}$$
$$= \sqrt{61} \approx 7.81$$

At a rate of 2 units per second, it will take $\frac{7.81}{2}$ seconds to travel 7.81 units. That's about 3.91.

36. C

Use the given angle to find the ratio of the legs:

$$\frac{BC}{AC} = \tan 38^\circ \approx 0.781$$

If you call AC (the leg you're looking for) x, then you can call leg BC $0.781x$. Half the product of the legs is the area, or 15:

$$\text{Area} = \frac{1}{2}(\text{leg}_1)(\text{leg}_2)$$
$$15 = \frac{1}{2}(x)(0.781x)$$
$$0.781x^2 = 30$$
$$x^2 = \frac{30}{0.781} \approx 38.4$$
$$x \approx 6.2$$

37. E

You should know what the ranges are of (A) $y = \sin x$ and (B) $y = \cos x$. They're both $-1 \leq y \leq 1$. They both include −1 and 1 themselves, so you can eliminate (A) and (B). The upper limit of (C) is 1, but (C) can be much smaller than −1, such as when $x = -0.99$. So you can eliminate (C). (D) can never be less than zero because the numerator and denominator are nonnegative for all possible x.

The answer is (E) because as x gets very large, $\frac{x}{\sqrt{1+x^2}}$ approaches $\frac{x}{\sqrt{x^2}}$ or $\frac{x}{x} = 1$, so the value of the whole expression approaches 1; and as x gets very small (that is, very negative), $\frac{x}{\sqrt{1+x^2}}$ approaches $\frac{x}{\sqrt{x^2}} = \frac{x}{|x|} = -1$, so the value of the whole expression approaches −1.

38. B

Use the formula for the sum of an infinite geometric series. Here the first term a is 1 and the ratio r is $-\frac{1}{3}$:

$$S = \frac{a}{1-r} = \frac{1}{1-\left(-\frac{1}{3}\right)} = \frac{1}{\left(\frac{4}{3}\right)} = \frac{3}{4}$$

39. B

Think about what happens to the indicated intercepts when all x's are doubled and all y's are quadrupled. The x-intercepts become 2 and –2, and the y-intercepts become 4 and –4. Choice (B) fits.

40. B

To find the limit of this expression as n approaches infinity, think about what happens as n gets extremely large. What happens is that the n^2 terms become so huge that they dwarf all other terms into insignificance. So you can think of the expression as, in effect,

$$\frac{-2n^2}{5n^2} = -\frac{2}{5}$$

41. E

If $\log_2(x^2 - 3) = 5$, then $x^2 - 3 = 2^5 = 32$:

$$x^2 - 3 = 32$$
$$x^2 = 35$$
$$x = \pm\sqrt{35} \approx \pm 5.92$$

42. C

If 2 is a zero, then $x - 2$ is a factor. Factor that out of the polynomial $6x^3 - 11x^2 - 3x + 2$:

$$\frac{6x^3 - 11x^2 - 3x + 2}{x-2} = 6x^2 + x - 1$$

Now you have a quadratic equation:

$$6x^2 + x - 1 = 0$$
$$(3x - 1)(2x + 1) = 0$$
$$3x - 1 = 0 \text{ or } 2x + 1 = 0$$
$$x = \frac{1}{3} \text{ or } -\frac{1}{2}$$

43. C

With 5 rotations, the circle covers a total distance equal to 5 times the circumference.

The circumference of a circle of radius 1 is 2π, so 5 times that is 10π. Of that total distance 10π, $5\sqrt{2}$ is on the incline.

The other $10\pi - 5\sqrt{2}$ is the distance traveled along the x-axis, so the x-coordinate of the endpoint is $10\pi - 5\sqrt{2} \approx 24.34$.

44. B

Sine is negative in the third and fourth quadrants. Cosine is negative in the second and third, so I is not necessarily true. Cosecant is 1 over sine, so where sine is negative, so is cosecant. Therefore II must be true. As for III: Absolute values are always nonnegative. The issue here is whether $\sin x + \cos x$ can equal 0.

In fact, if x is $\frac{7\pi}{4}$, then $\sin x = -\frac{\sqrt{2}}{2}$ and $\cos x = \frac{\sqrt{2}}{2}$;

$\sin x + \cos x = 0$ when $x = \frac{7\pi}{4}$.

45. A

Square the answer choices until you find one that gives you $8 - 6i$. Start with choice (A):

$$\begin{aligned}(3-i)^2 &= (3-i)(3-i)\\ &= 9 - 6i + i^2\\ &= 9 - 6i - 1\\ &= 8 - 6i\end{aligned}$$

46. E

To find the area of the rectangle, you need the base and height. Both parabolas are symmetric with respect to the y-axis, so if the x-coordinate of point B is –1.50, the length of segment BC, which is the base of the rectangle, is twice 1.50, or 3. To find the height of the rectangle, use the y-coordinate of point B and find the y-coordinate of point A. That's the value of y when you plug $x = -1.50$ into the equation $y = 2x^2 - 8$:

$$y = 2(-1.50)^2 - 8 = -3.5$$

So the height of the rectangle is $6.75 - (-3.5) = 6.75 + 3.5 = 10.25$, and the area of the rectangle is $3 \times 10.25 = 30.75$.

47. D

Take the sine of both sides:

$$\arcsin x = \arccos (2x)$$
$$\sin(\arcsin x) = \sin[\arccos (2x)]$$
$$x = \sin[\arccos (2x)]$$

The cosine and sine of an angle are related by the Pythagorean identity $\sin^2 x + \cos^2 x = 1$. Therefore, for an angle whose cosine is $2x$, the sine is $\sqrt{1-(2x)^2} = \sqrt{1-4x^2}$:

$$x = \sin[\arccos (2x)]$$
$$x = \sqrt{1-4x^2}$$
$$x^2 = 1 - 4x^2$$
$$5x^2 = 1$$
$$x^2 = \frac{1}{5}$$
$$x = \sqrt{\frac{1}{5}} \approx 0.447$$

48. D

If $f(g(x))=g(f(x))$, then it doesn't matter which function you apply first; you should get the same result. This will happen under two sets of circumstances. The first is when the functions are inverses, in which case both $f(g(x))$ and $g(f(x))$ will get you back to the x you started with. The second is when the two functions are identical, in which case $f(g(x))$ and $g(f(x))$ are identical. II is the inverse of $f(x)$ and III is the identical function $f(x)$, so II and III are possible.

49. B

Sketch or visualize the situation:

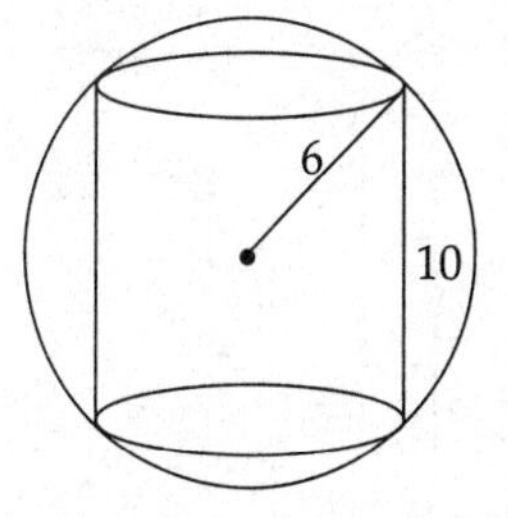

To get the volume of this cylinder, you need, in addition to the given height, the radius of the base. You can construct a right triangle and use the Pythagorean theorem to find that radius:

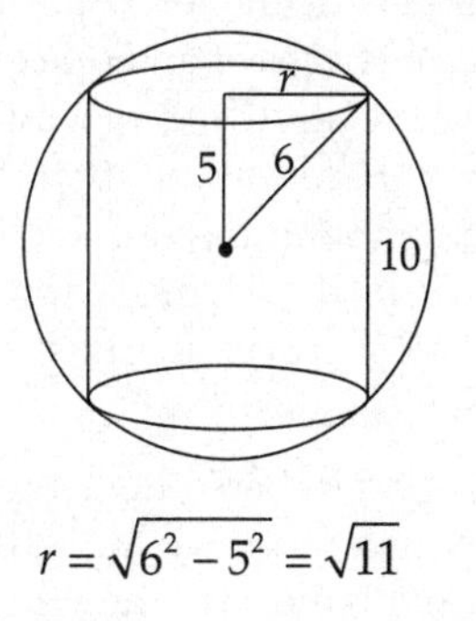

$$r = \sqrt{6^2 - 5^2} = \sqrt{11}$$

Now plug $r = \sqrt{11}$ and $h = 10$ into the cylinder volume formula:

$$\text{Volume} = \pi r^2 h = \pi\left(\sqrt{11}\right)^2 (10)$$
$$= 110\pi \approx 346$$

50. C

Look at the figure:

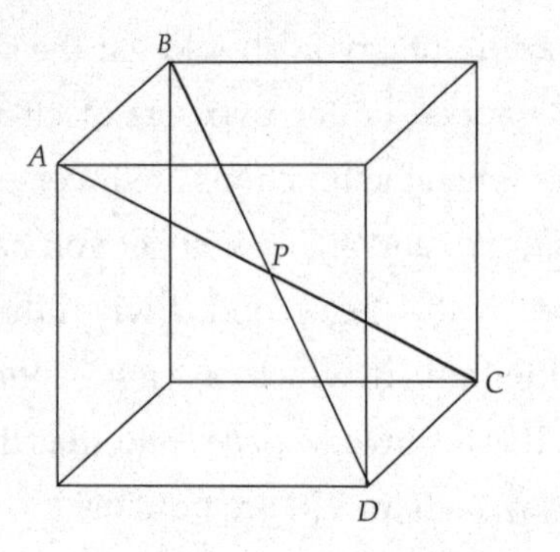

If we connect points A and D with a line segment and we connect points B and C with a line segment, then we will have a rectangle.

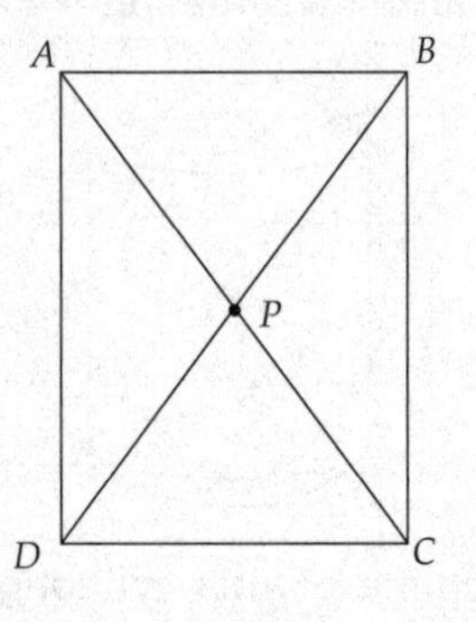

Side AB of rectangle $ABCD$ is an edge of the cube. Since all edges of a cube have the same length, we can pick a number for the length of the edge of the cube. Let's let the length of an edge of the cube be 1. So $AB = 1$. Diagonal AC of rectangle $ABCD$ connects opposite vertices A and C of the cube. Now in any rectangular solid having a length ℓ, a width w, and a height h, the formula for the distance d between opposite vertices is given by the formula $d = \sqrt{\ell^2 + w^2 + h^2}$. Here we have a cube, which is a rectangular solid where the length, width, and height are all equal. Since we are letting the edge of the cube be 1, we have that the length, width, and height are all equal to 1. So the length of AC is equal to $\sqrt{1^2 + 1^2 + 1^2} = \sqrt{1+1+1} = \sqrt{3}$.

We want the degree measure of angle APB. If we draw a perpendicular from point P to side AB of the rectangle and call the point where the perpendicular drawn from point P to side AB meets side AB point Q, then we will obtain two identical right triangles, triangles AQP and BQP.

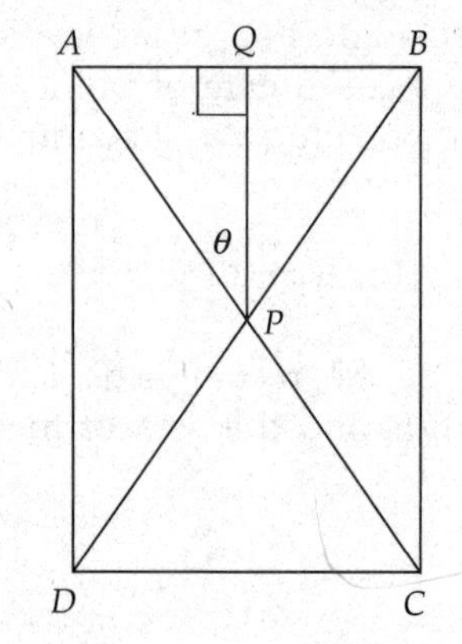

Point Q bisects AB. So the length of $\overline{AQ} = \frac{1}{2}$. Point P is the point of intersection of the diagonals, so point P is the midpoint of both diagonals. Therefore $AP = \frac{AC}{2} = \frac{\sqrt{3}}{2}$. Now angle APQ is $\frac{1}{2}$ of angle APB, whose degree measure we are seeking. Let's find the degree measure of angle APQ. Then we will take twice this number for the degree measure of angle APB.

If we refer to angle APQ by the letter θ, then we have that

$$\sin\theta = \frac{\text{opposite}}{\text{hypotenuse}}$$

$$= \frac{AQ}{AP}$$

$$= \frac{\left(\frac{1}{2}\right)}{\left(\frac{\sqrt{3}}{2}\right)}$$

$$= \frac{1}{2} \times \frac{2}{\sqrt{3}}$$

$$= \frac{1}{\sqrt{3}}$$

Thus, $\sin\theta = \frac{1}{\sqrt{3}}$. So $\theta = \arcsin\left(\frac{1}{\sqrt{3}}\right)$. Use your calculator to find that $\frac{1}{\sqrt{3}} \approx 0.57735$ and that $\theta = \arcsin\left(\frac{1}{\sqrt{3}}\right) \approx 35.26439°$. So the degree measure of angle APB is approximately $2(35.26439) = 70.52878$. This is closest to choice (C), 71. (To the nearest integer, 70.52878 is 71.)

HOW TO CALCULATE YOUR SCORE

Step 1: Figure out your raw score. Use the answer key to count the number of questions you answered correctly and the number of questions you answered incorrectly. (Do not count any questions you left blank.) Multiply the number wrong by 0.25 and subtract the result from the number correct. Round the result to the nearest whole number. This is your raw score.

SAT Subject Test: Mathematics Level 2 — Practice Test 3

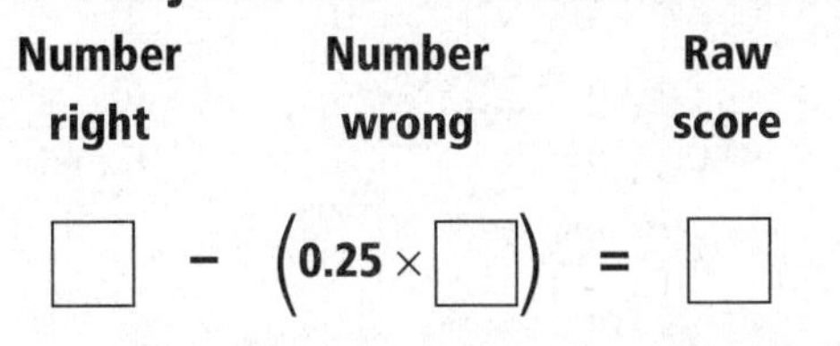

Step 2: Find your scaled score. In the Score Conversion Table below, find your raw score (rounded to the nearest whole number) in one of the columns to the left. The score directly to the right of that number will be your scaled score.

A note on your practice test scores: Don't take these scores too literally. Practice test conditions cannot precisely mirror real test conditions. Your actual SAT Subject Test: Mathematics Level 2 score will almost certainly vary from your practice test scores. However, your scores on the practice tests will give you a rough idea of your range on the actual exam.

Conversion Table

Raw	Scaled	Raw	Scaled	Raw	Scaled	Raw	Scaled
50	800	34	690	18	550	2	370
49	800	33	680	17	540	1	350
48	800	32	570	16	530	0	340
47	800	31	660	15	520	–1	330
46	800	30	650	14	520	–2	310
45	800	29	640	13	510	–3	300
44	790	28	630	12	500	–4	290
43	780	27	620	11	490	–5	280
42	770	26	610	10	480	–6	260
41	760	25	600	9	460	–7	250
40	750	24	590	8	450	–8	240
39	740	23	580	7	440	–9	220
38	730	22	580	6	420	–10	210
37	720	21	570	5	410	–11	200
36	710	20	560	4	400	–12	200
35	700	19	550	3	380		

Answer Grid
Practice Test 3

1. (A) (B) (C) (D) (E)
2. (A) (B) (C) (D) (E)
3. (A) (B) (C) (D) (E)
4. (A) (B) (C) (D) (E)
5. (A) (B) (C) (D) (E)
6. (A) (B) (C) (D) (E)
7. (A) (B) (C) (D) (E)
8. (A) (B) (C) (D) (E)
9. (A) (B) (C) (D) (E)
10. (A) (B) (C) (D) (E)
11. (A) (B) (C) (D) (E)
12. (A) (B) (C) (D) (E)
13. (A) (B) (C) (D) (E)
14. (A) (B) (C) (D) (E)
15. (A) (B) (C) (D) (E)
16. (A) (B) (C) (D) (E)
17. (A) (B) (C) (D) (E)
18. (A) (B) (C) (D) (E)
19. (A) (B) (C) (D) (E)
20. (A) (B) (C) (D) (E)
21. (A) (B) (C) (D) (E)
22. (A) (B) (C) (D) (E)
23. (A) (B) (C) (D) (E)
24. (A) (B) (C) (D) (E)
25. (A) (B) (C) (D) (E)
26. (A) (B) (C) (D) (E)
27. (A) (B) (C) (D) (E)
28. (A) (B) (C) (D) (E)
29. (A) (B) (C) (D) (E)
30. (A) (B) (C) (D) (E)
31. (A) (B) (C) (D) (E)
32. (A) (B) (C) (D) (E)
33. (A) (B) (C) (D) (E)
34. (A) (B) (C) (D) (E)
35. (A) (B) (C) (D) (E)
36. (A) (B) (C) (D) (E)
37. (A) (B) (C) (D) (E)
38. (A) (B) (C) (D) (E)
39. (A) (B) (C) (D) (E)
40. (A) (B) (C) (D) (E)
41. (A) (B) (C) (D) (E)
42. (A) (B) (C) (D) (E)
43. (A) (B) (C) (D) (E)
44. (A) (B) (C) (D) (E)
45. (A) (B) (C) (D) (E)
46. (A) (B) (C) (D) (E)
47. (A) (B) (C) (D) (E)
48. (A) (B) (C) (D) (E)
49. (A) (B) (C) (D) (E)
50. (A) (B) (C) (D) (E)

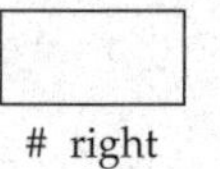

right

wrong

Use the answer key following the test to count up the number of questions you got right and the number you got wrong. (Remember not to count omitted questions as wrong.) "How to Calculate Your Score" on the back of this page will show you how to find your score.

Practice Test 3

50 Questions (1 hour)

Directions: For each question, choose the BEST answer from the choices given. If the precise answer is not among the choices, choose the one that best approximates the answer. Then fill in the corresponding oval on the answer sheet.

Notes:

(1) To answer some of these questions, you will need a calculator. You must use at least a scientific calculator, but programmable and graphing calculators are also allowed.

(2) Make sure your calculator is in the correct mode (degree or radian) for the question being asked.

(3) Figures in this test are drawn as accurately as possible UNLESS it is stated in a specific question that the figure is not drawn to scale. All figures are assumed to lie in a plane unless otherwise specified.

(4) The domain of any function f is assumed to be the set of all real numbers x for which $f(x)$ is a real number, unless otherwise indicated.

Reference Information: Use the following formulas as needed.

Right circular cone: If r = radius and h = height, then Volume = $\frac{1}{3}\pi r^2 h$; and if c = circumference of the base and ℓ = slant height, then Lateral Area = $\frac{1}{2}c\ell$.

Sphere: If r = radius, then Volume = $\frac{4}{3}\pi r^3$ and Surface Area = $4\pi r^2$.

Pyramid: If B = area of the base and h = height, then Volume = $\frac{1}{3}Bh$.

DO YOUR FIGURING HERE.

1. A certain type of account must be opened with an investment for a positive integer number of years. No other deposits or withdrawals are permitted, and the account earns 7 percent interest that is compounded annually. If \$400 were invested in the account for x years, what is the smallest possible value of x such that at the end of x years, the amount in the account will be at least 3 times the initial investment?

(A) 14
(B) 15
(C) 16
(D) 17
(E) 18

2. In Figure 1, point P is the endpoint of vector OP and point Q is the endpoint of vector OQ. When the vectors $\overline{OP}$ and $\overline{OQ}$ are added, what is the length of the resultant vector?

(A) 1.41
(B) 2.24
(C) 2.65
(D) 3.00
(E) 8.60

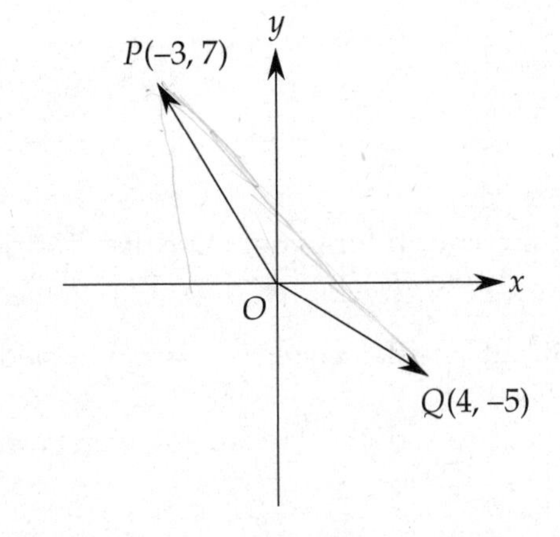

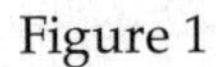

Figure 1

3. What is the area of a triangle whose vertices are $(0, 6\sqrt{3})$, $(\sqrt{35}, 7)$, and $(0, 3)$?

(A) 15.37
(B) 17.75
(C) 21.87
(D) 25.61
(E) 39.61

GO ON TO THE NEXT PAGE

4. The radius of right circular cone A is $\frac{1}{5}$ of the radius of right circular cone B, and the height of right circular cone A is $\frac{1}{4}$ of the height of right circular cone B. What is the ratio of the volume of right circular cone A to the volume of right circular cone B?

(A) $\frac{1}{16}$

(B) $\frac{1}{25}$

(C) $\frac{1}{64}$

(D) $\frac{1}{80}$

(E) $\frac{1}{100}$

5. The greatest possible distance between any two points on the surface of a right circular cylinder is $\sqrt{193}$ and the area of the circular base of the right circular cylinder is 36π. What is the volume of the right circular cylinder?

(A) 252π
(B) 294π
(C) 343π
(D) 386π
(E) $1{,}008\pi$

6. In Figure 2, the length of XY is 48. What is the length of YZ?

(A) 16.4
(B) 70.8
(C) 95.1
(D) 118.0
(E) 140.3

DO YOUR FIGURING HERE.

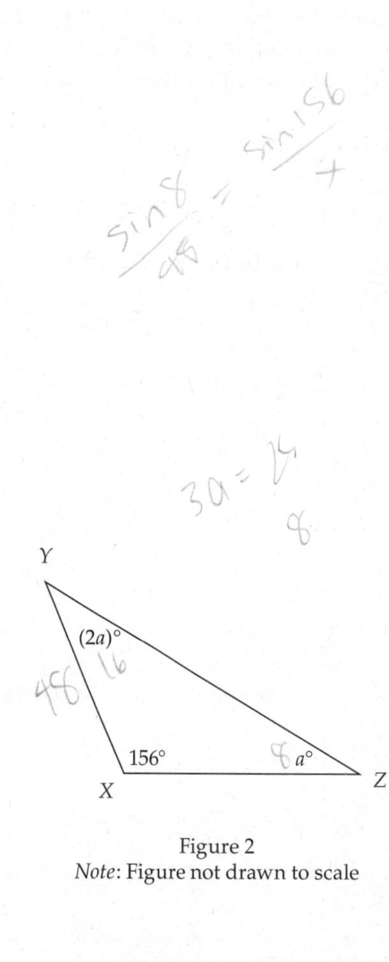

Figure 2
Note: Figure not drawn to scale

GO ON TO THE NEXT PAGE

7. In Figure 3, $STUV$ is a parallelogram with a perimeter of 14. What is the y-coordinate of point T?

(A) 1.26
(B) 1.89
(C) 3.26
(D) 3.89
(E) 4.26

8. The mean of a finite set S of numbers is 14, the median of this set of numbers is 12, and the standard deviation is 1.8. A new set T is formed by multiplying each member of the set S by 3. Which of the following statements must be true of the set T?

I. The mean of the numbers in set T is 42.
II. The median of the numbers in set T is 36.
III. The standard deviation of the numbers in set T is 5.4.

(A) I only
(B) II only
(C) I and II only
(D) I and III only
(E) I, II, and III

DO YOUR FIGURING HERE.

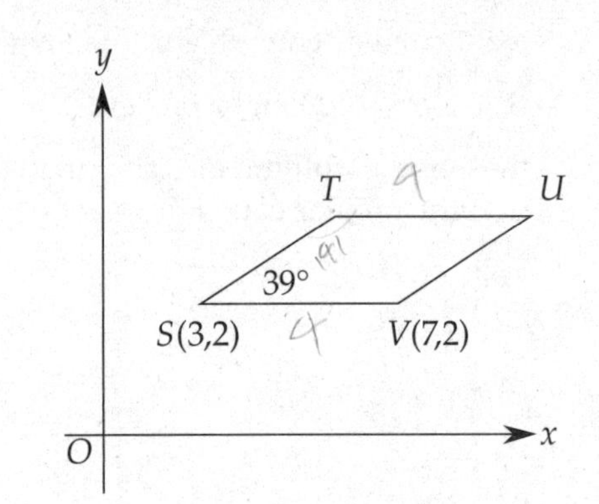

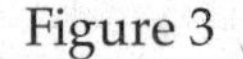
Figure 3

GO ON TO THE NEXT PAGE

DO YOUR FIGURING HERE.

9. There are 7 orange disks and 5 green disks in bag X and there are 5 orange disks and 15 green disks in bag Y. If one disk is selected at random from each bag, what is the probability that both disks selected are green?

(A) $\frac{5}{48}$

(B) $\frac{7}{48}$

(C) $\frac{5}{16}$

(D) $\frac{7}{16}$

(E) $\frac{3}{4}$

10. The terms of a sequence are defined by $a_n = 3a_{n-1} - a_{n-2}$ for $n > 2$. What is the value of a_5 if $a_1 = 4$ and $a_2 = 3$?

(A) 12
(B) 23
(C) 25
(D) 31
(E) 36

11. If $\frac{4}{y} + 4 = \frac{20}{y} + 20$, then what is the value of $\frac{4}{y} + 4$?

(A) −1
(B) 0
(C) 1
(D) 4
(E) 8

GO ON TO THE NEXT PAGE

DO YOUR FIGURING HERE.

12. $\dfrac{1}{\frac{x}{y}+\frac{y}{x}} =$

(A) $\dfrac{xy}{x^2+y^2}$

(B) $\dfrac{x^2+y^2}{xy}$

(C) $\dfrac{x^2+y^2}{2xy}$

(D) $\dfrac{xy}{x+y}$

(E) $\dfrac{(x+y)^2}{x^2+y^2}$

13. One complete cycle of the graph of $y = -\cos x$ is shown in the Figure 4. What are the coordinates of the point at which the maximum possible value of y occurs?

(A) $\left(\dfrac{\pi}{2}, 0\right)$

(B) $\left(\dfrac{\pi}{2}, \pi\right)$

(C) $(\pi, 1)$

(D) $\left(\dfrac{3\pi}{2}, -1\right)$

(E) $(2\pi, 1)$

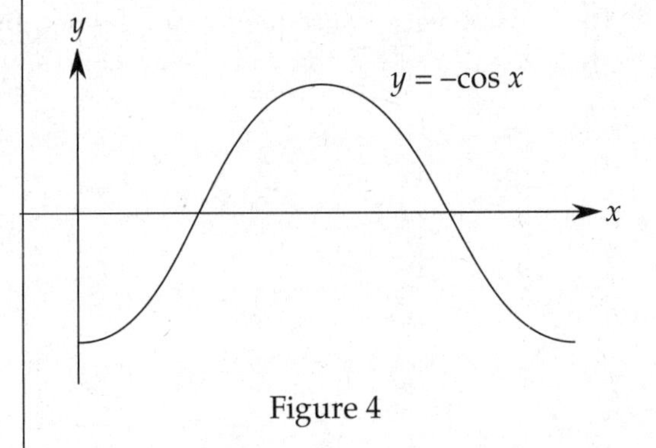

Figure 4

GO ON TO THE NEXT PAGE

14. Which of the following CANNOT occur when a line is in the same plane as a triangle?

    (A) The points of the line inside the triangle and on the perimeter of the triangle divide the triangle into a triangle and a quadrilateral.
    (B) The line has exactly three points in common with the perimeter of the triangle.
    (C) The line has exactly one point in common with the perimeter of the triangle.
    (D) The triangle and the line have infinitely many points in common.
    (E) The points of the line inside the triangle and on the perimeter of the triangle divide the triangle into two isosceles triangles.

DO YOUR FIGURING HERE.

15. What is the value of x if $\sqrt{\frac{x}{7}} = 2.74$?

    (A) 52.55
    (B) 57.54
    (C) 94.87
    (D) 105.11
    (E) 367.87

16. In Figure 5, which of the following is equal to $\csc \theta$?

    (A) $\frac{a}{\sqrt{a^2 + b^2}}$

    (B) $\frac{b}{\sqrt{a^2 + b^2}}$

    (C) $\frac{b}{a}$

    (D) $\frac{\sqrt{a^2 + b^2}}{a}$

    (E) $\frac{\sqrt{a^2 + b^2}}{b}$

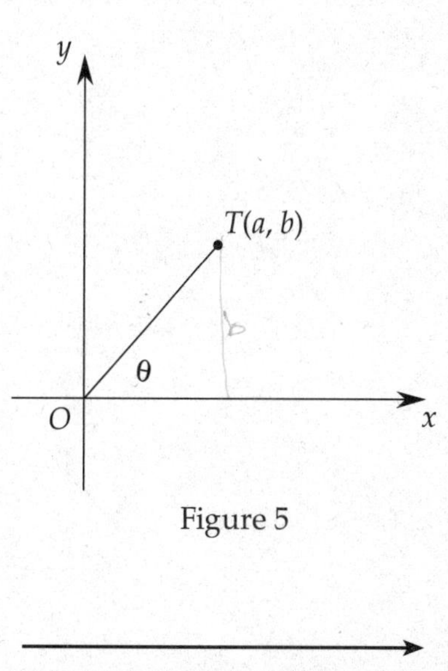

Figure 5

GO ON TO THE NEXT PAGE

17. If $f(x) = \sqrt{x^2 - 3x + 6}$ and $g(x) = \frac{156}{x+17}$, then what is the value of $g(f(4))$?

(A) 5.8
(B) 7.4
(C) 7.7
(D) 8.2
(E) 10.3

DO YOUR FIGURING HERE.

18. If $xyz \neq 0$ and $30x^{-5}y^{12}z^{-8} = 10x^{-6}y^{5}z^{4}$, then what is the value of x in terms of y and z ?

(A) $\frac{z^{12}}{3y^7}$

(B) $\frac{3z^{12}}{y^7}$

(C) $\frac{3y^{12}}{z^7}$

(D) $\frac{3}{y^7z^4}$

(E) $\frac{3y^{17}}{z^4}$

19. In Figure 6, $PQRS$ is a square. What is the slope of segment QR ?

(A) $-\frac{10}{3}$

(B) $-\frac{7}{3}$

(C) $-\frac{4}{3}$

(D) $\frac{3}{7}$

(E) $\frac{7}{3}$

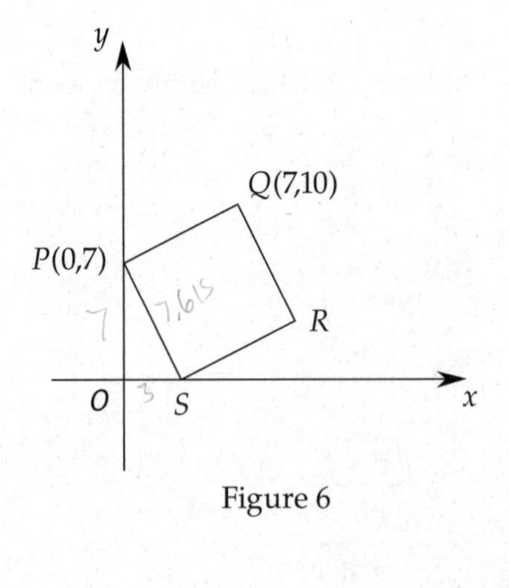

Figure 6

GO ON TO THE NEXT PAGE

20. When defined, $\tan(3x)\cot(3x) =$

(A) -1

(B) $\frac{\sqrt{3}}{3}$

(C) 1

(D) $\tan^3 3x$

(E) $\sec^2 3x - 1$

21. If 4 and 0 are both solutions to the equation $q(x) = 0$, where $q(x)$ is a polynomial, then it can be concluded that a factor of $q(x)$ is

(A) x^2

(B) $(x - 4)^2$

(C) $x^2 + 4x$

(D) $x^2 - 8x$

(E) $x^2 - 4x$

22. Which of the following could be $g(x)$ if $f(x) = 5x^2 + 4$ and $f(g(3)) = 84$?

(A) $3x - 10$

(B) $4x - 7$

(C) $6x - 17$

(D) $x^2 - 5$

(E) $x^2 - 3$

23. Figure 7 shows line n in a rectangular coordinate system. An equation of line n is

(A) $x = 4$

(B) $y = 4$

(C) $x = 5$

(D) $y = \frac{5}{4}x$

(E) $y = x + 1$

DO YOUR FIGURING HERE.

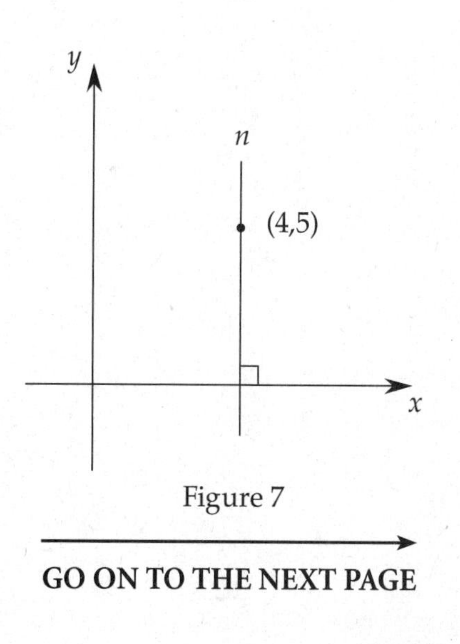

Figure 7

GO ON TO THE NEXT PAGE

DO YOUR FIGURING HERE.

24. A student's mean score on 5 tests was 84. The student's mean score on the first 4 of these tests was 87. What was the student's score on the fifth of these tests?

(A) 68
(B) 72
(C) 75
(D) 81
(E) 96

25. If $0 < x < \frac{\pi}{2}$ and $\cos x = 0.34$, what is the value of $\sin\left(\frac{x}{2}\right)$?

(A) 0.574
(B) 0.733
(C) 0.819
(D) 0.917
(E) 0.941

26. During a seven-day period, a company produced 49,812 items that it considered acceptable and 21,348 items that it considered unacceptable. If the company produced a total of 10,830 items on the first day and the percent of the items produced on the first day that the company considered acceptable was the same percent of items that the company considered acceptable for the entire seven-day period, how many items produced by the company on the first day did the company consider acceptable?

(A) 3,508
(B) 4,332
(C) 5,415
(D) 7,581
(E) 8,664

GO ON TO THE NEXT PAGE

DO YOUR FIGURING HERE.

27. For every pair (x,y) in the rectangular coordinate plane, $f: (x,y) \rightarrow (x, -8x + 3y)$. What is the set of points for which $f: (x,y) \rightarrow (x,y)$?

(A) The point $(-4,0)$
(B) The point $(4,12)$
(C) The set of points (x,y) that satisfy the equation $x = 4y$
(D) The set of points (x,y) that satisfy the equation $y = 4x$
(E) The set of points (x,y) that satisfy the equation $y = 8x$

28. When the number x is subtracted from each of the numbers 8, 16, and 40, the three numbers that result form a geometric progression. What is the value of x ?

(A) 3
(B) 4
(C) 6
(D) 12
(E) 18

29. If $f(x) = ax^2 + bx + c$, $f(-1) = -18$, and $f(1) = 10$, what is the value of b ?

(A) −12
(B) −4
(C) 14
(D) 21
(E) 28

30. What is the domain of the function $f(x) = \sqrt{x^2 + 3}$?

(A) $-1.73 \leq x \leq 1.73$
(B) $-1.32 \leq x \leq 1.32$
(C) $x > 1.32$
(D) $x > 1.73$
(E) All real numbers

GO ON TO THE NEXT PAGE

31. The function f is defined by $f(x) = \dfrac{180}{x+3}$ for $x \geq 0$, and $f(x) = 60$ for $x < 0$. Figure 8 shows the graph of $y = f(x)$. What is the sum of the areas of the three shaded rectangles?

(A) 111
(B) 135
(C) 141
(D) 180
(E) 195

32. Which of the following are the equations of lines that are asymptotes of the graph of $y = \dfrac{x^2 - 64}{(3x+4)(x-5)}$?

I. $x = -8$
II. $x = 5$
III. $y = \dfrac{1}{3}$

(A) I only
(B) II only
(C) I and II only
(D) II and III only
(E) I, II, and III

33. If $f(x) = 7x + 12$ and $f(g(x)) = 21x^2 + 40$, then which of the following is $g(x)$?

(A) $21x^2 + 28$
(B) $21x^2$
(C) $7x^2 + 4$
(D) $3x^2 + 28$
(E) $3x^2 + 4$

DO YOUR FIGURING HERE.

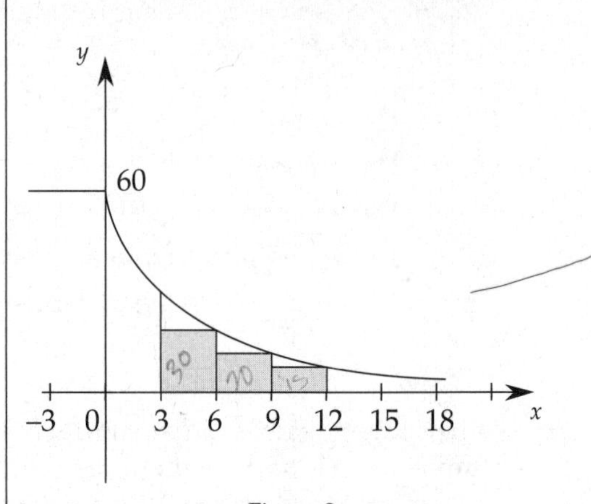

Figure 8
Note: Figure not drawn to scale

GO ON TO THE NEXT PAGE

34. A circle is tangent to the lines with the equations $x = 5$ and $y = 7$. Which of the following could be the coordinates of the center of the circle?

(A) (3,7)
(B) (8,4)
(C) (10,8)
(D) (10,14)
(E) (17,21)

35. What is the range of the function f that is defined by

$$f(x) = \begin{cases} 3^{\frac{1}{x^2+1}}, \text{ if } x \geq 0 \\ 5x+3, \text{ if } x < 0 \end{cases} ?$$

(A) $y \leq 0$
(B) $0 < y < 3$
(C) $y \leq 3$
(D) $3 \leq y \leq 5$
(E) All real numbers

36. If $x^2 - 7y = 8$, $x - y = 1$, and $y > 0$, what is the value of y?

(A) 1.64
(B) 4.78
(C) 5.64
(D) 6.14
(E) 7.28

DO YOUR FIGURING HERE.

GO ON TO THE NEXT PAGE

DO YOUR FIGURING HERE.

37. If $f(x) = \log_7 \frac{x}{8}$ for $x \geq 8$, then for the values x of its domain, $f^{-1}(x) =$

(A) $8(7^x)$
(B) $7(8^x)$
(C) $8(7^{x+1})$
(D) $\log_7 \frac{x}{8}$
(E) $\frac{x}{8}$

38. If $x_1 = 2$ and $x_{n+1} = \sqrt{x_n^2 + 8}$, then $x_4 =$

(A) 3.46
(B) 4.47
(C) 5.29
(D) 8.49
(E) 10.39

39. In Figure 9, C is the center of the semicircle, and the area of the semicircle is 8π. What is the area of triangle ABC in terms of θ?

(A) $2 \sin \theta \tan \theta$
(B) $4 \sin \theta$
(C) $8 \cos \theta$
(D) $8 \sin \theta$
(E) $8 \sin \theta \tan \theta$

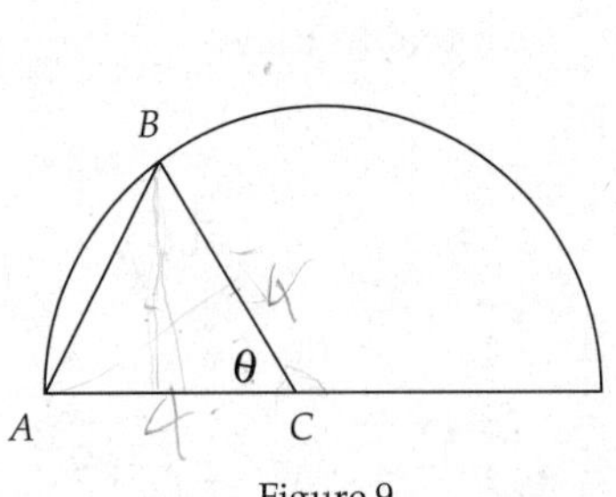

Figure 9

40. Exactly 70 percent of the people in each of 3 rooms are seniors at university X. If one person is selected at random from each of the 3 rooms, what is the probability that each of the 3 people selected is a senior at university X?

(A) 0.2401
(B) 0.343
(C) 0.49
(D) 0.64
(E) 0.7

GO ON TO THE NEXT PAGE

41. What is the value of x if $4.18^x = 36.54$?

(A) 0.86
(B) 1.43
(C) 1.80
(D) 2.17
(E) 2.52

42. If $\sin(\arcsin x) = \frac{\sqrt{2}}{4}$, then what is the value of x ?

(A) $\frac{\sqrt{2}}{4}$
(B) $\frac{\sqrt{7}}{7}$
(C) $\frac{\sqrt{2}}{2}$
(D) $\frac{\sqrt{14}}{4}$
(E) $\frac{2\sqrt{2}}{3}$

43. The sum of the first 25 terms of an arithmetic sequence is 1,400, and the 25th term is 104. If the first term of the sequence is a_1 and the second term is a_2, what is the value of $a_2 - a_1$?

(A) −3
(B) 2
(C) 4
(D) 5
(E) 8

DO YOUR FIGURING HERE.

GO ON TO THE NEXT PAGE

44. For all θ, $\sin \theta + \sin (\theta + \pi) + \sin (2\pi + \theta) =$

(A) $-\sin \theta$
(B) $\sin \theta$
(C) $2 \sin \theta$
(D) $3 \sin \theta$
(E) $2 \sin \theta + \cos \theta$

45. If n is a positive integer, then $\frac{(n+4)!}{(n+7)!} =$

(A) $\frac{1}{(n+1)(n+2)(n+3)}$

(B) $\frac{1}{n+5}$

(C) $\frac{1}{(n+5)(n+6)}$

(D) $\frac{1}{(n+6)(n+7)}$

(E) $\frac{1}{(n+5)(n+6)(n+7)}$

46. The graph of $g(x) = x^3 + 1$ was translated 4 units to the right and 2 units up, resulting in a new graph $h(x)$. What is the value of $h(3.7)$?

(A) 0.973
(B) 1.784
(C) 1.973
(D) 2.027
(E) 2.973

DO YOUR FIGURING HERE.

GO ON TO THE NEXT PAGE

47. A five-letter code is formed by selecting 5 different letters from the 12 letters A, B, C, D, E, F, G, H, I, J, K, and L and placing these 5 letters in the 5 spaces shown in Figure 10. Which of the following expressions is the number of different five-letter codes that are possible?

(A) $\frac{12!}{4!8!}$

(B) $\frac{12!}{(5!)(7!)}$

(C) $\frac{12!}{7!}$

(D) $\frac{12!}{5!}$

(E) $\frac{12!}{4!}$

Figure 10

48. Which of the following sets of real numbers is such that if x is an element of the set and y is an element of the set, then the sum of x and y is an element of the set?

I. The set of negative integers
II. The set of rational numbers
III. The set of irrational numbers

(A) None
(B) I only
(C) I and II only
(D) II and III only
(E) I, II, and II

49. If the length of the major axis of an ellipse with the equation $5x^2 + 24y^2 = 40$ is j and the length of the minor axis of the ellipse is n, then what is the value of $j + n$?

(A) 2.58
(B) 5.66
(C) 6.95
(D) 8.24
(E) 9.78

GO ON TO THE NEXT PAGE

50. Which of the following describes the values of x for which $\frac{1-5x}{x^2+1}$ is negative?

(A) $x > 0$

(B) $x > \frac{1}{5}$

(C) $x < \frac{1}{5}$

(D) $0 < x < \frac{1}{5}$

(E) None of the above

DO YOUR FIGURING HERE.

STOP!

If you finish before time is up, you may check your work.

Turn the page
for answers and explanations
to Practice Test 3.

Answer Key
Practice Test 3

1. D
2. B
3. C
4. E
5. A
6. E
7. D
8. E
9. C
10. D
11. B
12. A
13. C
14. B
15. A
16. E
17. C
18. A
19. B
20. C
21. E
22. D
23. A
24. B
25. A
26. D
27. D
28. B
29. C
30. E
31. C
32. D
33. E
34. B
35. C
36. D
37. A
38. C
39. D
40. B
41. E
42. A
43. C
44. B
45. E
46. E
47. C
48. C
49. D
50. B

ANSWERS AND EXPLANATIONS

1. D

The amount in the account at the end of the x years must be at least 3($400) = $1,200. At the end of x years, the amount in the account is ($400)($1.07^x$). You must find the smallest possible value of x such that ($400)($1.07^x$) ≥ $1,200. Thus, $400(1.07^x) \geq 1{,}200$. Solve this equation using natural logarithms.

$$400(1.07^x) \geq 1{,}200$$
$$1.07^x \geq 3$$
$$x(\ln 1.07) \geq \ln 3$$
$$x \geq \frac{\ln 3}{\ln 1.07}$$

Using the calculator, $\frac{\ln 3}{\ln 1.07} \approx 16.23757\ldots$

So $x \geq 16.23757\ldots$.

Since x must be an integer, 16 years will not be enough time for the initial amount invested to triple. You must have $x \geq 17$. The smallest possible value of x is 17.

2. B

The sum of vectors $\overrightarrow{OP}$ and $\overrightarrow{OQ}$ is equal to $(-3, 7) + (4, -5) = (-3 + 4, 7 + (-5)) = (1, 2)$.

Then $\left|\overrightarrow{OP} + \overrightarrow{OQ}\right| = \sqrt{1^2 + 2^2} = \sqrt{1+4} = \sqrt{5} \approx 2.24$.

3. C

Draw a picture showing the triangle.

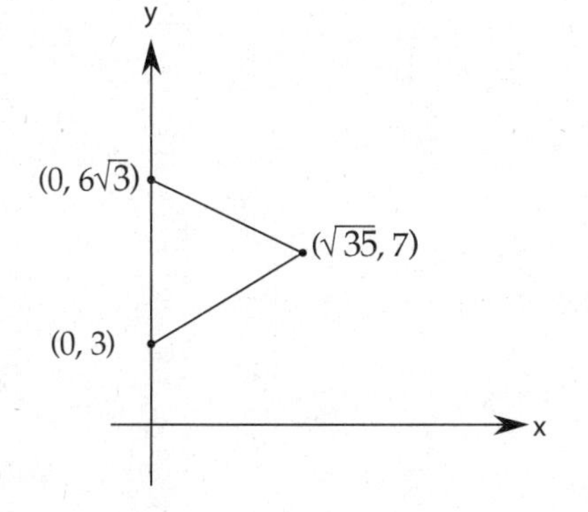

The area of a triangle is $\frac{1}{2} \times \text{base} \times \text{height}$. The base of this triangle can be considered to be the side along the y-axis. The length of this side is $6\sqrt{3} - 3$. The height of this triangle drawn to the side along the y-axis is the distance from the point $(\sqrt{35}, 7)$ to the y-axis. This distance is the x-coordinate of the point $(\sqrt{35}, 7)$, which is $\sqrt{35}$. The area of the triangle is $\frac{1}{2}(6\sqrt{3} - 3)(\sqrt{35}) \approx 21.87$.

4. E

The volume of a right circular cone is $\frac{1}{3}\pi r^2 h$. If the radius of cone B is x and the height is y, then the area is $\frac{1}{3}\pi x^2 y$. The radius of cone A will be $\frac{1}{5}x$ and the height will be $\frac{1}{4}y$, so the volume is $\frac{1}{3}\pi\left(\frac{1}{5}x\right)^2\left(\frac{1}{4}y\right) = \frac{1}{300}\pi x^2 y$. The ratio of the volume of cone A to that of cone B is therefore

$$\frac{\frac{1}{300}\pi x^2 y}{\frac{1}{3}\pi x^2 y} = \frac{1}{300} \div \frac{1}{3} = \frac{1}{100}$$

5. A

The volume of a right circular cylinder having a radius r and a height h is $\pi r^2 h$. To find the volume of this right circular cylinder, you need its radius and its height. Now the area of a circle with a radius r is πr^2. The area of the circular base is 36π. If the radius of the circular base is r, then $\pi r^2 = 36\pi$, $r^2 = 36$, and $r = 6$. You now have the radius. The diameter is $2(6) = 12$. Let's call the height of the right circular cylinder h and draw a picture.

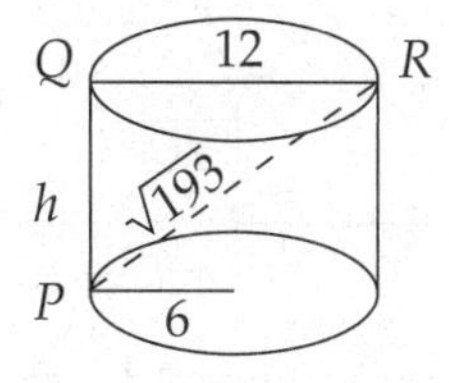

The greatest possible distance between two points on the surface of the right circular cylinder is the distance between points P and R. Use the Pythagorean theorem in right triangle PQR to find the height h.

$$12^2 + h^2 = (\sqrt{193})^2$$

$$144 + h^2 = 193$$

$$h^2 = 49$$

$$h = 7$$

The volume of the right circular cylinder is $\pi r^2 h = \pi(6^2)(7) = \pi(36)(7) = 252\pi$.

6. E

Find a first. $a + 2a + 156 = 180$, $3a + 156 = 180$, $3a = 24$, and $a = 8$. Now use the law of sines. $\frac{YZ}{\sin 156°} = \frac{48}{\sin 8°}$. Then $YZ \approx 140.3$.

7. D

The perimeter of any polygon is the sum of the lengths of its sides. Since points S and V have the same y-coordinate of 2, SV is parallel to the x-axis. The length of SV is the positive difference of the x-coordinates of points V and S. The length of SV is $7 - 3 = 4$. Opposite sides of a parallelogram are equal, so the length of TU is also 4. ST and UV are opposite sides of parallelogram $STUV$, so ST and UV also have equal lengths. The perimeter of the parallelogram is $2(4) + 2(ST)$. So $2(4) + 2(ST) = 14$, $8 + 2(ST) = 14$, $2(ST) = 6$, and $ST = 3$. Drop a perpendicular line from point T to SV that meets side SV at point P.

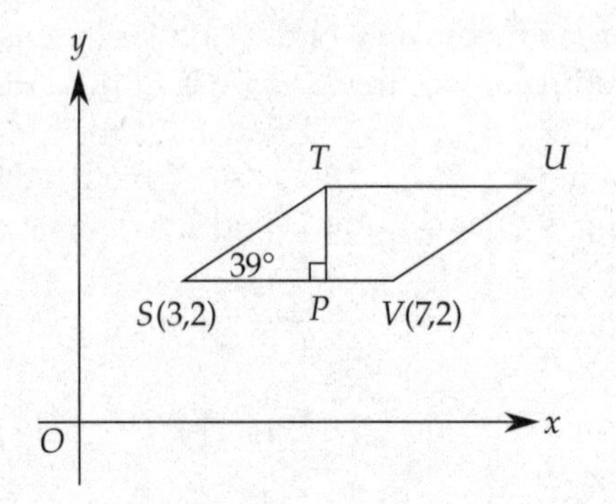

Since point P is on segment SV, the y-coordinate of point P is also 2. Then the y-coordinate of point T is equal to 2 plus the length of TP. Now $\sin 39° = \frac{TP}{ST}$. Since $ST = 3$, $\sin 39° = \frac{TP}{3}$, so $TP = 3 \sin 39°$. The y-coordinate of T is $2 + 3 \sin 39° \approx 3.89$.

8. E

Consider statement I. When every number in set T is multiplied by 3, the average is multiplied by 3. The average of the numbers in set T must be $3(14) = 42$. Statement I must be true. Eliminate (B), which does not contain I.

You can also show that statement I is true algebraically. If the members of the set S are $x_1, x_2, x_3, \cdots, x_n$, then the average of the numbers in set S is $\frac{x_1 + x_2 + \cdots + x_n}{n}$. The members of set T are $3x_1, 3x_2, 3x_3, \cdots, 3x_n$. The average of the numbers in set T is $\frac{3x_1 + 3x_2 + \cdots + 3x_n}{n} = \frac{3(x_1 + x_2 + \cdots + x_n)}{n} = 3\left(\frac{x_1 + x_2 + \cdots + x_n}{n}\right)$. The average $3\left(\frac{x_1 + x_2 + \cdots + x_n}{n}\right)$ of the numbers in set T is 3 times the average $\frac{x_1 + x_2 + \cdots + x_n}{n}$ of the numbers in set S. Thus, the average of the numbers in set T is $3(14) = 42$.

Now consider statement II. Suppose that the members of set S, in increasing order, are $y_1, y_2, y_3, \ldots, y_n$. Then the members of set T, in increasing order, are $3y_1, 3y_2, 3y_3, \ldots, 3y_n$. If there is an odd number of numbers in set S—that is, if n is odd—then the middle term among $y_1, y_2, y_3, \ldots, y_n$, which is the median of set S, was multiplied by 3 and is the middle term among $3y_1, 3y_2, 3y_3, \ldots, 3y_n$, which is the median of set T. If there is an even number of numbers in set S, then the median of the numbers in set S is the average of the two middle terms. Since the two middle terms among $3y_1, 3y_2, 3y_3, \ldots, 3y_n$ are the two corresponding middle terms among $y_1, y_2, y_3, \ldots, y_n$ that were multiplied by 3, the average of the two middle terms among $3y_1, 3y_2, 3y_3, \ldots, 3y_n$ must be the average of the two middle terms among $y_1, y_2, y_3, \ldots, y_n$ multiplied by 3. Thus, the median of set T must be 3(12) = 36. Statement II must be true, so eliminate (A) and (D) which do not contain II.

Consider statement III. If each number is multiplied by 3, the numbers are 3 times more dispersed. So the standard deviation of the numbers in set T must be 3 times the standard deviation of the numbers in set S. The standard deviation of the numbers in set T is 3(1.8) = 5.4. Statement III must be true. Statements I, II, and III must all be true, and (E) is correct.

You can also show that statement III is true algebraically. If the numbers in set S are $x_1, x_2, x_3, \ldots, x_n$, then the standard deviation of the numbers in set S is $\frac{\sqrt{(x_1-\overline{x})^2+(x_2-\overline{x})^2+\cdots+(x_n-\overline{x})^2}}{n}$, where $\overline{x}$ is the average of the numbers in set S.

Thus, $\overline{x}=\frac{x_1+x_2+\cdots+x_n}{n}$. You saw when considering statement I that the average of the numbers in set T is $3\overline{x}$, where $\overline{x}$ is the average of the numbers in set S. The standard deviation of the numbers in set T is

$$\frac{\sqrt{(3x_1-3\overline{x})^2+(3x_2-3\overline{x})^2+\cdots+(3x_n-3\overline{x})^2}}{n}$$
$$=\frac{\sqrt{[3(x_1-\overline{x})]^2+[3(x_2-\overline{x})]^2+\cdots+[3(x_n-\overline{x})]^2}}{n}$$
$$=\frac{\sqrt{9(x_1-\overline{x})^2+9(x_2-\overline{x})^2+\cdots+9(x_n-\overline{x})^2}}{n}$$
$$=\sqrt{9\left(\frac{(x_1-\overline{x})^2+(x_2-\overline{x})^2+\cdots+(x_n-\overline{x})^2}{n}\right)}$$
$$=\sqrt{9}\frac{\sqrt{(x_1-\overline{x})^2+(x_2-\overline{x})^2+\cdots+(x_n-\overline{x})^2}}{n}$$
$$=3\left(\frac{\sqrt{(x_1-\overline{x})^2+(x_2-\overline{x})^2+\cdots+(x_n-\overline{x})^2}}{n}\right)$$

So the standard deviation of the numbers in set T, $3\left(\frac{\sqrt{(x_1-\overline{x})^2+(x_2-\overline{x})^2+\cdots+(x_n-\overline{x})^2}}{n}\right)$, is 3 times the standard deviation of the numbers in set S, $\frac{\sqrt{(x_1-\overline{x})^2+(x_2-\overline{x})^2+\cdots+(x_n-\overline{x})^2}}{n}$

9. C

In bag X, there are 5 green disks among a total of 7 + 5 = 12 disks. The probability of choosing a green disk from bag X is $\frac{5}{12}$. In bag Y, there are 15 green disks among a total of 15 + 5 = 20 disks. The probability of choosing a green disk from bag Y is $\frac{15}{20}$, or $\frac{3}{4}$. Since the selections of the disks from the bags are independent, the probability that both disks selected are green is $\left(\frac{5}{12}\right)\left(\frac{3}{4}\right)=\frac{15}{48}=\frac{5}{16}$.

10. D

Letting $n = 3$ in the defining equation $a_n = 3a_{n-1} - a_{n-2}$, you have $a_3 = 3a_2 - a_1 = 3(3) - 4 = 9 - 4 = 5$. Letting $n = 4$ in the defining equation $a_n = 3a_{n-1} - a_{n-2}$, you have $a_4 = 3a_3 - a_2 = 3(5) - 3 = 15 - 3 = 12$. Letting $n = 5$ in the defining equation, you have $a_5 = 3a_4 - a_3 = 3(12) - 5 = 36 - 5 = 31$.

11. B

Solve the equation $\frac{4}{y} + 4 = \frac{20}{y} + 20$ for y. You have $4 = \frac{16}{y} + 20, -16 = \frac{16}{y}, -16y = 16$, and $y = -1$. Then $\frac{4}{y} + 4 = \frac{4}{-1} + 4 = -4 + 4 = 0$. Notice that incorrect choice (A), –1, is the value of y. The question requires finding the value of $\frac{4}{y} + 4$.

Here is a second way to solve this. The equation of the question stem is $\frac{4}{y} + 4 = \frac{20}{y} + 20$. The right side of the equation is 5 times the left side. Factoring 5 out

of the right side of this equation, you have $\frac{4}{y} + 4 = 5\left(\frac{4}{y} + 4\right)$. When a number equals 5 times itself, the number must be 0. Thus, $\frac{4}{y} + 4 = 0$.

12. A

$$\frac{1}{\frac{x}{y} + \frac{y}{x}} = \frac{1}{\left(\frac{x}{y}\right)\left(\frac{x}{x}\right) + \left(\frac{y}{x}\right)\left(\frac{y}{y}\right)} = \frac{1}{\frac{x^2}{xy} + \frac{y^2}{xy}} = \frac{1}{\left(\frac{x^2 + y^2}{xy}\right)} = \frac{xy}{x^2 + y^2}$$

13. C

The graph of the function $y = -\cos x$ has $y = -1$ at $x = 0$; y increases to 0 at $x = \frac{\pi}{2}$, y continues increasing until it is 1 at $x = \pi$, then y decreases to 0 at $x = \frac{3\pi}{2}$, and y continues decreasing until it reaches –1 again at the end of the cycle at $x = 2\pi$. The coordinates of the point where y is a maximum are $(\pi,1)$.

14. B

The following figure shows that choice (A) can occur.

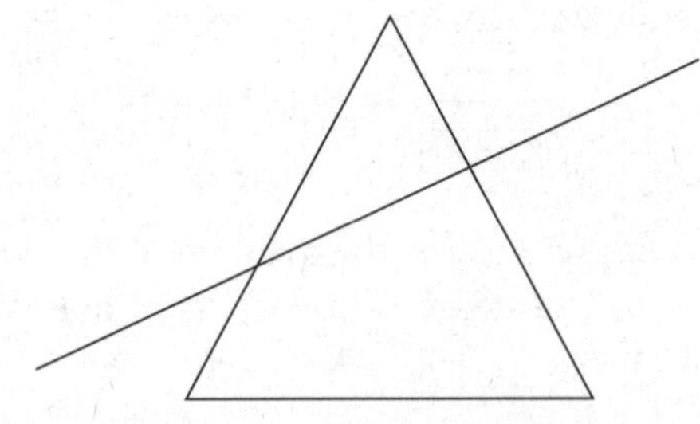

Choice (B) is correct. It cannot occur.

Drawing a line through the vertex of a triangle shows that choice (C) can occur. This is shown in this figure.

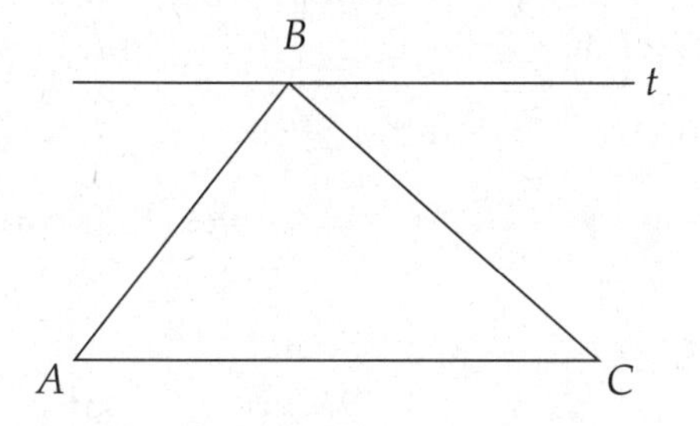

Choice (D) can occur when the line contains one side of the triangle. This is shown in this figure.

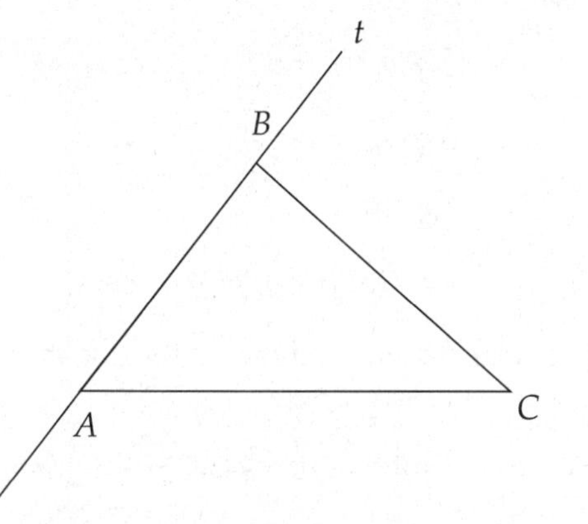

Line t contains side AB of the triangle.

Choice (E) can occur when a line divides an isosceles right triangle into two smaller identical isosceles right triangles. This is shown in this figure.

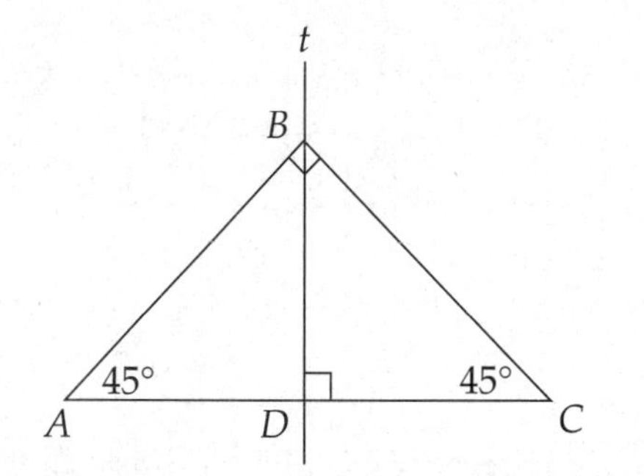

Line t divides isosceles right triangle ABC into smaller identical isosceles right triangles ABD and BCD.

It is also true that if the vertex opposite the hypotenuse of any right triangle is connected to the midpoint of the hypotenuse, the right triangle is divided into two isosceles triangles. Thus, a line containing the line segment whose endpoints are (1) the vertex of the right triangle that is opposite the hypotenuse of a right triangle and (2) the midpoint of the hypotenuse divides the right triangle into two isosceles triangles. However, if the original right triangle is not an isosceles right triangle, the two isosceles triangles formed from the original right triangle are not identical.

15. A

Solve the equation $\sqrt{\frac{x}{7}} = 2.74$ for x. Squaring both sides, $\frac{x}{7} = 2.74^2$. So $x = 7(2.74^2)$. Using the calculator, $x \approx 52.55$.

16. E

For any point (u,v), $\csc\theta = \frac{r}{v}$, where r is the distance from (u,v) to the origin. See the drawing below.

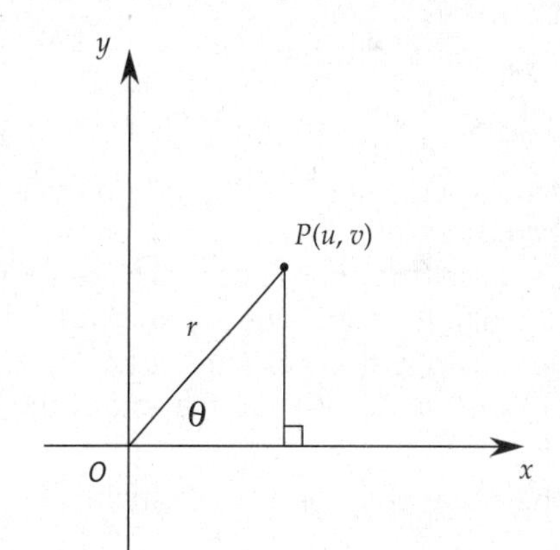

In this question, the distance r from point (a,b) to the origin is $\sqrt{a^2 + b^2}$. So $\csc\theta = \frac{\sqrt{a^2 + b^2}}{b}$.

17. C

$f(4)=\sqrt{4^2-3(4)+6}=\sqrt{16-12+6}=\sqrt{10}$.

$g(f(4))=g(\sqrt{10})=\dfrac{156}{\sqrt{10}+17}$. Use your calculator here. $g(f(4))=\dfrac{156}{\sqrt{10}+17}\approx 7.7$.

18. A

Try to solve the equation $30x^{-5}y^{12}z^{-8} = 10x^{-6}y^5z^4$ for x in terms of y and z. Dividing both sides by x^{-6}, you have $\dfrac{30x^{-5}y^{12}z^{-8}}{x^{-6}}=\dfrac{10x^{-6}y^5z^4}{x^{-6}}$,

$30x^{-5+6}y^{12}z^{-8} = 10x^{-6+6}y^5z^4$, and

$30xy^{12}z^{-8} = 10y^5z^4$. Dividing both sides by $30y^{12}z^{-8}$, you have $\dfrac{30xy^{12}z^{-8}}{30y^{12}z^{-8}}=\dfrac{10y^5z^4}{30y^{12}z^{-8}}$, and

then $x=\dfrac{1}{3}y^{5-12}z^{4-(-8)}$, $x=\dfrac{1}{3}y^{-7}z^{12}$, and $x=\dfrac{z^{12}}{3y^7}$.

19. B

The slope m of a line that goes through the points (x_1,y_1) and (x_2,y_2) is given by $m=\dfrac{y_2-y_1}{x_2-x_1}$. The slope of PQ is $\dfrac{10-7}{7-0}=\dfrac{3}{7}$. Since $PQRS$ is a square, QR is perpendicular to PQ. When two lines are perpendicular and the lines are not parallel to the coordinate axes (the coordinate axes are the x-axis and the y-axis), the slopes of the lines are negative reciprocals. Since QR is perpendicular to PQ, the slope of QR is the negative reciprocal of the slope of PQ. The slope of QR is $\dfrac{-1}{\left(\dfrac{3}{7}\right)}$, which is $-\dfrac{7}{3}$.

20. C

When the tangent and the cotangent of an angle are both defined, the tangent and the cotangent are reciprocals. So $\tan(3x)\cot(3x) = 1$.

21. E

Since 4 is a solution to $q(x) = 0$, $x - 4$ is a factor of $q(x)$. Since 0 is a solution to $q(x) = 0$, $x - 0 = x$ is a factor of $q(x)$.

So $x(x - 4)$ is a factor of $q(x)$.

Now $x(x - 4) = x^2 - 4x$. Choice (E) is correct.

22. D

Since $f(x) = 5x^2 + 4$, $f(g(3)) = 5(g(3))^2 + 4$. You know that $f(g(3)) = 84$. So $5(g(3))^2 + 4 = 84$. Solve this equation for $g(3)$.

$$5(g(3))^2 + 4 = 84$$

$$5(g(3))^2 = 80$$

$$(g(3))^2 = 16$$

$$g(3) = 4 \text{ or } g(3) = -4$$

The correct answer choice is the one that has a value of 4 or –4 when $x = 3$.

(A): $3x - 10 = 3(3) - 10 = 9 - 10 = -1$. This is not 4 or –4. Discard (A).

(B): $4x - 7 = 4(3) - 7 = 12 - 7 = 5$. This is not 4 or –4. Discard (B).

(C): $6x - 17 = 6(3) - 17 = 18 - 17 = 1$. This is not 4 or –4. Discard (C).

(D): $x^2 - 5 = 3^2 - 5 = 9 - 5 = 4$. Now you know that $x^2 - 5$ could be $g(x)$, so (D) is correct.

Checking (E) to be sure, $x^2 - 3 = 3^2 - 3 = 9 - 3 = 6$. This is not 4 or –4.

23. A

Since line n is perpendicular to the x-axis, and the y-axis is perpendicular to the x-axis, line n is parallel to the y-axis, and every point on line n has the same x-coordinate. You know that one point on line n has the coordinates (4,5). The x-coordinate of this point is 4. So the x-coordinate of every point on line n is 4. The equation of line n is $x = 4$.

24. B

The mean is the average. The average formula is Average = $\frac{\text{Sum of the terms}}{\text{Number of terms}}$. When you know the average and the number of terms, you can find the sum of the terms by rearranging the average formula in the form Sum of the terms = Average × Number of terms. So the sum of the scores on all 5 tests was 84 × 5 = 420. The sum of the scores on the first 4 tests was 87 × 4 = 348. So the score on the fifth test is the sum of the scores on all 5 tests minus the sum of the scores on the first 4 tests. Then the score on the fifth test is 420 – 348 = 72.

25. A

Use the calculator. If cos x = 0.34, then x = arccos 0.34 ≈ 70.123°. Then $\frac{x}{2} \approx \frac{70.123125}{2} = 35.06156°$. Then $\sin\frac{x}{2} \approx 0.574$.

You can also use the half-angle identity,

$$\sin\frac{x}{2} = \pm\sqrt{\frac{1-\cos x}{2}}.\text{ Since } 0 < x < \frac{\pi}{2}, 0 < \frac{x}{2} < \frac{\pi}{4}.$$

So you are in the first quadrant; work with

$$\sin\frac{x}{2} = +\sqrt{\frac{1-\cos x}{2}},\text{ or } \sin\frac{x}{2} = \sqrt{\frac{1-\cos x}{2}}.$$

$$\text{Here, } \sin\frac{x}{2} = \sqrt{\frac{1-\cos x}{2}} = \sqrt{\frac{1-0.34}{2}} = \sqrt{\frac{0.66}{2}} = \sqrt{0.33}.$$

Using the calculator, $\sqrt{0.33} \approx 0.574$. Again, $\sin\frac{x}{2} \approx$ 0.574.

26. D

Always remember that percents can be expressed as fractions. The fraction of all the items that were considered acceptable was

$$\frac{49,812}{49,812+21,348} = \frac{49,812}{71,160} = \frac{7}{10}.$$

Therefore, the fraction of the items produced on the first day that were acceptable must also have been $\frac{7}{10}$. To find the number of items that were acceptable, simply multiply the total amount produced on the first day by the fraction: $\frac{7}{10}$ (10,830) = 7,581. Thus, 7,581 items were produced on the first day that were considered acceptable.

27. D

The function f maps every point (x,y) to the point $(x, -8x + 3y)$. The goal is to find a point such that $f(x,y) = (x,y)$ or $(x, -8x + 3y) = (x,y)$.

In the Cartesian coordinate system, two points are the same if their x and y values are the same. Therefore, the solution of this equation can be simplified into two separate equations: $x = x$ and $y = -8x + 3y$.

These can be solved independently. Now x will always equal x, so only the second equation is used to define the points that satisfy the original equation. Solving for y, you have

$y = -8x + 3y$

$-2y = -8x$

$y = 4x$

This is the solution to the original equation, and it is the answer to the problem.

28. B

When x is subtracted from each of the numbers 8, 16, and 40, the resulting three numbers are $8 - x$, $16 - x$, and $40 - x$. In a geometric sequence $a_1, a_2, a_3, \ldots, a_n, \ldots$ for all integers $m \geq 1$, $\frac{a_{m+1}}{a_m} = r$, where r is a constant. Here, $8 - x$, $16 - x$, and $40 - x$ form a geometric progression. So $\frac{16-x}{8-x} = \frac{40-x}{16-x}$. Solve this equation for x. Cross multiplying, you have $(16 - x)(16 - x) = (8 - x)(40 - x)$. Multiplying out each side and solving for x, you have

$$(16)(16) - 16x - 16x + x^2 = (8)(40) - 8x - 40x + x^2$$

$$256 - 32x + x^2 = 320 - 48x + x^2$$

$$256 - 32x = 320 - 48x$$

$$16x = 64$$

$$x = 4$$

You can check that this is correct. The original numbers were 8, 16, and 40. When $x = 4$ is subtracted from each of these numbers, the resulting numbers are $8 - 4$, $16 - 4$, and $40 - 4$, which are the numbers 4, 12, and 36. Then $\frac{12}{4} = 3$ and $\frac{36}{12} = 3$. You have the same ratio, so you know that the value of x is 4 and (B) is correct.

29. C

Substitute –1 for x and 1 for x into $f(x) = ax^2 + bx + c$ and then set the resulting expressions equal to the values of the function f at $x = -1$ and $x = 1$.

When $x = -1$, $f(x) = f(-1) = ax^2 + bx + c = a((-1)^2) + b(-1) + c = a - b + c$. You know that $f(-1) = -18$, so $a - b + c = -18$.

When $x = 1$, $f(x) = f(1) = ax^2 + bx + c = a(1^2) + b(1) + c = a + b + c$. You know that $f(1) = 10$, so $a + b + c = 10$.

So you have the two equations $a - b + c = -18$ and $a + b + c = 10$. If you subtract the corresponding sides of the equation $a - b + c = -18$ from the corresponding sides of the equation $a + b + c = 10$, the a and c terms will cancel each other out, and you will be left with an equation with just the variable b.

$$(a + b + c) - (a - b + c) = 10 - (-18)$$

$$a + b + c - a + b - c = 28$$

$$2b = 28$$

$$b = 14$$

30. E

The square of any number is nonnegative. So $x^2 \geq 0$ for all real x. Then $x^2 + 3 > 0$ for all real x. The radical symbol $\sqrt{\ }$ is defined for any nonnegative number. So function $f(x)$ is defined for all real x. Remember that if $y > 0$, the mathematical convention is that $\sqrt{y}$ means the positive square root of y. Also, $\sqrt{0} = 0$.

31. C

The vertical dimension of the shaded rectangle whose horizontal dimension is on the x-axis for $3 \leq x \leq 6$ is $f(6) = \frac{180}{6+3} = \frac{180}{9} = 20$. The area of this shaded rectangle is $20(3) = 60$.

The vertical dimension of the shaded rectangle whose horizontal dimension is on the x-axis for $6 \leq x \leq 9$ is $f(9) = \frac{180}{9+3} = \frac{180}{12} = 15$. The area of this shaded rectangle is $15(3) = 45$.

The vertical dimension of the shaded rectangle whose horizontal dimension is on the x-axis for $9 \leq x \leq 12$ is $f(12) = \frac{180}{12+3} = \frac{180}{15} = 12$. The area of this shaded rectangle is $12(3) = 36$.

The sum of the areas of the three shaded rectangles is $60 + 45 + 36 = 141$.

32. D

The denominator $(3x + 4)(x - 5)$ is equal to 0 if $3x + 4 = 0$ or if $x - 5 = 0$. If $3x + 4 = 0$, then $3x = -4$, and $x = -\frac{4}{3}$. If $x - 5 = 0$, then $x = 5$. The vertical asymptotes are $x = -\frac{4}{3}$ and $x = 5$. Values of x with large absolute values can be large positive values of x and can also be negative values of x; both are far from 0 on the number line. For values of x with large absolute values, in $\frac{x^2 - 64}{(3x+4)(x-5)}$, x^2, x, and $3x$ have much greater absolute values than –64, 4, and –5. So when x approaches values that have a large absolute value, $\frac{x^2 - 64}{(3x+4)(x-5)}$ approaches $\frac{x^2}{(3x)(x)} = \frac{x^2}{3x^2} = \frac{1}{3}$. So the horizontal asymptote is $y = \frac{1}{3}$. Thus, $x = -8$ is not an asymptote, while $x = 5$ and $y = \frac{1}{3}$ are asymptotes. (D) is correct.

33. E

$$7(g(x)) + 12 = 21x^2 + 40$$

$$7(g(x)) = 21x^2 + 28$$

$$g(x) = \frac{21x^2 + 28}{7}$$

$$g(x) = \frac{21x^2}{7} + \frac{28}{7}$$

$$g(x) = 3x^2 + 4$$

34. B

The perpendicular distance from the center (h,k) of a circle to a line with the equation $x = x_1$ is $|x_1 - h|$. The perpendicular distance from the center (h,k) of a circle to a line with the equation $y = y_1$ is $|y_1 - k|$. If you draw the radii of a circle that are tangent to both lines, the distance from the center to the two lines will be the same because all radii of a circle are equal. So you must find the answer choice that is the same positive distance from the lines with the equations $x = 5$ and $y = 7$.

Look at choice (A). The distance from (3,7) to the line with the equation $x = 5$ is $|5 - 3| = |2| = 2$. The distance from (3,7) to the line with the equation $y = 7$ is $|7 - 7| = 0$. The distances from (3,7) to the lines with the equations $x = 5$ and $y = 7$ are not equal; (A) is incorrect.

Look at choice (B). The distance from (8,4) to the line with the equation $x = 5$ is $|5 - 8| = |-3| = 3$. The distance from (8,4) to the line with the equation $y = 7$ is $|7 - 4| = |3| = 3$. The distances from (3,7) to the lines with the equations $x = 5$ and $y = 7$ are equal; (B) is correct.

35. C

You should determine the range of each part of the piecewise function. Begin with $x < 0$. If this is true, then $5x < 0$ and $5x + 3 < 3$. As x becomes more and more negative, $5x + 3$ will also become more and more negative, so the range for this part of the function is all real values less than 3.

Now consider $x \geq 0$. Because the square of a number is nonnegative, $x^2 \geq 0$ and $x^2 + 1 \geq 1$. Dividing both sides by $x^2 + 1$ results in $1 \geq \frac{1}{x^2+1}$. The range of $\frac{1}{x^2+1}$ is (0, 1], since $x = 0$ results in 1 and as x increases, the fraction will decrease but always remain positive. Therefore, the range of the original function is $3^{\circ} < \frac{1}{3^{x^2+1}} \leq 3^1$, or (1, 3]. Combining these two ranges gives the range of the full function as all real values less than or equal to 3, or $y \leq 3$.

36. D

Since you want the value of y, solve the equation $x - y = 1$ for x in terms of y. Then substitute for x the expression containing y that equals x in the equation $x^2 - 7y = 8$. Adding y to both sides of the equation $x - y = 1$, you have $x = y + 1$. Now substitute $y + 1$ for x in the equation $x^2 - 7y = 8$. You have $(y + 1)^2 - 7y = 8$. Solve this equation for y.

$$(y + 1)^2 - 7y = 8$$

$$y^2 + 2y + 1 - 7y = 8$$

$$y^2 - 5y - 7 = 0$$

The general solution to the quadratic equation $ax^2 + bx + c = 0$ is

$$x = \frac{-b \pm \sqrt{b^2 - 4ac}}{2a}$$

Here, you have the equation $y^2 - 5y - 7 = 0$ in the variable y. This equation is also a quadratic equation where the right side is 0. You have the quadratic equation $ay^2 + by + c = 0$ where $a = 1$, $b = -5$, and $c = 7$. The solution to this equation is

$$y = \frac{-(-5) \pm \sqrt{(-5)^2 - 4(1)(-7)}}{2(1)} = \frac{5 \pm \sqrt{25 - (-28)}}{2} =$$

$$\frac{5 \pm \sqrt{53}}{2}$$

Now, the question stem says that $y > 0$. Since $\sqrt{53} > \sqrt{49} = 7$, $5 - \sqrt{53}$ is negative, so $\frac{5 - \sqrt{53}}{2}$ is negative. So the value of y is $\frac{5 + \sqrt{53}}{2}$. Using the calculator, $\frac{5 + \sqrt{53}}{2} \approx 6.14$. Thus, $y \approx 6.14$.

37. A

Set $y = \log_7 \frac{x}{8}$. To find the inverse, switch the values of x and y and then solve for y:

$$x = \log_7 \frac{y}{8}$$

This is equivalent to

$$7^x = \frac{y}{8}$$

$$8(7^x) = y$$

Therefore, $f^{-1}(x) = 8(7^x)$.

38. C

Using the defining equation $x_{n+1} = \sqrt{x_n^2 + 8}$ with $n = 1$, you have $x_2 = \sqrt{x_1^2 + 8} = \sqrt{2^2 + 8} = \sqrt{4 + 8} = \sqrt{12}$. Thus, $x_2 = \sqrt{12}$.

Using the defining equation $x_{n+1} = \sqrt{x_n^2 + 8}$ with $n = 2$, you have $x_3 = \sqrt{x_2^2 + 8} = \sqrt{(\sqrt{12})^2 + 8} = \sqrt{12 + 8} = \sqrt{20}$. Thus, $x_3 = \sqrt{20}$.

Using the defining equation $x_{n+1} = \sqrt{x_n^2 + 8}$ with $n = 3$, you have $x_4 = \sqrt{x_3^2 + 8} = \sqrt{(\sqrt{20})^2 + 8} = \sqrt{20 + 8} = \sqrt{28}$. Thus, $x_4 = \sqrt{28}$.

Use the calculator: $\sqrt{28} \approx 5.29$. Thus, $x_4 \approx 5.29$.

39. D

The area of any triangle is $\frac{1}{2} \times b \times h$. Call the base of triangle ABC side AC and say that the height drawn from point B to base AC meets base AC at point D.

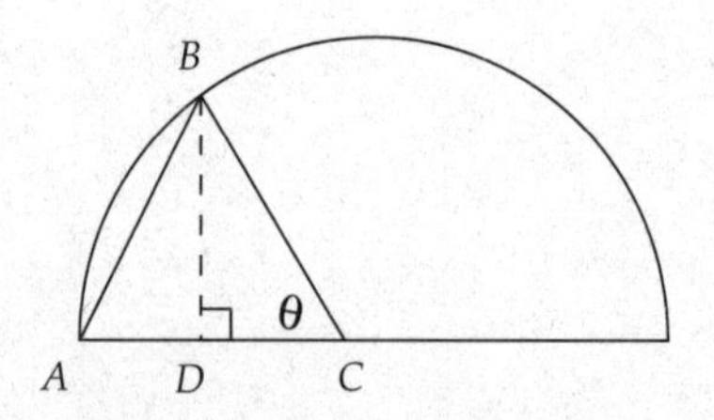

Base AC of triangle ABC is a radius of the circle. The area of the semicircle is 8π. So if the radius of the semicircle is r (in particular, $r = AC = BC$), then

$\frac{1}{2}\pi r^2 = 8\pi$, $\pi r^2 = 16\pi$, $r^2 = 16$, and $r = 4$. So $AC = 4$ and $BC = 4$.

You can find the height BD of the triangle in terms of θ by saying that $\frac{BD}{BC} = \sin\theta$, so $BD = BC \sin\theta = 4 \sin\theta$. Thus the base AC is 4, and the height BD is $4 \sin\theta$. The area of the triangle is $\frac{1}{2} \times AC \times BD = \frac{1}{2} \times 4 \times (4 \sin\theta) = 8 \sin\theta$.

40. B

When two or more events are independent, the probability that all the events occur is equal to the product of the probabilities that each event occurs. The decimal equivalent of 70% is 0.7. In this question, the probability that a person selected at random from each room is a senior at university X is 0.7. So the probability that each of the 3 people selected is a senior at university X is (0.7)(0.7)(0.7) = 0.343.

41. E

Begin to solve the equation $4.18^x = 36.54$ by taking the natural logarithm of both sides. Use the calculator when it is time to find the approximate value of x.

$$\ln(4.18^x) = \ln 36.54$$

$$x \ln 4.18 = \ln 36.54$$

$$x = \frac{\ln 36.54}{\ln 4.18} \approx 2.52$$

42. A

Before solving the question, be sure you understand the difference between the meanings of f and $f(x)$. The letter f can refer to a function. Other letters, like g, u, and t can also refer to functions. Then $f(x)$, which is read "f of x," means the value of the function f at x.

The idea in the solution to this question is that the function of the inverse function of x is x. If f^{-1} is the inverse function of the function f, then $f(f^{-1}(x)) = x$. Here you are working with the sine function and the inverse sine function. If you say that $f(x) = \sin x$, then $f^{-1}(x) = \arcsin x$. So $\sin(\arcsin x) = x$. If $\sin(\arcsin x) = \frac{\sqrt{2}}{4}$, then since $\sin(\arcsin x) = x$, $x = \frac{\sqrt{2}}{4}$; (A) is correct.

Let's briefly review the definition of the inverse sine function. The definition of the inverse sine function, which is called $\arcsin x$, is this. If $-1 \le x \le 1$, then $\arcsin x$ is that value y such that $-\frac{\pi}{2} \le y \le \frac{\pi}{2}$ and $\sin y = x$. In order for g to be a function, to each x of the domain of g, g must specify exactly one value. That is, for each specified x, $g(x)$ is not ambiguous. Now when x increases from $-\frac{\pi}{2}$ to $\frac{\pi}{2}$, $\sin x$ increases from -1 to 1, so the inverse sine function $\arcsin x$ assigns exactly one value y such that $-\frac{\pi}{2} \le y \le \frac{\pi}{2}$ to each x such that $-1 \le x \le 1$.

43. C

In this question, you want the constant d, where if n is an integer and $n \ge 1$, $d = a_{n+1} - a_n$.
In an arithmetic sequence, if a_1 is the first term and a_n is the nth term, then the sum S_n of the first n terms can be found by using the formula $S_n = \frac{a_1 + a_n}{2} \times n$. Using the formula with $a_{25} = 104$, $S_{25} = 1{,}400$, and $n = 25$, you can find a_1: $1{,}400 =$

$$\frac{a_1 + 104}{2} \times 25, \frac{1{,}400}{25} = \frac{a_1 + 104}{2}, 112 = a_1 + 104,$$

and $a_1 = 8$.

Now if a_1 is the first term of an arithmetic sequence, a_n is the nth term of the sequence, and for each term other than the first term, d is the value of that term minus the term right before it, then $a_n = a_1 + d(n - 1)$.

You know that $a_1 = 8$ and $a_{25} = 104$. So you have $n = 25$. Using the formula $a_n = a_1 + d(n - 1)$ with $a_1 = 8$, $a_{25} = 104$, and $n = 25$, you have

$104 = 8 + d(25 - 1)$. Then $104 = 8 + 24d$, $96 = 24d$, and $d = \frac{96}{24} = 4$. Now d is $a_2 - a_1$. Thus, $a_2 - a_1 = 4$.

44. B

For all θ, $\sin(\theta + \pi) = -\sin\theta$.
For all θ, $\sin(\theta + 2\pi) = \sin\theta$.
So $\sin\theta + \sin(\theta + \pi) + \sin(2\pi + \theta) = \sin\theta - \sin\theta + \sin\theta = \sin\theta$.

45. E

The definition of $x!$ for positive integers x is $x! = x(x - 1)(x - 2)\ldots(3)(2)(1)$.
$(n + 4)! = (n + 4)(n + 3)(n + 2)(n + 1)(n)\ldots(3)(2)(1)$
$(n + 7)! = (n + 7)(n + 6)(n + 5)(n + 4)(n + 3)(n + 2)(n + 1)(n)\ldots(3)(2)(1)$
To solve this question, write

$$(n + 7)! = (n + 7)(n + 6)(n + 5)(n + 4)!$$

Then

$$\frac{(n+4)!}{(n+7)!} = \frac{(n+4)!}{(n+7)(n+6)(n+5)(n+4)!}$$
$$= \frac{1}{(n+7)(n+6)(n+5)}$$

46. E

When a function $n(x)$ is translated to the right v units, where $v > 0$, the resulting function, say $r(x)$, can be described by $r(x) = n(x - v)$.

When a function $q(x)$ is translated up w units, where $w > 0$, the resulting function, say $t(x)$, can be described by $t(x) = q(x) + w$.

Here, when $g(x) = x^3 + 1$ is translated to the right 4 units, the resulting function, say $v(x)$, can be described by $v(x) = g(x - 4) = (x - 4)^3 + 1$. When the function $v(x)$ is translated 2 units up, the resulting function, which the question stem says is $h(x)$, can be described by $h(x) = v(x) + 2 = [(x - 4)^3 + 1] + 2 = (x - 4)^3 + 3$.

Thus, $h(x) = (x - 4)^3 + 3$.

Then $h(3.7) = (3.7 - 4)^3 + 3 = (-0.3)^3 + 3 = -0.027 + 3 = 2.973$.

We can draw graphs of both $g(x)$ and $h(x)$.

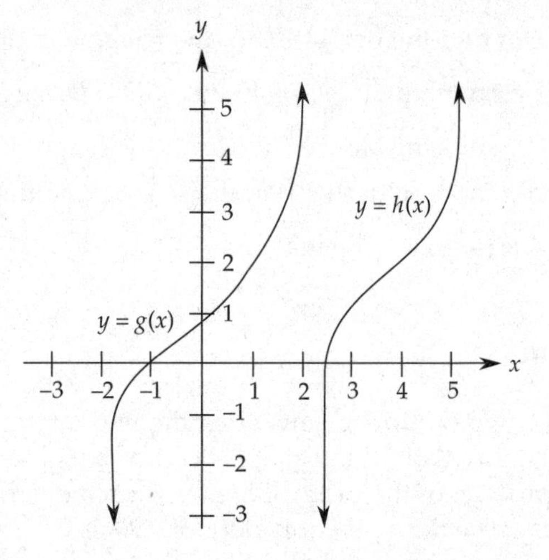

47. C

The number of five-letter codes possible is the number of ways of permuting 5 different objects from 12 different objects. Say that ${}_nP_k$ is the number of ways to permute k different objects from n different objects, where n is a positive integer and k is an integer such that $0 \le k \le n$. Then ${}_nP_k = \frac{n!}{(n-k)!}$. In this question, $n = 12$ and $k = 5$. The number of possible five-letter codes is $\frac{12!}{(12-5)!} = \frac{12!}{7!}$.

You can solve the question in this way also: Call the space furthest to the left space I; the space adjacent to the space furthest to the left space II; the space in the center space III; the space adjacent to the space furthest to the right space IV; and the space furthest to the right space V. You can place any of the 12 letters in space I. For each of these 12 letters in space I, you can place $12 - 1 = 11$ letters in space II. For any pair of letters in spaces I and II, you can place $12 - 2 = 10$ letters in space III. For any triplet of letters in spaces I, II, and III, you can place 9 letters in space IV. For any set of 4 letters in spaces I, II, III, and IV, you can place 8 letters in space V. The number of possible five-letter codes is (12)(11)(10)(9)(8).This can be rewritten as $\frac{(12)(11)(10)(9)(8)7!}{7!}$ which equals $\frac{12!}{7!}$, matching (C).

48. C

Consider option I. The sum of two negative integers is a negative integer. So option I will be part of the correct answer; eliminate (A) and (D).

Consider option II. The sum of two rational numbers is a rational number; eliminate (B).

Consider option III. The sum of two irrational numbers is not necessarily an irrational number. For example, the sum of the irrational number $3 - \sqrt{7}$ and the irrational number $\sqrt{7}$ is 3, which is a rational number. Option III will not be part of the correct answer. (C) is correct.

49. D

Begin by writing the equation in the form $\frac{x^2}{a^2} + \frac{y^2}{b^2} = 1$. Dividing both sides of the equation $5x^2 + 24y^2 = 40$ by 40 will leave 1 on the right side of the equation, and you have

$$\frac{5x^2 + 24y^2}{40} = \frac{40}{40}, \text{ or}$$

$$\frac{5x^2}{40} + \frac{24y^2}{40} = 1, \text{ or } \frac{x^2}{8} + \frac{3y^2}{5} = 1.$$

$$\frac{3y^2}{5} = \frac{y^2}{\left(\frac{5}{3}\right)}, \text{ so } \frac{x^2}{8} + \frac{y^2}{\left(\frac{5}{3}\right)} = 1. \text{ Now our equation is}$$

in the form $\frac{x^2}{a^2} + \frac{y^2}{b^2} = 1$. Thus, $a^2 = 8$ and $b^2 = \frac{5}{3}$, and the two axes will have lengths $2a = 2\sqrt{8}$ and $2b = 2\sqrt{\frac{5}{3}}$. Therefore, the sum of the axes is

$$2\sqrt{8} + 2\sqrt{\frac{5}{3}} \approx 8.24.$$

50. B

Since the square of any real number is nonnegative, x^2 is nonnegative. Then $x^2 + 1$ is the sum of the nonnegative quantity x^2 and the positive quantity 1. So $x^2 + 1$ is positive. The fraction $\frac{1-5x}{x^2+1}$ will be negative when the numerator $1 - 5x$ is negative. Solve the inequality $1 - 5x < 0$:

$$1 - 5x < 0$$

$$1 < 5x$$

$$\frac{1}{5} < x$$

Thus, the fraction $\frac{1-5x}{x^2+1}$ will be negative when and only when $x > \frac{1}{5}$.

HOW TO CALCULATE YOUR SCORE

Step 1: Figure out your raw score. Use the answer key to count the number of questions you answered correctly and the number of questions you answered incorrectly. (Do not count any questions you left blank.) Multiply the number wrong by 0.25 and subtract the result from the number correct. Round the result to the nearest whole number. This is your raw score.

SAT Subject Test: Mathematics Level 2 — Practice Test 4

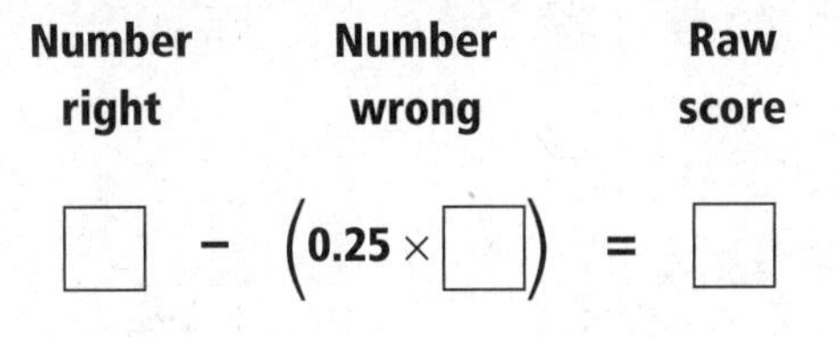

Step 2: Find your scaled score. In the Score Conversion Table below, find your raw score (rounded to the nearest whole number) in one of the columns to the left. The score directly to the right of that number will be your scaled score.

A note on your practice test scores: Don't take these scores too literally. Practice test conditions cannot precisely mirror real test conditions. Your actual SAT Subject Test: Mathematics Level 2 score will almost certainly vary from your practice test scores. However, your scores on the practice tests will give you a rough idea of your range on the actual exam.

Conversion Table

Raw	Scaled	Raw	Scaled	Raw	Scaled	Raw	Scaled
50	800	34	690	18	550	2	370
49	800	33	680	17	540	1	350
48	800	32	570	16	530	0	340
47	800	31	660	15	520	−1	330
46	800	30	650	14	520	−2	310
45	800	29	640	13	510	−3	300
44	790	28	630	12	500	−4	290
43	780	27	620	11	490	−5	280
42	770	26	610	10	480	−6	260
41	760	25	600	9	460	−7	250
40	750	24	590	8	450	−8	240
39	740	23	580	7	440	−9	220
38	730	22	580	6	420	−10	210
37	720	21	570	5	410	−11	200
36	710	20	560	4	400	−12	200
35	700	19	550	3	380		

Answer Grid
Practice Test 4

1. (A) (B) (C) (D) (E)
2. (A) (B) (C) (D) (E)
3. (A) (B) (C) (D) (E)
4. (A) (B) (C) (D) (E)
5. (A) (B) (C) (D) (E)
6. (A) (B) (C) (D) (E)
7. (A) (B) (C) (D) (E)
8. (A) (B) (C) (D) (E)
9. (A) (B) (C) (D) (E)
10. (A) (B) (C) (D) (E)
11. (A) (B) (C) (D) (E)
12. (A) (B) (C) (D) (E)
13. (A) (B) (C) (D) (E)
14. (A) (B) (C) (D) (E)
15. (A) (B) (C) (D) (E)
16. (A) (B) (C) (D) (E)
17. (A) (B) (C) (D) (E)
18. (A) (B) (C) (D) (E)
19. (A) (B) (C) (D) (E)
20. (A) (B) (C) (D) (E)
21. (A) (B) (C) (D) (E)
22. (A) (B) (C) (D) (E)
23. (A) (B) (C) (D) (E)
24. (A) (B) (C) (D) (E)
25. (A) (B) (C) (D) (E)
26. (A) (B) (C) (D) (E)
27. (A) (B) (C) (D) (E)
28. (A) (B) (C) (D) (E)
29. (A) (B) (C) (D) (E)
30. (A) (B) (C) (D) (E)
31. (A) (B) (C) (D) (E)
32. (A) (B) (C) (D) (E)
33. (A) (B) (C) (D) (E)
34. (A) (B) (C) (D) (E)
35. (A) (B) (C) (D) (E)
36. (A) (B) (C) (D) (E)
37. (A) (B) (C) (D) (E)
38. (A) (B) (C) (D) (E)
39. (A) (B) (C) (D) (E)
40. (A) (B) (C) (D) (E)
41. (A) (B) (C) (D) (E)
42. (A) (B) (C) (D) (E)
43. (A) (B) (C) (D) (E)
44. (A) (B) (C) (D) (E)
45. (A) (B) (C) (D) (E)
46. (A) (B) (C) (D) (E)
47. (A) (B) (C) (D) (E)
48. (A) (B) (C) (D) (E)
49. (A) (B) (C) (D) (E)
50. (A) (B) (C) (D) (E)

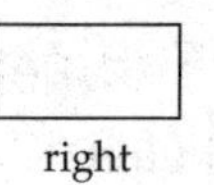

right

#wrong

Use the answer key following the test to count up the number of questions you got right and the number you got wrong. (Remember not to count omitted questions as wrong.) "How to Calculate Your Score" on the back of this page will show you how to find your score.

Practice Test 4

50 Questions (1 hour)

Directions: For each question, choose the BEST answer from the choices given. If the precise answer is not among the choices, choose the one that best approximates the answer. Then fill in the corresponding oval on the answer sheet.

Notes:

(1) To answer some of these questions, you will need a calculator. You must use at least a scientific calculator, but programmable and graphing calculators are also allowed.

(2) Make sure your calculator is in the correct mode (degree or radian) for the question being asked.

(3) Figures in this test are drawn as accurately as possible UNLESS it is stated in a specific question that the figure is not drawn to scale. All figures are assumed to lie in a plane unless otherwise specified.

(4) The domain of any function f is assumed to be the set of all real numbers x for which $f(x)$ is a real number, unless otherwise indicated.

Reference Information: Use the following formulas as needed.

Right circular cone: If r = radius and h = height, then Volume = $\frac{1}{3}\pi r^2 h$; and if c = circumference of the base and ℓ = slant height, then Lateral Area = $\frac{1}{2}c\ell$.

Sphere: If r = radius, then Volume = $\frac{4}{3}\pi r^3$ and Surface Area = $4\pi r^2$.

Pyramid: If B = area of the base and h = height, then Volume = $\frac{1}{3}Bh$.

1. If $|2x - 4| \geq \frac{x}{4}$, which of the following statements must be true?

   (A) $x \geq \frac{9}{16}$ or $x = \frac{16}{7}$

   (B) $x \geq \frac{9}{16}$ or $x \leq \frac{7}{16}$

   (C) $\frac{16}{9} < x < \frac{16}{7}$

   (D) $\frac{7}{16} \leq x \leq \frac{9}{16}$

   (E) $x \geq \frac{16}{7}$ or $x \leq \frac{16}{9}$

2. $f(x) = |4x| - 2x^3$. If $f(a) = 66$, which of the following could be the value of a?

   (A) –6
   (B) –4
   (C) –3
   (D) 3
   (E) 6

3. If $x \geq 4$, $A^2 = x^2 + 12x + 36$, and $B^2 = 4x^2 - 28x + 49$, then $(A + B)^2 =$

   (A) $2x^2 + 5x + -42$
   (B) $3x^2 - 9x - 1$
   (C) $4x^2 - x + 13$
   (D) $5x^2 - 26x + 85$
   (E) $9x^2 - 6x + 1$

DO YOUR FIGURING HERE.

GO ON TO THE NEXT PAGE

4. $f(x) = 3x^{\frac{2}{3}}$

   $f(64) =$

   (A) 48
   (B) 128
   (C) 256
   (D) 1,204
   (E) 2,304

5. A circle with center (3,8) contains the point (2,–1). Which of the following is also a point on the circle?

   (A) (1,–10)
   (B) (4,17)
   (C) (5,–9)
   (D) (7,15)
   (E) (9,6)

6. For all $y \neq 5$, $\dfrac{y^3 - 6y^2 + 3y + 10}{y^2 - 10y + 25} =$

   (A) $\dfrac{y^2 - y + 2}{y + 5}$

   (B) $\dfrac{y^2 - y - 2}{y - 5}$

   (C) $\dfrac{y^2 + y - 2}{y + 5}$

   (D) $\dfrac{y^2 + y - 2}{y - 5}$

   (E) $\dfrac{y^2 - y + 2}{y - 5}$

DO YOUR FIGURING HERE.

GO ON TO THE NEXT PAGE

7. Which of the following functions has a domain of $x \leq 3$?

(A) $f(x) = (3 - x)^{\frac{1}{4}} + \frac{x}{2}$

(B) $f(x) = (x - 3)^{\frac{1}{2}} + \frac{x}{2}$

(C) $f(x) = (x - 2)^{\frac{1}{3}} + \frac{x}{4}$

(D) $f(x) = (2 - x)^{\frac{1}{4}} + \frac{x}{3}$

(E) $f(x) = (3 - x)^{\frac{1}{3}} + \frac{x}{4}$

8. If $x^2 - 8x + 13 = 0$, $x =$

(A) $-4 \pm \sqrt{3}$
(B) $-4 \pm 2\sqrt{3}$
(C) $4 \pm \sqrt{3}$
(D) $4 \pm 3\sqrt{2}$
(E) $4 \pm 2\sqrt{2}$

9. What is the perimeter of a triangle with vertices at coordinates (–2,3), (4,3), and (6,–3)?

(A) $4\sqrt{11}$
(B) $18\sqrt{10}$
(C) $10 + 4\sqrt{5}$
(D) $16 + 2\sqrt{10}$
(E) $16 + 3\sqrt{7}$

10. $f(x) = x + 4$
$g(x) = 6 - x^2$
What is the maximum value of $g(f(x))$?

(A) –6
(B) –4
(C) 2
(D) 4
(E) 6

DO YOUR FIGURING HERE.

GO ON TO THE NEXT PAGE

DO YOUR FIGURING HERE.

11. If $x^{\frac{3}{2}} = 27$, $x^{\frac{5}{2}} =$

(A) 27
(B) 81
(C) 243
(D) 729
(E) 2,181

12. Which of the following lines intersects $y = \frac{x}{3} + 5$ at (9,8) and does not intersect the line $y = -3x - 7$?

(A) $y = -3x + 13$
(B) $y = -3x + 35$
(C) $y = -\frac{x}{3} + 8$
(D) $y = -\frac{x}{3} + 11$
(E) $y = \frac{x}{3} - 7$

13. In the right triangle in Figure 1, if $\theta = 67°$, what is the value of x?

(A) 2.9
(B) 7.6
(C) 16.5
(D) 17.9
(E) 18.2

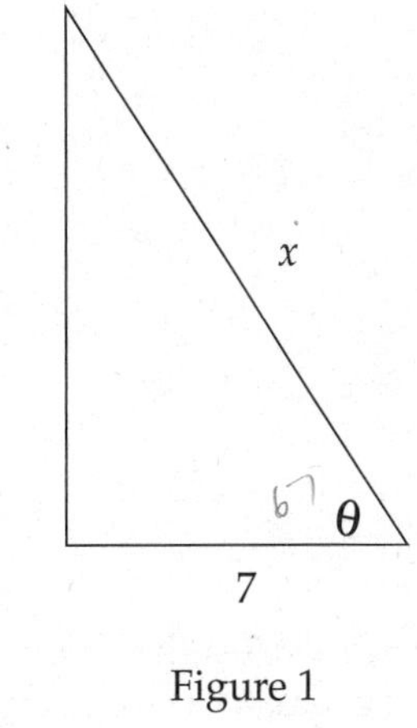

Figure 1

14. What is the minimum value of $f(x) = |2x - 5| + 6$?

(A) 2
(B) 3
(C) 5
(D) 6
(E) 8

GO ON TO THE NEXT PAGE

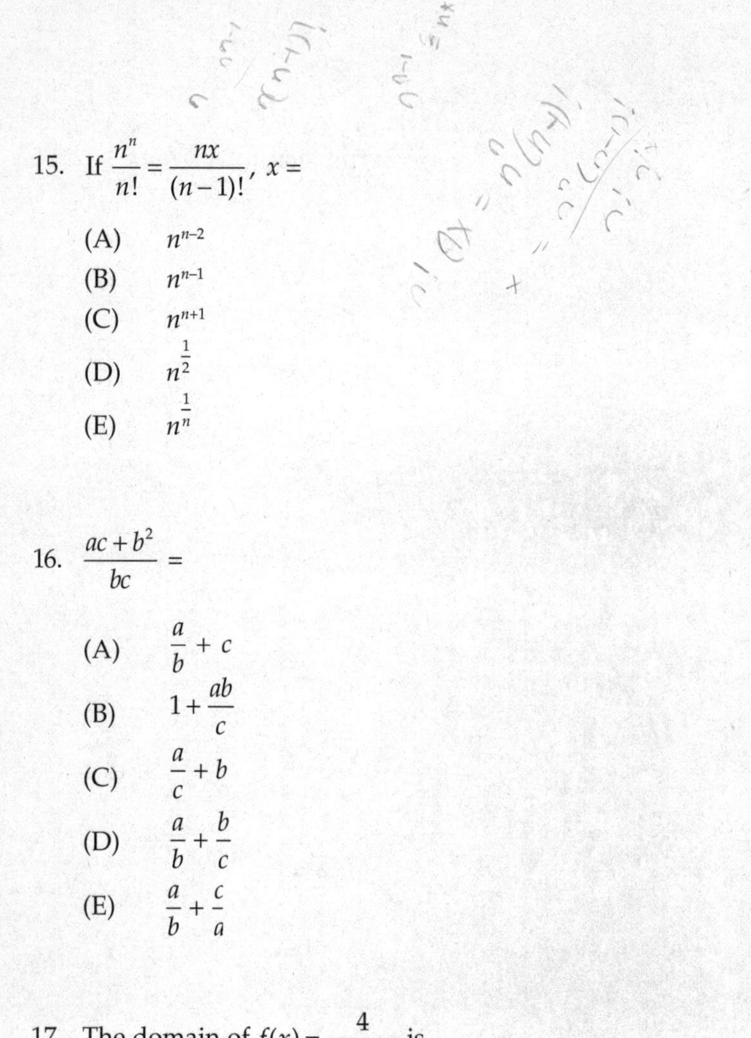

DO YOUR FIGURING HERE.

15. If $\frac{n^n}{n!} = \frac{nx}{(n-1)!}$, $x =$

(A) n^{n-2}
(B) n^{n-1}
(C) n^{n+1}
(D) $n^{\frac{1}{2}}$
(E) $n^{\frac{1}{n}}$

16. $\frac{ac+b^2}{bc} =$

(A) $\frac{a}{b} + c$
(B) $1 + \frac{ab}{c}$
(C) $\frac{a}{c} + b$
(D) $\frac{a}{b} + \frac{b}{c}$
(E) $\frac{a}{b} + \frac{c}{a}$

17. The domain of $f(x) = \frac{4}{|x| - x}$ is

(A) $x < -4$
(B) $x > 0$
(C) $x < 0$
(D) $x > 1$
(E) $x > 4$

18. A square is formed by the points (4,5), (12,5), (12,–3), and (4,–3). The diagonals of the square intersect at which of the following points?

(A) (8,5)
(B) (9,6)
(C) (8,1)
(D) (12,1)
(E) (16,2)

GO ON TO THE NEXT PAGE

19. If $s = t + \sqrt{\frac{r^3}{q}}$, what is the value of r in terms of q, s, and t?

(A) $\sqrt[3]{qs^2 - qt}$

(B) $\sqrt[3]{\frac{s^2 - 2st - t^2}{q}}$

(C) $\sqrt[3]{qs^2 - 2qst + qt^2}$

(D) $\sqrt{\frac{qs^2 + 2qst - t^2}{3}}$

(E) $\frac{qs^2 - 2st + t^2}{3}$

20. The hyberbola $\frac{x^2 + 4x + 4}{25} - \frac{y^2 - 6x + 9}{16} = 1$ is centered at which of the following points?

(A) (–4,–9)
(B) (–2,3)
(C) (2,–3)
(D) (5,4)
(E) (25,16)

21. If $3^n = n^6$, $n^{18} =$

(A) $3^n n^3$
(B) $3^n n^{12}$
(C) 9^n
(D) 3^{12n}
(E) 3^{n+12}

22. What is the domain of $f(x) = \sqrt{(4 - x)^2 - 5}$?

(A) $x \leq -3$ or $x \geq 7$
(B) $2 - \sqrt{5} \leq x \leq 2 + \sqrt{5}$
(C) $x \leq 4 - \sqrt{5}$ or $x \geq 4 + \sqrt{5}$
(D) $4 - \sqrt{5} \leq x \leq 4 + \sqrt{5}$
(E) $x \leq 21$ or $x \geq 29$

DO YOUR FIGURING HERE.

GO ON TO THE NEXT PAGE

23. If one solution of $x^2 - 22x + d = 0$ is 6, which of the following could be the value of d?

(A) 36
(B) 48
(C) 72
(D) 96
(E) 108

DO YOUR FIGURING HERE.

24. Which line has a slope of $\frac{5}{3}$?

(A) $3x - 5y + 2 = 0$
(B) $3x + 5y + 6 = 0$
(C) $3x - 4y + 5 = 0$
(D) $5x + 3y + 8 = 0$
(E) $5x - 3y + 4 = 0$

25. What is the range of $f(x) = x^3 - \sqrt{-x-6}$?

(A) All real numbers less than or equal to –222
(B) All real numbers less than or equal to –216
(C) All real numbers greater than or equal to –216
(D) All real numbers greater than or equal to 216
(E) All real numbers greater than or equal to 222

26. A cone has a slant height of 8 and a lateral area of 48π. What is the radius of the base of the cone?

(A) 3
(B) 6
(C) 12
(D) 16
(E) 24

GO ON TO THE NEXT PAGE

27. If $f(x) = 4x - 3$ and $g(x) = x - 4$, which of the following has a value of –11?

(A) $f(g(2))$
(B) $g(f(2))$
(C) $g(f(3))$
(D) $f(g(3))$
(E) $f(g(4))$

28. What is the length of side AC in Figure 2?

(A) 10.77
(B) 10.83
(C) 10.89
(D) 13.16
(E) 17.90

29. What is the range of $f(x) = -7\sin\frac{x}{8}$?

(A) All real numbers greater than or equal to –7 and less than or equal to 7
(B) All real numbers greater than or equal to –7 and less than or equal to 0
(C) All real numbers greater than or equal to 0 and less than or equal to 8
(D) All real numbers greater than or equal to 0 and less than or equal to $\frac{1}{8}$
(E) All real numbers greater than or equal to $-\frac{1}{8}$ and less than or equal to $\frac{1}{8}$

DO YOUR FIGURING HERE.

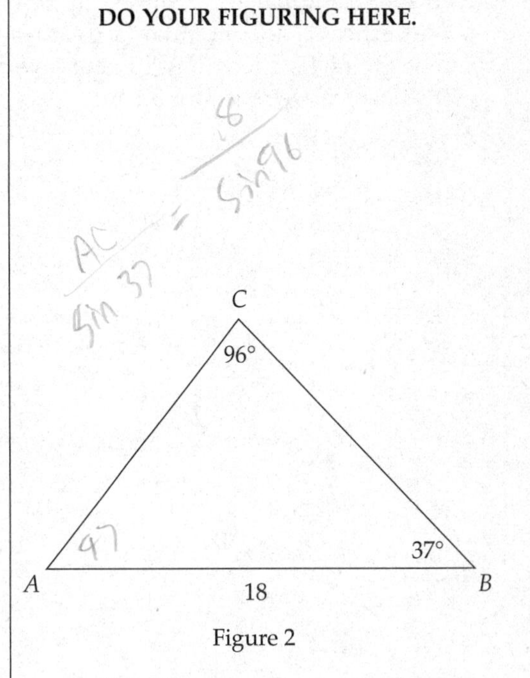

Figure 2

Note: Figure not drawn to scale.

GO ON TO THE NEXT PAGE

DO YOUR FIGURING HERE.

30. In the triangular solid shown in Figure 3, points A and B are vertices. The triangular faces are isosceles. The solid has a height of 12, a length of 21, and a width of 18. What is the distance between A and B?

(A) $9\sqrt{5}$
(B) $3\sqrt{58}$
(C) $3\sqrt{74}$
(D) $3\sqrt{85}$
(E) $6\sqrt{11}$

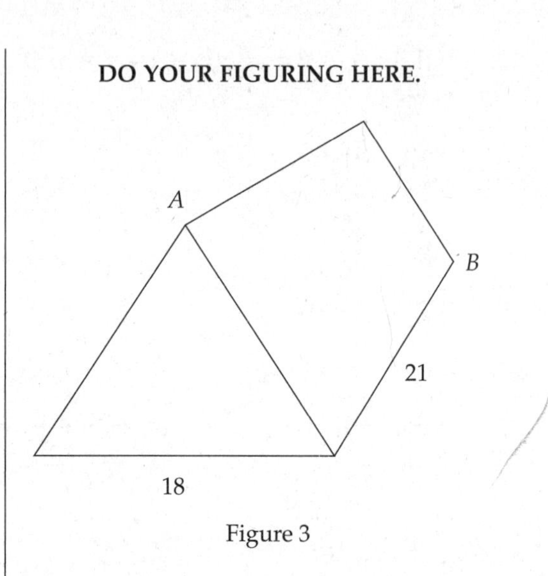

Figure 3

31. If $0 \le x \le \frac{\pi}{2}$ and $\sin^2 x = a$ and $\cos^2 x = b$, then $\sin 2x + \cos 2x =$

(A) $2\sqrt{ab} + b - a$
(B) $2ab + a^2 - b^2$
(C) $2\sqrt{ab} + 2a - 1$
(D) $2ab + 2a - 1$
(E) $2\sqrt{ab} - 2b + 1$

32. The mean number of tickets sold daily by a theater over a seven-day period was 52. The theater sold 46 tickets on the last day of that period. What was the mean number of tickets sold daily over the first six days?

(A) 53
(B) 54
(C) 55
(D) 56
(E) 57

33. The ratio of the surface area of sphere A to the surface area of sphere B is 729:1. What is ratio of the volume of sphere A to sphere B?

(A) 27:1
(B) 81:1
(C) 19,683:1
(D) 26,224:1
(E) 531,441:1

GO ON TO THE NEXT PAGE

34. If $f(x) = \frac{2x}{5} + \frac{7}{3}$, $f^{-1}(x) =$

(A) $\frac{15}{6x+35}$

(B) $\frac{6}{15x+35}$

(C) $\frac{15x-6}{35}$

(D) $\frac{15x-35}{6}$

(E) $\frac{15-6x}{35}$

35. The graph of $y = \cos\frac{x}{2}$ is shown in Figure 4. Point a has coordinates $(t,0)$. What is the value of t in radians?

(A) $\frac{2\pi}{3}$

(B) $\frac{3\pi}{4}$

(C) $\frac{3\pi}{2}$

(D) 2π

(E) 3π

36. The domain of which function does NOT include 2?

(A) $f(x) = \frac{x^2 - 2x}{x^2 - 2x^3}$

(B) $f(x) = \frac{x^{-2} - \frac{2}{x}}{x^2 - 2x^2}$

(C) $f(x) = \frac{2x^2}{x^3 - 2x^2}$

(D) $f(x) = \frac{x^2 - 2x}{x^2 - x}$

(E) $f(x) = \frac{4x^2}{x^{-2} + 2x}$

DO YOUR FIGURING HERE.

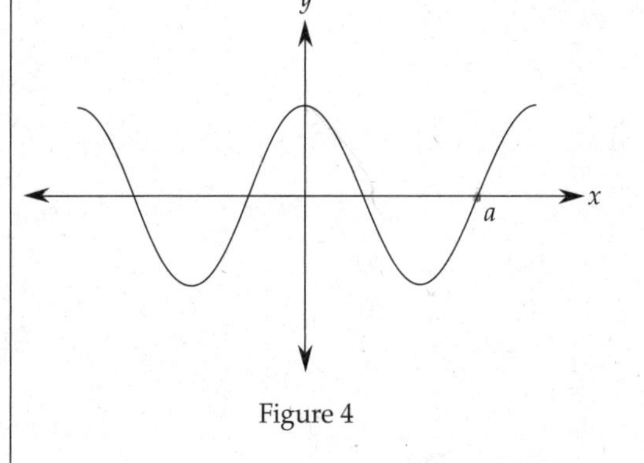

Figure 4

GO ON TO THE NEXT PAGE

37. The arithmetic sequence $s = \{4, 7, 10, 13, n_5, n_6, \ldots\}$. Which step in the sequence involves a 12% increase over the immediately preceding term?

(A) n_5 to n_6
(B) n_6 to n_7
(C) n_7 to n_8
(D) n_8 to n_9
(E) n_9 to n_{10}

38. To what sum does the geometric series 2.8 + 2.1 + 1.575 + ... converge?

(A) 3.73
(B) 4.97
(C) 6.47
(D) 11.20
(E) 44.80

39. $\dfrac{1 - \cos 40^\circ}{2} =$

(A) $\cos^2 20^\circ$
(B) $\sin^2 20^\circ$
(C) $\tan 20^\circ$
(D) $\cos 80^\circ$
(E) $\tan^2 80^\circ$

DO YOUR FIGURING HERE.

GO ON TO THE NEXT PAGE

40. Which of the following graphs is NOT a function?

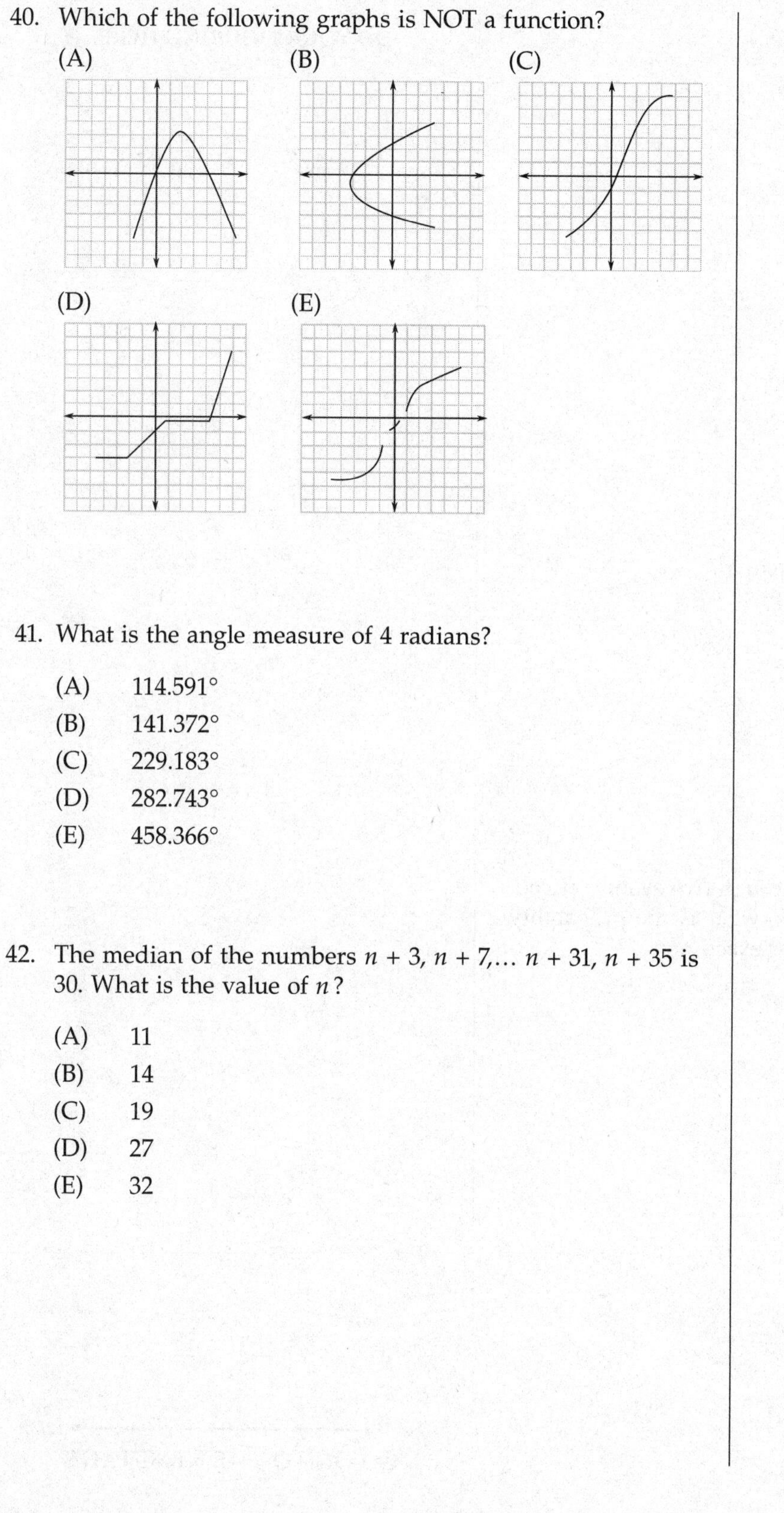

DO YOUR FIGURING HERE.

41. What is the angle measure of 4 radians?

(A) 114.591°
(B) 141.372°
(C) 229.183°
(D) 282.743°
(E) 458.366°

42. The median of the numbers $n + 3, n + 7, \ldots n + 31, n + 35$ is 30. What is the value of n?

(A) 11
(B) 14
(C) 19
(D) 27
(E) 32

GO ON TO THE NEXT PAGE

43. If $g(x) = (2x + 3)^2$, then $g\left(\frac{x}{3} - 1\right) =$

(A) $\frac{4x^2}{9} + \frac{4x}{3} + 4$

(B) $\frac{4x^2}{3} + 2x + 2$

(C) $\frac{4x^2 + 2x}{3} + 1$

(D) $\frac{4x^2}{9} + \frac{4x}{3} + 1$

(E) $\frac{4x^2 + 2x}{3} + 2$

DO YOUR FIGURING HERE.

44. What is the amplitude of $y = 3 - 6\sin^2 x$?

(A) 2
(B) 3
(C) 4
(D) 6
(E) 12

45. A spinner has the numbers one through five evenly spaced. If the spinner is used three times, what is the probability that it will land on an odd number exactly once?

(A) $\frac{12}{125}$

(B) $\frac{18}{125}$

(C) $\frac{27}{125}$

(D) $\frac{36}{125}$

(E) $\frac{54}{125}$

GO ON TO THE NEXT PAGE

46. What is the arithmetic mean of $\frac{1}{2}, \frac{1}{3}, 2n$, and m?

(A) $\frac{5+12n+6m}{24}$

(B) $\frac{5+8n+4m}{24}$

(C) $\frac{5+2n+m}{4}$

(D) $\frac{5+12n+6m}{6}$

(E) $\frac{5+2n+m}{12}$

47. Four letters mailed today each have a $\frac{2}{3}$ probability of arriving in two days or sooner. What is the probability that exactly two of the four letters will arrive in two days or sooner?

(A) $\frac{4}{81}$

(B) $\frac{16}{81}$

(C) $\frac{6}{27}$

(D) $\frac{8}{27}$

(E) $\frac{4}{9}$

48. Paula jogged for a total of 30 minutes. Her average speed for the first 10 minutes was 5 miles per hour. During the remainder of her time, she jogged 2.5 miles. What was Paula's average speed for her entire jog?

(A) $6\frac{1}{2}$ mph

(B) $6\frac{2}{3}$ mph

(C) $7\frac{1}{2}$ mph

(D) $7\frac{1}{3}$ mph

(E) $7\frac{2}{3}$ mph

DO YOUR FIGURING HERE.

GO ON TO THE NEXT PAGE

49. The front row of an auditorium has 28 seats. Each of the remaining rows has 3 more seats than the row in front of it. If the auditorium has 32 rows, how many seats does it have?

(A) 1,440
(B) 2,384
(C) 2,688
(D) 2,784
(E) 3,552

50. Charles put $1,000 in a savings account that earns 2% compound interest each year. The account would earn approximately $126 in how many years?

(A) 5
(B) 6
(C) 7
(D) 8
(E) 9

DO YOUR FIGURING HERE.

STOP!

If you finish before time is up, you may check your work.

**Turn the page
for answers and explanations
to Practice Test 4.**

Answer Key
Practice Test 4

1. E
2. C
3. E
4. A
5. B
6. B
7. A
8. C
9. D
10. E
11. C
12. B
13. D
14. D
15. A
16. D
17. C
18. C
19. C
20. B
21. B
22. C
23. D
24. E
25. B
26. B
27. A
28. C
29. A
30. C
31. A
32. A
33. C
34. D
35. E
36. C
37. D
38. D
39. B
40. B
41. C
42. A
43. D
44. B
45. D
46. A
47. D
48. B
49. B
50. B

ANSWERS AND EXPLANATIONS

1. E

Since this is an absolute value inequality, you must solve two related inequalities, $2x - 4 \geq \frac{x}{4}$ and $-(2x - 4) \geq \frac{x}{4}$. Both of these inequalities hold when $|2x - 4| \geq \frac{x}{4}$.

Inequality 1: $2x - 4 \geq \frac{x}{4}$

$$\frac{7x}{4} \geq 4$$

$$x \geq \frac{16}{7}$$

Inequality 2: $-(2x - 4) \geq \frac{x}{4}$

$$4 - 2x \geq \frac{x}{4}$$

$$\frac{9x}{4} \leq 4$$

$$x \leq \frac{16}{9}$$

So the solution to the inequality is $x \geq \frac{16}{7}$ or $x \leq \frac{16}{9}$.

2. C

$f(a) = |4a| - 2a^3$. To find the possible values of a, solve the absolute value equation $|4a| - 2a^3 = 66$. If $|4a| - 2a^3 = 66$, then $4a - 2a^3 = 66$ and $-4a - 2a^3 = 66$. Plug the values of the answer options into each of the equations until you find one that works, starting with the middle value: $-4(-3) - 2(-3)^3 = 12 + 54 = 66$. So $a = -3$ is one possible value.

3. E

Since each polynomial is expressed in terms of the square of an expression, find the square root of each polynomial.

$$A^2 = x^2 + 12x + 36 = (x + 6)^2$$

$$A = x + 6$$

$$B^2 = 4x^2 - 28x + 49 = (2x - 7)^2$$

$$B = 2x - 7$$

Now add the two binomial square roots.

$$A + B = (x + 6) + (2x - 7) = 3x - 1$$

$$(A + B)^2 = (3x - 1)^2 = 9x^2 - 6x + 1$$

4. A

$x^{\frac{1}{3}}$ is the cube root of x. $x^{\frac{2}{3}}$ is the square of $x^{\frac{1}{3}}$. $\left(x^{\frac{1}{3}}\right)^2 = x^{\frac{1}{3}} \times x^{\frac{1}{3}} = x^{\left(\frac{1}{3}+\frac{1}{3}\right)} = x^{\frac{2}{3}}$. The cube root of 64 is 4; $4^2 = 16$; $3 \times 16 = 48$. So $f(64) = 48$.

5. B

Every point on a circle is the same distance from the center. We can use the distance formula to find the distance between (3,8) and (2,–1).

$$D = \sqrt{(3-2)^2 + [8-(-1)]^2}$$

$$= \sqrt{1^2 + 9^2} = \sqrt{1 + 81} = \sqrt{82}$$

Only a point that has a distance of $\sqrt{82}$ from (3, 8) lies on the circle.

$$(1,-10): D = \sqrt{(3-1)^2 + [8-(-10)]^2} = \sqrt{2^2 + 18^2}$$

$$= \sqrt{4 + 324} = \sqrt{328}$$

$$(4,17): D = \sqrt{(3-4)^2 + (8-17)^2}$$

$$= \sqrt{(-1)^2 + (-9)^2} = \sqrt{1 + 81} = \sqrt{82}$$

(4,17) has a distance of $\sqrt{82}$ from (3, 8).

6. B

Since the denominator has a lower power, factor that expression first.

$$y^2 - 10y + 25 = (y - 5)(y - 5)$$

Check each numerator in the answer options to determine which polynomial gives a product of $y^3 - 6y^2 + 3y + 10$ when multiplied by $(y - 5)$.

$y^3 - 6y^2 + 3y + 10 = (y - 5)(y^2 - y - 2)$.

Then $\frac{y^3 - 6y^2 + 3y + 10}{y^2 - 10y + 25} = \frac{(y-5)(y^2 - y - 2)}{(y-5)(y-5)}$

$= \frac{y^2 - y - 2}{y - 5}$.

7. A

The domain of a function cannot include numbers that figure into negative square roots. This means that any expression raised to a power of $\frac{1}{2}$ or $\frac{1}{4}$ must be greater than 0. x must be less than or equal to 3 when $(3 - x)$ is raised to the $\frac{1}{4}$ power. Any number greater than 3 would mean that a negative number is being raised to that power. So $f(x) = (3-x)^{\frac{1}{4}} + \frac{x}{2}$ has a domain of $x \leq 3$.

8. C

The solution to the quadratic equation $ax^2 + bx + c = 0$ is

$$x = \frac{-b \pm \sqrt{b^2 - 4ac}}{2a}$$

Here, $a = 1$, $b = -8$, and $c = 13$.

$$x = \frac{-(-8) \pm \sqrt{(-8)^2 - 4(1)(13)}}{2(1)} =$$

$$\frac{8 \pm \sqrt{64 - 4(1)(13)}}{2}$$

$$= \frac{8 \pm \sqrt{64 - 52}}{2} = \frac{8 \pm \sqrt{12}}{2} = \frac{8 \pm \sqrt{(4)(3)}}{2}$$

$$= \frac{8 \pm 2\sqrt{3}}{2} = 4 \pm \sqrt{3}$$

9. D

Use the distance formula to calculate the lengths between each pair of points. The perimeter of the triangle is the sum of these lengths.

$$D_1 = \sqrt{(-2-4)^2 + (3-3)^2} = \sqrt{(-6)^2 + (0)^2} = \sqrt{36} = 6$$

$$D_2 = \sqrt{(-2-6)^2 + (3-(-3))^2} = \sqrt{(-8)^2 + (6)^2} =$$

$$\sqrt{64 + 36} = \sqrt{100} = 10$$

$$D_3 = \sqrt{(4-6)^2 + (3-(-3))^2} = \sqrt{(-2)^2 + (6)^2} =$$

$$\sqrt{4 + 36} = \sqrt{40} = 2\sqrt{10}$$

$$D_1 + D_2 + D_3 = 16 + 2\sqrt{10}$$

10. E

$g(f(x))$ is a compound function. Since $f(x) = x + 4$, $g(f(x)) = 6 - (x + 4)^2$

Once you have figured out the compound function, you answer this question with the help of a graphing calculator, which will indicate the highest point of the graph.

Alternatively, you can find the maximum of the function by first finding the minimum value of the squared binomial. $(x + 4)^2$ has a minimum value of zero, since a square never has a value less than 0. Six minus 0 equals 6, and that is the maximum value of the function $g(f(x))$, which occurs when $x = -4$. For any value of x other than -4, $(x + 4)^2$ will be greater than 0, so the function will have a value less than 6.

11. C

$$x^{\frac{3}{2}} = x^{1+\frac{1}{2}} = x \times x^{\frac{1}{2}}$$

So $x^{\frac{3}{2}}$ equals a number multiplied by its square root. We know that 9 multiplied by its square root, 3, equals 27. So $x = 9$.

$$x^{\frac{5}{2}} = x^{2+\frac{1}{2}} = x^2 \times x^{\frac{1}{2}} = 9^2 \times 9^{\frac{1}{2}} = 81 \times 3 = 243$$

12. B

If two lines do not intersect on the coordinate plane, then they are parallel. Parallel lines have the same slope, so the slope of the line is -3. Since the line passes through the point (9,8), we can plug those x and y coordinate values into the slope-intercept formula:

$$y = mx + b$$

$$8 = -3(9) + b = -27 + b$$

$$b = 35$$

So the equation of the line is $y = -3x + 35$.

13. D

The value x here is the length of the hypotenuse of the right triangle. Remember that the cosine of an angle of a right triangle equals the ratio of the angle's adjacent side to the hypotenuse. Use this ratio to make an equation and then solve it for the variable. Use your calculator to calculate cos 67°.

$$\cos 67° = \frac{7}{x}$$

$$0.391 = \frac{7}{x}$$

$$x = \frac{7}{0.391} \approx 17.9$$

14. D

The minimum value of the absolute value expression is 0 $\left(\text{when } x = \frac{5}{2}\right)$ since it cannot have a negative value. $f\left(\frac{5}{2}\right) = \left|2\left(\frac{5}{2}\right) - 5\right| + 6 = 0 + 6 = 6$. So 6 is the minimum value of the function.

15. A

$$\frac{n^n}{n!} = \frac{n \times n^{n-1}}{n \times (n-1)!} = \frac{n}{n} \times \frac{n^{n-1}}{(n-1)!}$$

$$= 1 \times \frac{n^{n-1}}{(n-1)!} = \frac{n^{n-1}}{(n-1)!}$$

So $nx = n^{n-1}$. Dividing both sides by n gets you $x = \frac{n^{n-1}}{n} = n^{n-2}$. Remember that you must subtract exponents when dividing powers. Since the exponent of n is 1, you subtract that value from $n - 1$.

16. D

$\frac{ac + b^2}{bc}$ is the sum of two fractions, $\frac{ac}{bc}$ and $\frac{b^2}{bc}$. Each of these can be simplified:

$$\frac{ac}{bc} = \frac{a}{b} \times \frac{c}{c} = \frac{a}{b} \times 1 = \frac{a}{b}$$

$$\frac{b^2}{bc} = \frac{b \times b}{bc} = \frac{b}{c} \times \frac{b}{b} = \frac{b}{c} \times 1 = \frac{b}{c}$$

So $\frac{ac + b^2}{bc} = \frac{a}{b} + \frac{b}{c}$.

17. C

Any fraction with a denominator equal to 0 has an undefined value. So the domain of $f(x)$ includes numbers such that $|x| - x \neq 0$. If $x \geq 0$, then $|x| - x = x - x = 0$.

If $x < 0$, $|x| - x = -x - x = -2x$. So the domain of $f(x)$ includes all real numbers less than 0.

18. C

The diagonals of a square intersect at their midpoints. To find the point of intersection, find the midpoint of one of the diagonals. Take the points (4,5) and (12,–3):

$$\text{Midpoint} = \left(\frac{x_1 + x_2}{2}, \frac{y_1 + y_2}{2}\right)$$

$$\left(\frac{4 + 12}{2}, \frac{5 + (-3)}{2}\right) = (8,1)$$

19. C

Your goal is to get an equation with just the variable r on one side. The first variable you should move is t. Subtract it from both sides:

$$s - t = \sqrt{\frac{r^3}{q}}$$

Next, get rid of the radical sign on the left side by squaring both sides:

$$(s-t)^2 = \left(\sqrt{\frac{r^3}{q}}\right)^2$$

$$s^2 - 2st + t^2 = \frac{r^3}{q}$$

You can get rid of the denominator on the right side of the equation by multiplying both sides by q:

$$q(s^2 - 2st + t^2) = q\left(\frac{r^3}{q}\right)$$

$$qs^2 - 2qst + qt^2 = r^3$$

Finally, to get r alone on the right side, take the cube root of each side:

$$\sqrt[3]{qs^2 - 2qst + qt^2} = \sqrt[3]{r^3}$$

$$\sqrt[3]{qs^2 - 2qst + qt^2} = r$$

20. B

The hyperbola $\frac{(x-h)^2}{a^2} - \frac{(y-k)^2}{b^2} = 1$ is centered at the point (h,k).

$$\frac{x^2+4x+4}{25} - \frac{y^2-6x+9}{16} = \frac{(x+2)^2}{25} - \frac{(y-3)^2}{16} = 1$$

So $h = -2$ and $k = 3$; the hyperbola is centered at $(-2,3)$.

21. B

To find the value of n^{18}, determine what operation was performed on n^6. Then see whether performing the same operation on 3^n gives you one of the results found in the answer options.

$$n^{18} = n^{6+12} = n^6 \times n^{12}$$

Since $3^n = n^6$, you can substitute in the equation:

$$n^{18} = n^6 \times n^{12} = 3^n \times n^{12} = 3^n n^{12}$$

22. C

Radicals of negative numbers are undefined, so $(4-x)^2 - 5 \geq 0$.

To find the domain of the function, solve the inequality for x.

$$(4-x)^2 \geq 5$$

Since $(4-x)^2 \geq 5$, $4 - x \geq \sqrt{5}$ or $4 - x \leq -\sqrt{5}$.

$$4 - x \geq \sqrt{5}$$

$$-x \geq \sqrt{5} - 4$$

$$x \leq -(\sqrt{5} - 4)$$

$$x \leq 4 - \sqrt{5}$$

or

$$4 - x \leq -\sqrt{5}$$

$$4 + \sqrt{5} - x \leq 0$$

$$4 + \sqrt{5} \leq x$$

So, $x \leq 4 - \sqrt{5}$ or $x \geq 4 + \sqrt{5}$.

23. D

Since one solution is $x = 6$, let's substitute 6 for x in the equation $x^2 - 22x + d = 0$. Then $6^2 - 22(6) + d = 0$, $36 - 132 + d = 0$, $-96 + d = 0$, and $d = 96$.

24. E

The line with slope $\frac{5}{3}$ has an equation of the form $y = \frac{5}{3}x + b$. Find the equation that fits this form when solved for y.

(A): $3x - 5y + 2 = 0 : y = \frac{3}{5}x + \frac{2}{5}$

(B): $3x + 5y + 6 = 0 : y = -\frac{3}{5}x - \frac{6}{5}$

(C): $3x - 4y + 5 = 0 : y = \frac{3}{4}x + \frac{5}{2}$

(D): $5x + 3y + 8 = 0 : y = -\frac{5}{3}x - \frac{8}{3}$

(E): $5x - 3y + 4 = 0 : y = \frac{5}{3}x + \frac{4}{3}$

(E) is the only equation with a slope of $\frac{5}{3}$ and therefore is the correct answer.

25. B

The expression under the radical must have a value greater than or equal to 0. Therefore,

$$-x - 6 \geq 0$$
$$-x \geq 6$$
$$x \leq -6$$
$$f(-6) = (-6)^3 - \sqrt{-(-6) - 6} = -216$$

For any x less than or equal to -6, $f(x) \leq -216$.

26. B

The lateral area of a cone is determined with the formula $A = \frac{1}{2}c\ell$, where c is the circumference and ℓ is the slant height.

So to find the circumference of the cone, solve the formula for c.

$$c = \frac{2A}{\ell}$$
$$c = \frac{2(48\pi)}{8} = \frac{96\pi}{8} = 12\pi$$

The circumference of the circular base of a cone is $2\pi r$, where r = radius, so the radius of the base of the cone is 6.

27. A

Since this question involves compound functions, you can combine them.

$$f(g(x)) = 4(x - 4) - 3 = 4x - 16 - 3 = 4x - 19$$
$$4x - 19 = -11$$
$$4x = 8$$
$$x = 2$$

So $f(g(x)) = -11$ when $x = 2$ and $f(g(2))$ is one of the choices.

This problem can also be solved easily by backsolving.

28. C

With the information provided, you can use the Law of Sines to solve for the unknown.

$$\frac{b}{\sin B} = \frac{c}{\sin C}$$

where b and c represent the lengths of the sides opposite the angles B and C, respectively.

$$\frac{b}{\sin 37^\circ} = \frac{18}{\sin 96^\circ}$$
$$b = \frac{18 \sin 37^\circ}{\sin 96^\circ}$$
$$b = 10.89$$

29. A

The first number in the function gives the amplitude of the function. A sine curve like the one in the function moves up and down between this value and its inverse. So the range of values is the real numbers between –7 and 7.

You could use a graphing calculator to answer this question by observing the range of y values for the function.

30. C

You can find the distance by using the distance formula twice. First, calculate the distance from A to the vertex on the bottom right side of the same face. The vertical distance is 12 (the height), and the horizontal distance is 9, since A is above the midpoint of the edge of the solid.

$$D_1 = \sqrt{12^2 + 9^2} = \sqrt{81 + 144} = \sqrt{225} = 15$$

The distance between A and B is the hypotenuse of the right triangle with sides of length 15 and 21. So you can use the distance formula again.

$$D_{21} = \sqrt{15^2 + 21^2} = \sqrt{225 + 441} = \sqrt{666}$$
$$\sqrt{666} = 3\sqrt{74}$$

31. A

Start by finding the positive square roots of $\sin^2 x$ and $\cos^2 x$:

$$\sin x = \sqrt{a}$$

$$\cos x = \sqrt{b}$$

Next, use the double angle formulas to simplify $\sin 2x$ and $\cos 2x$:

$$\sin 2x = 2\sin x \cos x = 2\sqrt{ab}$$

$$\cos 2x = \cos^2 x - \sin^2 x = b - a$$

Therefore, $\sin 2x + \cos 2x = 2\sqrt{ab} + b - a$.

32. A

To find the mean number for the six-day period, begin by figuring the total number of tickets sold:

$$52 \times 7 = 364$$

Since 46 tickets were sold on the last day, 318 tickets were sold before that day:

$$364 - 46 = 318$$

To find the mean for the six-day period, divide this total by six:

$$\frac{318}{6} = 53$$

33. C

To solve this problem, you need the formulas for the volumes and surface areas of spheres.

$$\text{Surface area} = 4\pi r^2$$

Since the surface area of sphere A is 729 times greater than that of sphere B, the radius is $\sqrt{729}$ or 27 times greater.

$$\text{Volume} = \frac{4}{3}\pi r^3$$

The volume of sphere A, therefore, is 27^3 or 19,683 times greater than the volume of sphere B. So the ratio of the volumes is 19,683:1.

34. D

To find the inverse function $f^{-1}(x)$ of $f(x) = \frac{2x}{5} + \frac{7}{3}$, write the inverse function as an equation $x = \frac{2y}{5} + \frac{7}{3}$, and solve for y.

$$15x = 15\left(\frac{2y}{5} + \frac{7}{3}\right)$$

$$15x = 6y + 35$$

$$15x - 35 = 6y$$

$$y = \frac{15x - 35}{6}$$

$$\text{So } f^{-1}(x) = \frac{15x - 35}{6}.$$

35. E

The graph of $y = \cos\frac{x}{2}$ has a period of 4π, since the period of $y = a\cos bx$ is $\frac{2\pi}{b}$.

$$\frac{2\pi}{\frac{1}{2}} = 4\pi$$

The curve has completed $\frac{3}{4}$ of a period from the origin to point a. So the value of t is $\frac{3}{4}$ of 4π, or 3π.

36. C

Since a defined fraction cannot have a denominator with a value of 0, the domain of $f(x)$ cannot include any number that gives the denominator a value of 0. Find the value of each denominator for $x = 2$.

When $x = 2$

$$x^2 - 2x^3 = -12$$

$$x^2 - 2x^2 = -4$$

$$x^3 - 2x^2 = 0$$

$$x^2 - x = 2$$

$$x^{-2} + 2x = \frac{1}{4} + 4 = 4\frac{1}{4}$$

So $f(x) = \frac{2x^2}{x^3 - 2x^2}$ cannot include 2 in its domain.

37. D

The value of each number in an arithmetic sequence increases by a fixed increment. In this sequence, each term is three greater than the previous one. Use this information to find the values of the other terms of the sequence through n_{10}.

$n_5 = 16$

$n_6 = 19$

$n_7 = 22$

$n_8 = 25$

$n_9 = 28$

$n_{10} = 31$

You can find the percent increase over the last number in each step by dividing the increase, 3, by the previous number:

n_5 to n_6 : $\frac{3}{16} = 0.1875$

n_6 to n_7 : $\frac{3}{19} = 0.1578$

n_7 to n_8 : $\frac{3}{22} = 0.1363$

n_8 to n_9 : $\frac{3}{25} = 0.12$

n_9 to n_{10} : $\frac{3}{28} = 0.1071$

So the percent increase from n_8 to n_9 is 12%

38. D

You can use the formula $S_\infty = \frac{a_1}{1-r}$ to find the sum that an infinite geometric series converges to, where a_1 is the first term and r is the ratio of one term to the previous term.

Here, $a_1 = 2.8$ and $r = \frac{2.1}{2.8} = 0.75$.

$$S_\infty = \frac{a_1}{1-r} = \frac{2.8}{1-0.75} = \frac{2.8}{0.25} = 11.2$$

So the sum that this infinite series converges to is 11.2.

39. B

You will need an identity equation into which the expression $\frac{1-\cos 40°}{2}$ can fit. One of the half-angle identities is

$$\sqrt{\frac{1-\cos A}{2}} = \sin\frac{1}{2}a$$

It follows that $\left(\sqrt{\frac{1-\cos A}{2}}\right)^2 = \left(\sin\frac{1}{2}a\right)^2$. Therefore,

$$\frac{1-\cos A}{2} = \sin^2\frac{1}{2}a$$

So $\frac{1-\cos 40°}{2} = \sin^2\frac{1}{2}(40°) = \sin^2 20°$.

40. B

In every function, there can be only one value in the range (only one y value) for each value in the domain (each x value). In the graph shown in (B), there are two y values for each value of x (that is greater than –3).

41. C

π radians = 180 degrees

To find the measure of 4 radians, you can solve the equation $\frac{\pi}{180} = \frac{4}{x}$ by cross multiplying:

$\pi x = 720$

$x = \frac{720}{\pi} \approx 229.183$

42. A

To find the value of n, you can get the median of the numbers in terms of n. The median is the middle value in a set of numbers. Fill the gap in the sequence of numbers:

$$n+3, n+7, n+11, n+15, n+19, n+23,$$
$$n+27, n+31, n+35$$

There are nine numbers in the set, so the middle number is the fifth, which is $n + 19$.

Since $n + 19 = 30$, $n = 30 - 19 = 11$.

43. D

Insert the whole expression $\frac{x}{3}-1$ into the function $g(x)=(2x+3)^2$:

$$g\left(\frac{x}{3}-1\right)=\left(2\left(\frac{x}{3}-1\right)+3\right)^2=\left(\left(\frac{2x}{3}-2\right)+3\right)^2$$
$$=\left(\frac{2x}{3}+1\right)^2$$
$$=\frac{4x^2}{9}+\frac{4x}{3}+1$$

44. B

Use the identity rule $\cos 2x = 1 - 2\sin^2 x$. First factor the 3 out of the expression $3 - 6\sin^2 x$. You should get $y = 3(1 - 2\sin^2 x)$. The identity rule allows you to substitute $\cos 2x$ for $1 - 2\sin^2 x$, so $y = 3(1 - 2\sin^2 x) = 3(\cos 2x)$. The amplitude of $y = 3\cos 2x$ is 3.

45. D

To find the probability of this outcome, begin by counting the number of ways it can occur. The spinner can land on an odd number on the first, second, or third spin. So there are three ways the outcome can occur.

On any given spin, the probability of getting an odd number (1, 3, 5) is $\frac{3}{5}$. The probability of getting an even number is $\frac{2}{5}$ $\left(\text{equal to } 1 - \frac{3}{5}\right)$. The probability of the spinner landing on an odd number on the first spin only is

$$\frac{3}{5}\times\frac{2}{5}\times\frac{2}{5}=\frac{12}{125}$$

The probability of getting an odd number on just the second spin or just the third spin is the same. So the probability of getting one of these three outcomes is

$$\frac{12}{125}\times 3=\frac{36}{125}$$

46. A

You can find the mean by adding these terms and dividing the sum by four, the number of terms in the set.

$$\frac{1}{2}+\frac{1}{3}=\frac{3}{6}+\frac{2}{6}=\frac{5}{6}$$
$$\frac{5}{6}+2n=\frac{5}{6}+\frac{12n}{6}=\frac{5+12n}{6}$$
$$\frac{5+12n}{6}+m=\frac{5+12n}{6}+\frac{6m}{6}=\frac{5+12n+6m}{6}$$

Now divide this sum by 4:

$$\frac{\frac{5+12n+6m}{6}}{4}=\frac{5+12n+6m}{6}\times\frac{1}{4}=\frac{5+12n+6m}{24}$$

47. D

You can start by counting the number of ways this outcome can occur. There are $\frac{4!}{2!(4-2)!} = \frac{4!}{2!2!} = \frac{(4)(3)(2)(1)}{(2)(1)(2)(1)} = \frac{12}{2} = 6$ ways that exactly two of four letters will arrive in two days or sooner (the first and second letters, the first and third, the second and fourth, etc.).

The probability of a letter failing to arrive in two days or sooner is $1 - \frac{2}{3} = \frac{1}{3}$.

So the probability of the first two letters arriving in two days or sooner, and the other two arriving later than that is:

$$\frac{2}{3} \times \frac{2}{3} \times \frac{1}{3} \times \frac{1}{3} = \frac{8}{81}$$

The probability of any other pair of letters arriving in two days or sooner and the other two arriving later than that is the same. The probability of any of those outcomes occurring is the sum of all of them. Since they are the same, the probability of exactly two arriving in two days or sooner is $\frac{4}{81} \times 6 = \frac{24}{81} = \frac{8}{27}$.

48. B

You need to figure out the total distance traveled so that you can compute her overall average rate. Since her jog lasted 30 minutes overall, we can already plug total time into the formula: $\text{rate} = \frac{\text{distance}}{\text{time}}$. Since the question asks for a rate in miles per hour, we need to convert minutes to hours; 30 minutes equals $\frac{1}{2}$ hour.

During the first 10 minutes, or the first $\frac{1}{6}$ hours, Paula jogged at five miles per hour. That means that she jogged $\frac{5}{6}$ miles. She then jogged 2.5, or $\frac{5}{2}$, miles.

$$\frac{5}{6} + \frac{5}{2} = \frac{5}{6} + \frac{15}{6} = \frac{20}{6}$$

So Paula jogged a total of $\frac{20}{6}\left(\text{or } 3\frac{1}{3}\right)$ miles.

Now divide this distance by the time.

$$\frac{\frac{20}{6}}{\frac{1}{2}} = \frac{20}{6} \times 2 = \frac{40}{6}$$

So Paula jogged at an overall average rate of $\frac{40}{6}$, or $6\frac{2}{3}$, miles per hour.

49. B

This question requires you to find the sum of an arithmetic sequence. One formula you can use to find the sum is $S_n = \frac{n}{2}(2a_1 + (n-1)d)$, where n is the number of terms in the sequence, a_1 is the first term, and d is the difference between one term and the next. Since there are 32 rows, and the number of seats in a row increases by 3:

$$S_n = \frac{32}{2}(2(28) + (32-1)3) = 16(56 + 93) = 2,384$$

50. B

You can use logarithms to solve this problem. After x years, the bank account will have $1,126 ($1,000 + $126). You need to solve an equation where rate of increase due to interest is raised to the x power:

$$1,000(1.02)^x = 1,126$$

Divide both sides by 1,000 to get the exponential expression alone on the left side:

$$1.02^x = 1.126$$

Solve this equation for x using logarithms:

$$\log 1.02^x = \log 1.126$$

Once you have converted an exponential expression to a logarithm, you can move the exponent to the front:

$$x \log 1.02 = \log 1.126$$

Now you can solve the equation for x by dividing:

$$x = \frac{\log 1.126}{\log 1.020} \approx 5.99$$

Since the question asks for the number of years needed to reach approximately $1,126, you can round the number up to 6.

100 Essential Math Concepts

The math on the SAT subject tests covers a lot of ground—from arithmetic to algebra to geometry.

Don't let yourself be intimidated. We've highlighted the 100 most important concepts that you'll need for Math 2 and listed them in this chapter.

Use this list to remind yourself of the key areas you'll need to know. Do four concepts a day, and you'll be ready within a month. If a concept continually causes you trouble, circle it and come back to it as you try to do the questions.

You've probably been taught most of these concepts in school already, so this list is a great way to refresh your memory.

NUMBER PROPERTIES

1. Number Categories

Integers are **whole numbers**; they include negative whole numbers and zero.

A **rational number** is a number that can be expressed as a **ratio of two integers**. **Irrational numbers** are real numbers—they have locations on the number line—but they can't be expressed precisely as a fraction or decimal. The most important irrational numbers are $\sqrt{2}$, $\sqrt{3}$, and π.

2. Adding/Subtracting Signed Numbers

To **add a positive and a negative number**, first ignore the signs and find the positive difference between the number parts. Then attach the sign of the original number with the larger number part. For example, to add 23 and –34, first ignore the minus sign and find the positive difference between 23 and 34—that's 11. Then attach the sign of the number with the larger number part—in this case it's the minus sign from the –34. So 23 + (–34) = –11.

Make **subtraction** situations simpler by turning them into addition. For example, you can think of –17 – (–21) as –17 + (+21) or –17 – 21 as –17 + (–21).

To **add or subtract a string of positives and negatives**, first turn everything into addition. Then combine the positives and negatives so that the string is reduced to the sum of a single positive number and a single negative number.

3. Multiplying/Dividing Signed Numbers

To multiply and/or divide positives and negatives, treat the number parts as usual and **attach a minus sign if there were originally an odd number of negatives**. For example, to multiply –2, –3, and –5, first multiply the number parts: 2 × 3 × 5 = 30. Then go back and note that there were *three*—an *odd* number—of negatives, so the product is negative: (–2) × (–3) × (–5) = –30.

4. PEMDAS

When performing multiple operations, remember to perform them in the right order.

PEMDAS, which means **Parentheses** first, then **Exponents**, then **Multiplication** and **Division** (left to right), and lastly **Addition** and **Subtraction** (left to right). In the expression $9 - 2 \times (5 - 3)^2 + 6 \div 3$, begin with the parentheses: (5 – 3) = 2. Then do the exponent: $2^2 = 4$. Now the expression is: 9 – 2 × 4 + 6 ÷ 3. Next do the multiplication and division to get: 9 – 8 + 2, which equals 3. If you have difficulty remembering PEMDAS, use this sentence to recall it: Please Excuse My Dear Aunt Sally.

5. Counting Consecutive Integers

To count consecutive integers, **subtract the smallest from the largest and add 1**. To count the number of integers from 13 through 31, subtract: 31 – 13 = 18. Then add 1: 18 + 1 = 19.

NUMBER OPERATIONS AND CONCEPTS

6. Exponential Growth

If r is the ratio between consecutive terms, a_1 is the first term, a_n is the nth term, and S_n is the sum of the first n terms, then $a_n = a_1 r^{n-1}$ and $S_n = \frac{a_1 - a_1 r^n}{1 - r}$.

7. Union and Intersection of Sets

The things in a set are called elements or members. The **union** of Set A and Set B, sometimes expressed as $A \cup B$, is the set of elements that are in either or both of Set A and Set B. If Set $A = \{1, 2\}$ and Set $B = \{3, 4\}$, then $A \cup B = \{1, 2, 3, 4\}$. The **intersection** of Set A and Set B, sometimes expressed as $A \cap B$, is the set of elements common to both Set A and Set B. If Set $A = \{1, 2, 3\}$ and Set $B = \{3, 4, 5\}$, then $A \cap B = \{3\}$.

DIVISIBILITY

8. Factor/Multiple

The **factors** of integer *n* are the positive integers that divide into *n* with no remainder. The **multiples** of *n* are the integers that *n* divides into with no remainder. For example, 6 is a factor of 12, and 24 is a multiple of 12. 12 is both a factor and a multiple of itself, since 12 × 1 = 12 and 12 ÷ 12 = 1.

9. Prime Factorization

To find the prime factorization of an integer, continue factoring until **all the factors are prime**. For example, to factor 36: 36 = 4 × 9 = 2 × 2 × 3 × 3.

10. Relative Primes

Relative primes are integers that have no common factor other than 1. To determine whether two integers are relative primes, break them both down to their prime factorizations. For example, 35 = 5 × 7, and 54 = 2 × 3 × 3 × 3. They have **no prime factors in common**, so 35 and 54 are relative primes.

11. Common Multiple

A common multiple of two or more integers is a number that is a multiple of all of these integers. You can always get a common multiple of two integers by **multiplying** them, but, unless the two numbers are relative primes, the product will not be the *least* common multiple. For example, to find a common multiple for 12 and 15, you could just multiply: 12 × 15 = 180.

To find the **least common multiple** (LCM), test the **multiples of the larger integer** until you find one that's **also a multiple of the smaller**. To find the LCM of 12 and 15, begin by taking the multiples of 15: 15 is not divisible by 12; 30 is not; nor is 45. But the next multiple of 15, 60, *is* divisible by 12, so it's the LCM.

12. Greatest Common Factor (GCF)

To find the greatest common factor of two or more integers, break down the integers into their prime factorizations and multiply **all the prime factors they have in common**. For example, 36 = 2 × 2 × 3 × 3, and 48 = 2 × 2 × 2 × 2 × 3. These integers have a 2 × 2 and a 3 in common, so the GCF is 2 × 2 × 3 = 12.

13. Even/Odd

To predict whether a sum, difference, or product will be even or odd, just **take simple numbers like 1 and 2 and see what happens**. There are rules—"odd times even is even," for example—but there's no need to memorize them. What happens with one set of numbers generally happens with all similar sets.

14. Multiples of 2 and 4

An integer is divisible by 2 (even) if the **last digit** is even. An integer is divisible by 4 if the **last two digits form a multiple of 4**. The last digit of 562 is 2, which is even, so 562 is a multiple of 2. The last two digits form 62, which is *not* divisible by 4, so 562 is not a multiple of 4. The integer 512, however, is divisible by 4 because the last two digits form 12, which is a multiple of 4.

15. Multiples of 3 and 9

An integer is divisible by 3 if the **sum of its digits is divisible by 3**. An integer is divisible by 9 if the **sum of its digits is divisible by 9**. The sum of the digits in 957 is 21, which is divisible by 3 but not by 9, so 957 is divisible by 3 but not by 9.

16. Multiples of 5 and 10

An integer is divisible by 5 if the **last digit is 5 or 0**. An integer is divisible by 10 if the **last digit is 0**. The last digit of 665 is 5, so 665 is a multiple of 5 but *not* a multiple of 10.

17. Remainders

The remainder is the **whole number left over after division**. 487 is 2 more than 485, which is a multiple of 5, so when 487 is divided by 5, the remainder is 2.

FRACTIONS AND DECIMALS

18. Reducing Fractions

To reduce a fraction to lowest terms, **factor out and cancel** all factors the numerator and denominator have in common.

$$\frac{28}{36}=\frac{4\times7}{4\times9}=\frac{7}{9}$$

19. Adding/Subtracting Fractions

To add or subtract fractions, first find a **common denominator**, then add or subtract the numerators.

$$\frac{2}{15}+\frac{3}{10}=\frac{4}{30}+\frac{9}{30}=\frac{4+9}{30}=\frac{13}{30}$$

20. Multiplying Fractions

To multiply fractions, **multiply** the numerators and **multiply** the denominators.

$$\frac{5}{7}\times\frac{3}{4}=\frac{5\times3}{7\times4}=\frac{15}{28}$$

21. Dividing Fractions

To divide fractions, **invert** the second one and **multiply**.

$$\frac{1}{2}\div\frac{3}{5}=\frac{1}{2}\times\frac{5}{3}=\frac{1\times5}{2\times3}=\frac{5}{6}$$

22. Mixed Numbers and Improper Fractions

To convert a mixed number to an improper fraction, **multiply** the whole number part by the denominator, then **add** the numerator. The result is the new numerator (over the same denominator). To convert $7\frac{1}{3}$, first multiply 7 by 3, then add 1, to get the new numerator of 22. Put that over the same denominator, 3, to get $\frac{22}{3}$.

To convert an improper fraction to a mixed number, divide the denominator into the numerator to get a **whole number quotient with a remainder**. The quotient becomes the whole number part of the mixed number, and the remainder becomes the new numerator—with the same denominator. For example, to convert $\frac{108}{5}$, first divide 5 into 108, which yields 21 with a remainder of 3. Therefore, $\frac{108}{5} = 21\frac{3}{5}$.

23. Reciprocal

To find the reciprocal of a fraction, **switch the numerator and the denominator**. The reciprocal of $\frac{3}{7}$ is $\frac{7}{3}$. The reciprocal of 5 is $\frac{1}{5}$. The product of reciprocals is 1.

24. Comparing Fractions

One way to compare fractions is to **re-express them with a common denominator**. $\frac{3}{4} = \frac{21}{28}$ and $\frac{5}{7} = \frac{20}{28}$. $\frac{21}{28}$ is greater than $\frac{20}{28}$, so $\frac{3}{4}$ is greater than $\frac{5}{7}$. Another method is to **convert them both to decimals**: $\frac{3}{4}$ converts to 0.75, and $\frac{5}{7}$ converts to approximately 0.714.

25. Converting Fractions and Decimals

To convert a fraction to a decimal, **divide the bottom into the top**. To convert $\frac{5}{8}$, divide 8 into 5, yielding 0.625.

To convert a decimal to a fraction, set the decimal over 1 and **multiply the numerator and denominator by 10** raised to the number of digits which are to the right of the decimal point.

To convert 0.625 to a fraction, you would multiply $\frac{0.625}{1}$ by $\frac{10^3}{10^3}$ or $\frac{1{,}000}{1{,}000}$.

Then simplify: $\frac{625}{1{,}000} = \frac{5 \times 125}{8 \times 125} = \frac{5}{8}$.

26. Repeating Decimal

To find a particular digit in a repeating decimal, note the **number of digits in the cluster that repeats**. If there are 2 digits in that cluster, then every second digit is the same. If there are 3 digits in that cluster, then every third digit is the same. And so on. For example, the decimal equivalent of $\frac{1}{27}$ is 0.037037037..., which is best written $0.\overline{037}$. There are 3 digits in the repeating cluster, so every third digit is the same. To find the 50th digit, look for the multiple of 3 just less than 50—that's 48. The 48th digit is 7, and with the 49th digit, the pattern repeats with 0. The 50th digit is 3.

27. Identifying the Parts and the Whole

The key to solving most fraction and percent word problems is to identify the **part** and the **whole**. Usually you'll find the **part** associated with the verb ***is/are*** and the **whole** associated with the word ***of***. In the sentence "Half of the boys are blonds," the whole is the boys ("*of* the boys"), and the part is the blonds ("*are* blonds").

PERCENTS

28. Percent Formula

Whether you need to find the part, the whole, or the percent, use the same formula:

$$\textbf{Part} = \textbf{Percent} \times \textbf{Whole}$$

Example: What is 12 percent of 25?

Setup: Part = 0.12×25.

Example: 15 is 3 percent of what number?

Setup: $15 = 0.03 \times$ Whole.

Example: 45 is what percent of 9?

Setup: $45 =$ Percent $\times 9$.

29. Percent Increase and Decrease

To increase a number by a percent, **add the percent to 100 percent**, convert to a decimal, and multiply. To increase 40 by 25 percent, add 25 percent to 100 percent, convert 125 percent to 1.25, and multiply by 40. $1.25 \times 40 = 50$.

30. Finding the Original Whole

To find the **original whole before a percent increase or decrease**, set up an equation. Think of the result of a 15 percent increase over x as $1.15x$.

Example: After a 5 percent increase, the population was 59,346. What was the population before the increase?

Setup: $1.05x = 59{,}346$

31. Combined Percent Increase and Decrease

To determine the combined effect of multiple percent increases and/or decreases, **start with 100 and see what happens**.

Example: A price went up 10 percent one year, and the new price went up 20 percent the next year. What was the combined percent increase?

Setup: First year: 100 + (10 percent of 100) = 110. Second year: 110 + (20 percent of 110) = 132. That's a combined 32 percent increase.

RATIOS, PROPORTIONS, AND RATES

32. Setting Up a Ratio

To find a ratio, put the number associated with the word ***of* on top** and the quantity associated with the word ***to* on the bottom** and reduce. The ratio of 20 oranges to 12 apples is $\frac{20}{12}$, which reduces to $\frac{5}{3}$.

33. Part-to-Part Ratios and Part-to-Whole Ratios

If the parts add up to the whole, a part-to-part ratio can be turned into two part-to-whole ratios by putting **each number in the original ratio over the sum of the numbers**. If the ratio of males to females is 1 to 2, then the males-to-people ratio is $\frac{1}{1+2}=\frac{1}{3}$, and the females-to-people ratio is $\frac{2}{1+2}=\frac{2}{3}$. In other words, $\frac{2}{3}$ of all the people are female.

34. Solving a Proportion

To solve a proportion, cross multiply:

$$\frac{x}{5}=\frac{3}{4}$$

$$4x=3\times5$$

$$x=\frac{15}{4}=3.75$$

35. Rate

To solve a rate problem, **use the units** to keep things straight.

Example: If snow is falling at the rate of one foot every four hours, how many inches of snow will fall in seven hours?

Setup:

$$\frac{1\text{ foot}}{4\text{ hours}}=\frac{x\text{ inches}}{7\text{ hours}}$$

$$\frac{12\text{ inches}}{4\text{ hours}}=\frac{x\text{ inches}}{7\text{ hours}}$$

$$4x=12\times7$$

$$x=21$$

36. Average Rate

Average rate is *not* simply the average of the rates.

$$\text{Average } A \text{ per } B=\frac{\text{Total } A}{\text{Total } B}$$

$$\text{Average Speed}=\frac{\text{Total distance}}{\text{Total time}}$$

To find the average speed for 120 miles at 40 mph and 120 miles at 60 mph, **don't just average the two speeds**. First, figure out the total distance and the total time. The total distance is 120 + 120 = 240 miles. The times are 3 hours for the first leg and 2 hours for the second leg, or 5 hours total. The average speed, then, is $\frac{240}{5}$ = 48 miles per hour.

AVERAGES

37. Average Formula

To find the average of a set of numbers, **add them up and divide by the number of numbers**.

$$\text{Average} = \frac{\text{Sum of the terms}}{\text{Number of terms}}$$

To find the average of the 5 numbers 12, 15, 23, 40, and 40, first add them: 12 + 15 + 23 + 40 + 40 = 130. Then, divide the sum by 5: 130 ÷ 5 = 26.

38. Average of Evenly Spaced Numbers

To find the average of evenly spaced numbers, just **average the smallest and the largest**. The average of all the integers from 13 through 77 is the same as the average of 13 and 77:

$$\frac{13+77}{2} = \frac{90}{2} = 45$$

39. Using the Average to Find the Sum

$$\text{Sum} = (\text{Average}) \times (\text{Number of terms})$$

If the average of 10 numbers is 50, then they add up to 10 × 50, or 500.

40. Finding the Missing Number

To find a missing number when you're given the average, **use the sum**. If the average of 4 numbers is 7, then the sum of those 4 numbers is 4 × 7, or 28. Suppose that 3 of the numbers are 3, 5, and 8. These 3 numbers add up to 16 of that 28, which leaves 12 for the fourth number.

41. Median and Mode

The median of a set of numbers is the **value that falls in the middle of the set**. If you have 5 test scores and they are 88, 86, 57, 94, and 73, you must first list the scores in increasing or decreasing order: 57, 73, 86, 88, 94.

The median is the middle number, or 86. If there is an even number of values in a set (6 test scores, for instance), simply take the average of the two middle numbers.

The mode of a set of numbers is the **value that appears most often**. If your test scores were 88, 57, 68, 85, 99, 93, 93, 84, and 81, the mode of the scores would be 93 because it appears more often than any other score. If there is a tie for the most common value in a set, the set has more than one mode.

POSSIBILITIES AND PROBABILITY

42. Counting the Possibilities

The fundamental counting principle: If there are ***m* ways** one event can happen and ***n* ways** a second event can happen, then there are ***m* × *n* ways** for the two events to happen. For example, with 5 shirts and 7 pairs of pants to choose from, you can have 5 × 7 = 35 different outfits.

43. Probability

$$\textbf{Probability} = \frac{\textbf{Favorable Outcomes}}{\textbf{Total Possible Outcomes}}$$

For example, if you have 12 shirts in a drawer and 9 of them are white, the probability of picking a white shirt at random is $\frac{9}{12} = \frac{3}{4}$. This probability can also be expressed as 0.75 or 75%.

POWERS AND ROOTS

44. Multiplying and Dividing Powers

To multiply powers with the same base, **add the exponents and keep the same base:**

$$x^3 \times x^4 = x^{3+4} = x^7$$

To divide powers with the same base, **subtract the exponents and keep the same base:**

$$y^{13} \div y^8 = y^{13-8} = y^5$$

45. Raising Powers to Powers

To raise a power to a power, **multiply the exponents:**

$$(x^3)^4 = x^{3\times 4} = x^{12}$$

46. Simplifying Square Roots

To simplify a square root, **factor out the perfect squares** under the radical, unsquare them, and put the result in front.

$$\sqrt{12} = \sqrt{4\times 3} = \sqrt{4}\times\sqrt{3} = 2\sqrt{3}$$

47. Adding and Subtracting Roots

You can add or subtract radical expressions **when the part under the radicals is the same:**

$$2\sqrt{3}+3\sqrt{3}=5\sqrt{3}$$

Don't try to add or subtract when the radical parts are different. There's not much you can do with an expression like:

$$3\sqrt{5}+3\sqrt{7}$$

48. Multiplying and Dividing Roots

The product of square roots is equal to the **square root of the product:**

$$\sqrt{3}\times\sqrt{5}=\sqrt{3\times5}=\sqrt{15}$$

The quotient of square roots is equal to the **square root of the quotient**:

$$\frac{\sqrt{6}}{\sqrt{3}}=\sqrt{\frac{6}{3}}=\sqrt{2}$$

49. Negative Exponent and Rational Exponent

To find the value of a number raised to a negative exponent, simply rewrite the number, without the negative sign, as the bottom of a fraction with 1 as the numerator of the fraction: $3^{-2}=\frac{1}{3^2}=\frac{1}{9}$. If x is a positive number and a is a nonzero number, then $x^{\frac{1}{a}}=\sqrt[a]{x}$. So $4^{\frac{1}{2}}=\sqrt[2]{4}=2$. If p and q are integers, then $x^{\frac{p}{q}}=\sqrt[q]{x^p}$. So $4^{\frac{3}{2}}=\sqrt[2]{4^3}=\sqrt{64}=8$.

ABSOLUTE VALUE

50. Determining Absolute Value

The absolute value of a number is the distance of the number from zero on the number line. Because absolute value is a distance, it is always positive. The absolute value of 7 is 7; this is expressed $|7|=7$. Similarly, the absolute value of –7 is 7: $|-7|=7$. Every positive number is the absolute value of two numbers: itself and its negative.

ALGEBRAIC EXPRESSIONS

51. Evaluating an Expression

To evaluate an algebraic expression, **plug in** the given values for the unknowns and calculate according to **PEMDAS**. To find the value of $x^2 + 5x - 6$ when $x = -2$, plug in -2 for x: $(-2)^2 + 5(-2) - 6 = -12$.

52. Adding and Subtracting Monomials

To combine like terms, **keep the variable part unchanged while adding or subtracting the coefficients:**

$$2a + 3a = (2 + 3)a = 5a$$

53. Adding and Subtracting Polynomials

To add or subtract polynomials, **combine like terms**.

$$(3x^2 + 5x - 7) - (x^2 + 12) =$$
$$(3x^2 - x^2) + 5x + (-7 - 12) =$$
$$2x^2 + 5x - 19$$

54. Multiplying Monomials

To multiply monomials, **multiply the coefficients and the variables separately:**

$$2a \times 3a = (2 \times 3)(a \times a) = 6a^2$$

55. Multiplying Binomials—FOIL

To multiply binomials, use **FOIL**. To multiply $(x + 3)$ by $(x + 4)$, first multiply the **F**irst terms: $x \times x = x^2$. Next the **O**uter terms: $x \times 4 = 4x$. Then the **I**nner terms: $3 \times x = 3x$. And finally the **L**ast terms: $3 \times 4 = 12$. Then add and combine like terms:

$$x^2 + 4x + 3x + 12 = x^2 + 7x + 12$$

56. Multiplying Other Polynomials

FOIL works only when you want to multiply two binomials. If you want to multiply polynomials with more than two terms, make sure you **multiply each term in the first polynomial by each term in the second**.

$$(x^2+3x+4)(x+5)=$$
$$x^2(x+5)+3x(x+5)+4(x+5)=$$
$$x^3+5x^2+3x^2+15x+4x+20=$$
$$x^3+8x^2+19x+20$$

After multiplying two polynomials together, the number of terms in your expression before simplifying should equal the number of terms in one polynomial multiplied by the number of terms in the second. In the example, you should have $3 \times 2 = 6$ terms in the product before you simplify like terms.

FACTORING ALGEBRAIC EXPRESSIONS

57. Factoring out a Common Divisor

A factor common to all terms of a polynomial can be **factored out**. All three terms in the polynomial $3x^3 + 12x^2 - 6x$ contain a factor of $3x$. Pulling out the common factor yields $3x(x^2 + 4x - 2)$.

58. Factoring the Difference of Squares

One of the test maker's favorite factorables is the **difference of squares**.

$$a^2 - b^2 = (a - b)(a + b)$$

$x^2 - 9$, for example, factors to $(x - 3)(x + 3)$.

59. Factoring the Square of a Binomial

Recognize polynomials that are squares of binomials:

$$a^2 + 2ab + b^2 = (a + b)^2$$
$$a^2 - 2ab + b^2 = (a - b)^2$$

For example, $4x^2 + 12x + 9$ factors to $(2x + 3)^2$, and $n^2 - 10n + 25$ factors to $(n - 5)^2$.

60. Factoring Other Polynomials—FOIL in Reverse

To factor a quadratic expression, **think about what binomials you could use FOIL on to get that quadratic expression**. To factor $x^2 - 5x + 6$, think about what **F**irst terms will produce x^2, what **L**ast terms will produce +6, and what **O**uter and **I**nner terms will produce $-5x$. Some common sense—and a little trial and error—lead you to $(x - 2)(x - 3)$.

61. Simplifying an Algebraic Fraction

Simplifying an algebraic fraction is a lot like simplifying a numerical fraction. The general idea is to **find factors common to the numerator and denominator and cancel them**. Thus, simplifying an algebraic fraction begins with factoring.

For example, to simplify $\frac{x^2 - x - 12}{x^2 - 9}$, first factor the numerator and denominator:

$$\frac{x^2 - x - 12}{x^2 - 9} = \frac{(x-4)(x+3)}{(x-3)(x+3)}$$

Canceling $x + 3$ from the numerator and denominator leaves you with $\frac{x-4}{x-3}$.

SOLVING EQUATIONS

62. Solving a Linear Equation

To solve an equation, do whatever is necessary to both sides to **isolate the variable**. To solve the equation $5x - 12 = -2x + 9$, first get all the x's on one side by adding $2x$ to both sides: $7x - 12 = 9$. Then add 12 to both sides: $7x = 21$. Then divide both sides by 7: $x = 3$.

63. Solving "In Terms Of"

To solve an equation for one variable **in terms of** another means to **isolate the one variable on one side of the equation**, leaving an expression containing the other variable on the other side of the equation. To solve the equation $3x - 10y = -5x + 6y$ for x in terms of y, isolate x:

$$3x - 10y = -5x + 6y$$
$$3x + 5x = 6y + 10y$$
$$8x = 16y$$
$$x = 2y$$

64. Translating from English into Algebra

To translate from English into algebra, look for the key words and systematically turn phrases into algebraic expressions and sentences into equations. Be careful about order, especially when subtraction is called for.

Example: Celine and Remi play tennis. Last year, Celine won 3 more than twice the number of matches that Remi won. If Celine won 11 more matches than Remi, how many matches did Celine win?

Setup: You are given two sets of information. One way to solve this is to write a system of equations–one equation for each set of information. Use variables that relate well with what they represent. For example, use r to represent Remi's winning matches. Use c to represent Celine's winning matches. The phrase "Celine won 3 more than twice Remi" can be written as $c = 2r + 3$. The phrase "Celine won 11 more matches than Remi" can be written as $c = r + 11$.

65. Solving a Quadratic Equation

To solve a quadratic equation, put it in the "$ax^2 + bx + c = 0$" form, **factor** the left side (if you can), and set each factor equal to 0 separately to get the two solutions. To solve $x^2 + 12 = 7x$, first rewrite it as $x^2 - 7x + 12 = 0$. Then factor the left side:

$$(x-3)(x-4) = 0$$
$$x - 3 = 0 \text{ or } x - 4 = 0$$
$$x = 3 \text{ or } 4$$

66. Solving a System of Equations

You can solve for two variables only if you have two distinct equations. Two forms of the same equation will not be adequate. **Combine the equations** in such a way that **one of the variables cancels out**. To solve the two equations $4x + 3y = 8$ and $x + y = 3$, multiply both sides of the second equation by -3 to get: $-3x - 3y = -9$. Now add the two equations; the $3y$ and the $-3y$ cancel out, leaving $x = -1$. Plug that back into either one of the original equations, and you'll find that $y = 4$.

67. Solving an Inequality

To solve an inequality, do whatever is necessary to both sides to **isolate the variable**. Just remember that when you **multiply or divide both sides by a negative number**, you must **reverse the sign**. To solve $-5x + 7 < -3$, subtract 7 from both sides to get $-5x < -10$. Now divide both sides by -5, remembering to reverse the sign: $x > 2$.

68. Radical Equations

A radical equation is one that contains at least one radical expression. Solve radical equations by using standard rules of algebra. If $5\sqrt{x} - 2 = 13$, then $5\sqrt{x} = 15$ and $\sqrt{x} = 3$, so $x = 9$.

FUNCTIONS

69. Function Notation and Evaluation

Standard function notation is written $f(x)$ and read "f of x." To evaluate the function $f(x) = 2x + 3$ for $f(4)$, replace x with 4 and simplify: $f(4) = 2(4) + 3 = 11$.

70. Direct and Inverse Variation

In direct variation, $y = kx$, where k is a nonzero constant. In direct variation, the variable y changes directly as x does. If a unit of Currency A is worth 2 units of Currency B, then $A = 2B$. If the number of units of B were to double, the number of units of A would double, and so on for halving, tripling, etc. In inverse variation, $xy = k$, where x and y are variables and k is a constant. A famous inverse relationship is *rate* × *time* = *distance*, where distance is constant. Imagine having to cover a distance of 24 miles. If you were to travel at 12 miles per hour, you'd need 2 hours. But if you were to halve your rate, you would have to double your time. This is just another way of saying that rate and time vary inversely.

71. Domain and Range of a Function

The domain of a function is the set of values for which the function is defined. For example, the domain of $f(x) = \frac{1}{1 - x^2}$ is all values of x except 1 and -1, because for those values the denominator has a value of 0 and the fraction is therefore undefined. The range of a function is the set of outputs or results of the function. For example, the range of $f(x) = x^2$ is all numbers greater than or equal to zero, because x^2 cannot be negative.

COORDINATE GEOMETRY

72. Finding the Distance Between Two Points

To find the distance between points, **use the Pythagorean theorem** or **special right triangles**. The difference between the x's is one leg and the difference between the y's is the other.

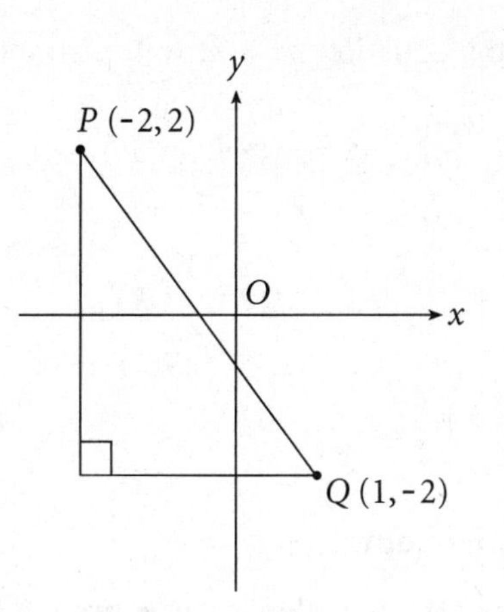

In the figure above, PQ is the hypotenuse of a 3-4-5 triangle, so $PQ = 5$.

You can also use the **distance formula**:

$$d=\sqrt{(x_1-x_2)^2+(y_1-y_2)^2}$$

To find the distance between $R(3,6)$ and $S(5,-2)$:

$$\begin{aligned} d &= \sqrt{(3-5)^2+[6-(-2)]^2} \\ &= \sqrt{(-2)^2+(8)^2} \\ &= \sqrt{68}=2\sqrt{17} \end{aligned}$$

73. Using Two Points to Find the Slope

$$\textbf{Slope} = \frac{\textbf{Change in } y}{\textbf{Change in } x} = \frac{\textbf{Rise}}{\textbf{Run}}$$

The slope of the line that contains the points $A(2,3)$ and $B(0,-1)$ is

$$\frac{y_1-y_2}{x_1-x_2}=\frac{3-(-1)}{2-0}=\frac{4}{2}=2$$

74. Using an Equation to Find the Slope

To find the slope of a line from an equation, put the equation into the **slope-intercept** form:

$$y = mx + b$$

The **slope is *m***. To find the slope of the equation $3x + 2y = 4$, rearrange it:

$$3x+2y=4$$
$$2y=-3x+4$$
$$y=-\frac{3}{2}x+2$$

The slope is $-\frac{3}{2}$.

75. Using an Equation to Find an Intercept

To find the *y*-intercept, you can either put the equation into $\boldsymbol{y = mx + b}$ **(slope-intercept)** form—in which case ***b* is the *y*-intercept**—or you can just **plug $x = 0$** into the equation and **solve for *y***. To find the *x*-intercept, plug $\boldsymbol{y = 0}$ into the equation and **solve for *x***.

76. Finding the Midpoint

The midpoint of two points on a line segment is the average of the *x*-coordinates of the endpoints and the average of the *y*-coordinates of the endpoints. If the endpoints are (x_1,y_1) and (x_2,y_2), the midpoint is $\left(\frac{x_1+x_2}{2},\frac{y_1+y_2}{2}\right)$. The midpoint of (3,5) and (9,1) is $\left(\frac{3+9}{2},\frac{5+1}{2}\right)$ (or (6, 3).

LINES AND ANGLES

77. Intersecting Lines

When two lines intersect, **adjacent angles are supplementary, and vertical angles are equal**.

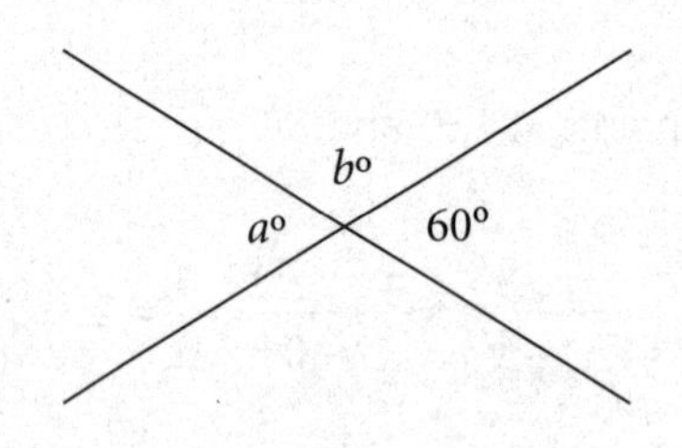

In the figure above, the angles marked $a°$ and $b°$ are adjacent and supplementary, so $a + b = 180$. Furthermore, the angles marked $a°$ and 60° are vertical and equal, so $a = 60$.

78. Parallel Lines and Transversals

A transversal across parallel lines forms **four equal acute angles and four equal obtuse angles**, unless the transversal meets the lines at a right angle; then all eight angles are right angles.

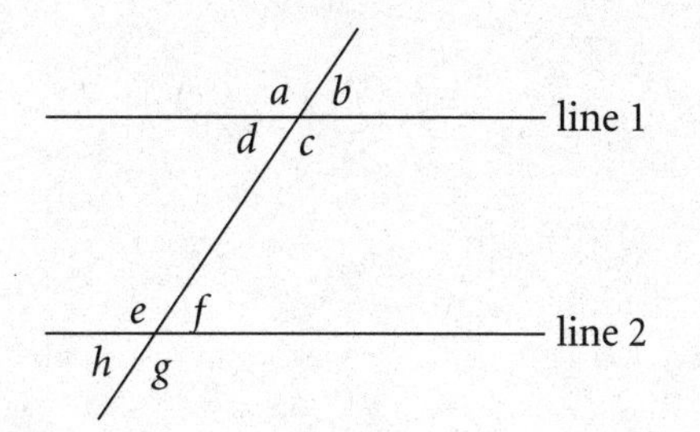

In the figure above, line 1 is parallel to line 2. Angles *a, c, e,* and *g* are obtuse, so they are all equal. Angles *b, d, f,* and *h* are acute, so they are all equal.

Furthermore, **any of the acute angles is supplementary to any of the obtuse angles**. Angles *a* and *h* are supplementary, as are *b* and *e, c* and *f,* and so on.

TRIANGLES—GENERAL

79. Interior and Exterior Angles of a Triangle

The three angles of any triangle **add up to 180 degrees**.

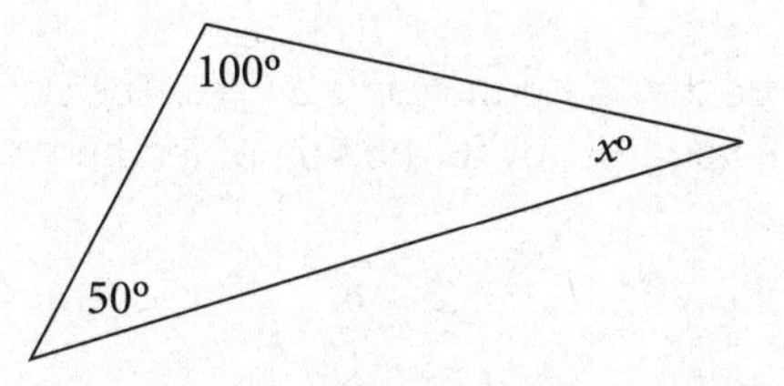

In the figure above, $x + 50 + 100 = 180$, so $x = 30$.

An exterior angle of a triangle is equal to the **sum of the remote interior angles**.

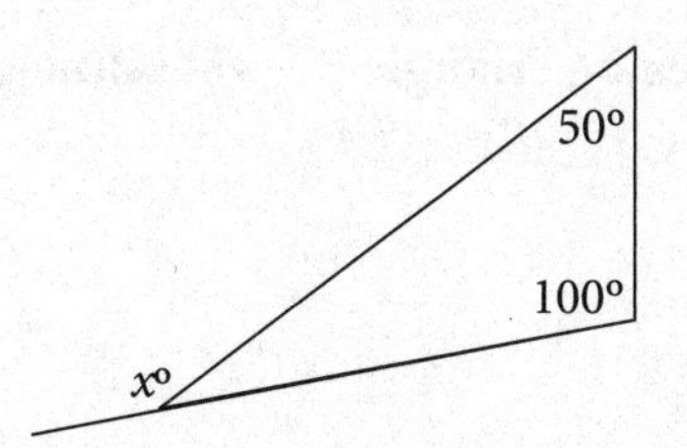

In the figure above, the exterior angle labeled $x°$ is equal to the sum of the remote angles: $x = 50 + 100 = 150$.

The three exterior angles of a triangle **add up to 360 degrees**.

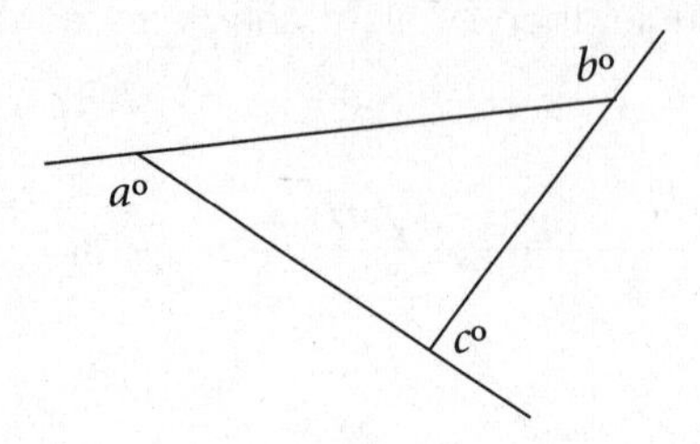

In the figure above, $a + b + c = 360$.

80. Similar Triangles

Similar triangles have the same shape: **corresponding angles are equal and corresponding sides are proportional**.

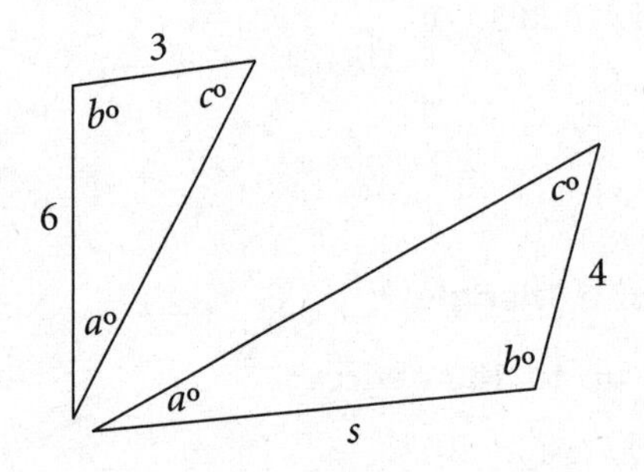

The triangles above are similar because they have the same angles. The side of length 3 corresponds to the side of length 4, and the side of length 6 corresponds to the side of length *s*.

$$\frac{3}{4} = \frac{6}{s}$$
$$3s = 24$$
$$s = 8$$

81. Area of a Triangle

$$\textbf{Area of Triangle} = \frac{1}{2}\ \textbf{(base)(height)}$$

The height is the perpendicular distance between the side that's chosen as the base and the opposite vertex.

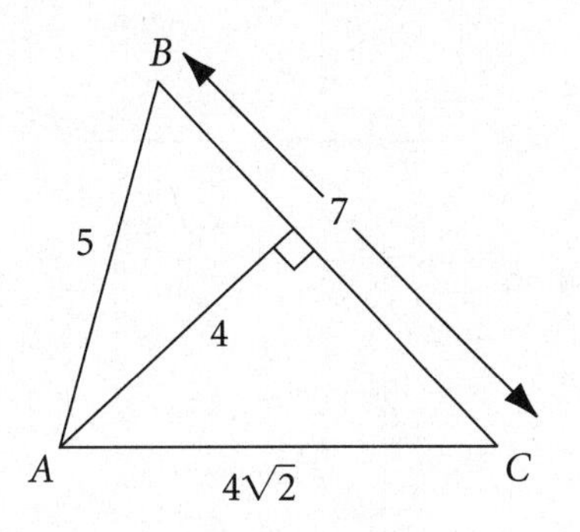

In the triangle above, 4 is the height when 7 is chosen as the base.

$$\text{Area} = \frac{1}{2}bh = \frac{1}{2}(7)(4) = 14$$

82. Triangle Inequality Theorem

The length of one side of a triangle must be **greater than the difference between and less than the sum** of the lengths of the other two sides. For example, if it is given that the length of one side is 3 and the length of another side is 7, then you know that the length of the third side must be greater than 7 – 3 = 4 and less than 7 + 3 = 10.

83. Isosceles and Equilateral Triangles

An isosceles triangle is a triangle that has **two equal sides**. Not only are two sides equal, but the angles opposite the equal sides, called **base angles**, are also equal.

Equilateral triangles are triangles in which **all three sides are equal**. Since all the sides are equal, all the angles are also equal. All three angles in an equilateral triangle measure 60 degrees, regardless of the lengths of the sides.

RIGHT TRIANGLES

84. Pythagorean Theorem

For all right triangles:

$$(\text{leg}_1)^2 + (\text{leg}_2)^2 = (\text{hypotenuse})^2$$

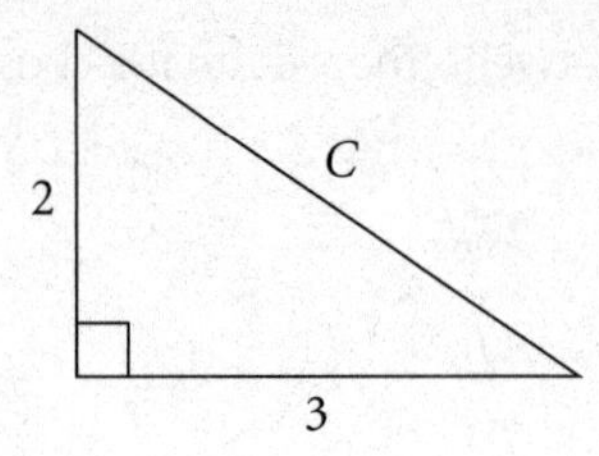

If one leg is 2 and the other leg is 3, then:

$$2^2 + 3^2 = c^2$$
$$c^2 = 4 + 9$$
$$c = \sqrt{13}$$

85. The 3-4-5 Triangle

If a right triangle's leg-to-leg ratio is 3:4, or if the leg-to-hypotenuse ratio is 3:5 or 4:5, it's a 3-4-5 triangle, and you don't need to use the Pythagorean theorem to find the third side. Just figure out what multiple of 3-4-5 it is.

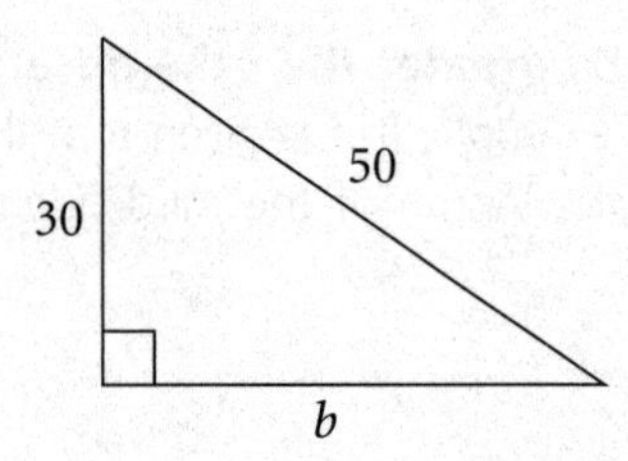

In the right triangle shown, one leg is 30, and the hypotenuse is 50. This is 10 times 3-4-5. The other leg is 40.

86. The 5-12-13 Triangle

If a right triangle's leg-to-leg ratio is 5:12, or if the leg-to-hypotenuse ratio is 5:13 or 12:13, then it's a 5-12-13 triangle, and you don't need to use the Pythagorean theorem to find the third side. Just figure out what multiple of 5-12-13 it is.

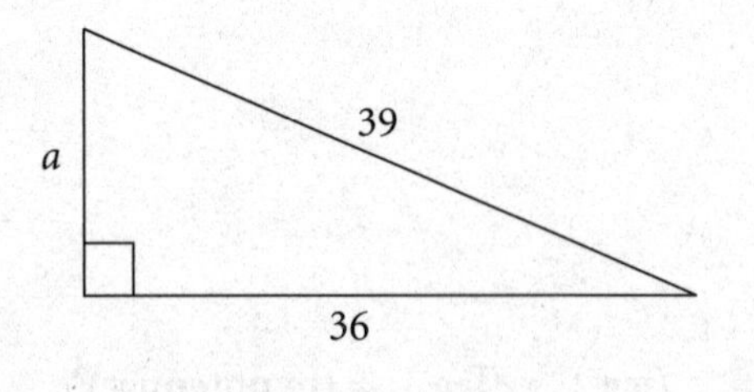

Here one leg is 36, and the hypotenuse is 39. This is 3 times 5-12-13. The other leg is 15.

87. The 30-60-90 Triangle

The sides of a 30-60-90 triangle are in a ratio of $\boldsymbol{x : x\sqrt{3} : 2x}$. You don't need the Pythagorean theorem.

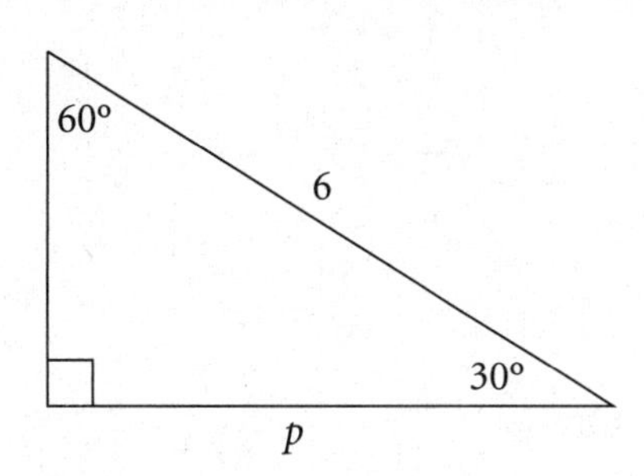

If the hypotenuse is 6, then the shorter leg is half that, or 3, and then the longer leg is equal to the short leg times $\sqrt{3}$, or $p = 3\sqrt{3}$.

88. The 45-45-90 Triangle

The sides of a 45-45-90 triangle are in a ratio of $\boldsymbol{x : x : x\sqrt{2}}$.

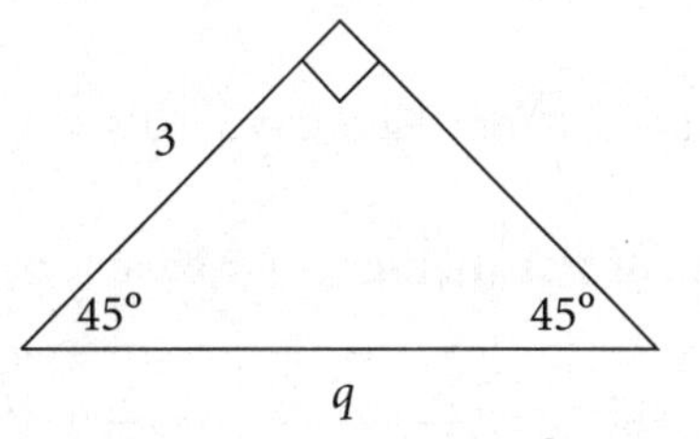

If one leg has a length of 3, then the other leg also has a length of 3, and the hypotenuse is equal to a leg times $\sqrt{2}$, or $q = 3\sqrt{2}$.

OTHER POLYGONS

89. Characteristics of a Rectangle

A rectangle is a **four-sided figure with four right angles**. Opposite sides are equal. Diagonals are equal.

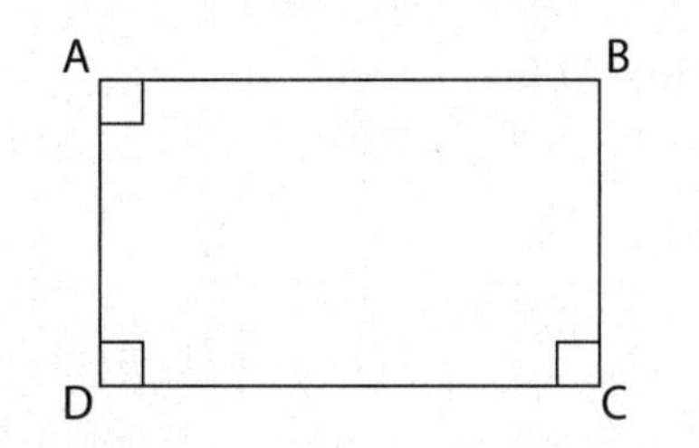

Quadrilateral *ABCD* above is shown to have three right angles. The fourth angle therefore also measures 90 degrees, and *ABCD* is a rectangle. The **perimeter** of a rectangle is equal to the sum of the lengths of the four sides, which is equivalent to 2(length + width).

Area of Rectangle = length × width

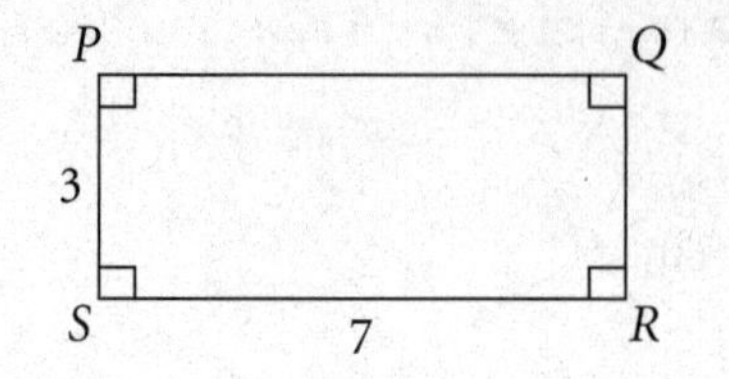

The area of a 7-by-3 rectangle is $7 \times 3 = 21$.

90. Characteristics of a Parallelogram

A parallelogram has **two pairs of parallel sides**. Opposite sides are equal. Opposite angles are equal. Consecutive angles add up to 180 degrees.

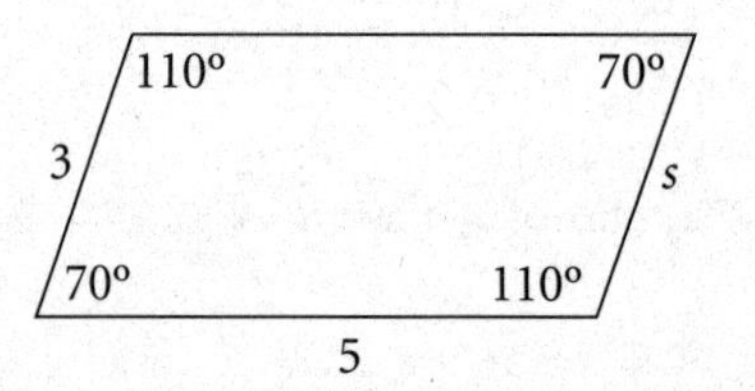

In the figure above, s is the length of the side opposite the 3, so $s = 3$.

Area of Parallelogram = base × height

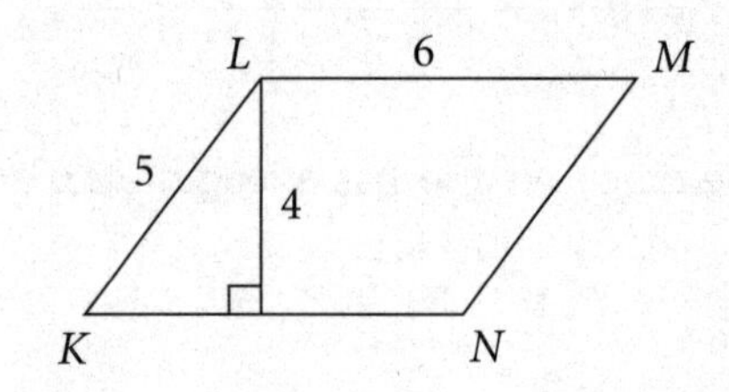

In parallelogram *KLMN* above, 4 is the height when *LM* or *KN* is used as the base.

Base × height = $6 \times 4 = 24$.

91. Characteristics of a Square

A square is a **rectangle with four equal sides**.

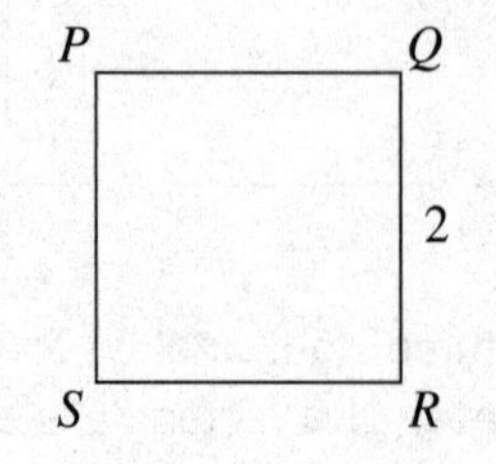

If *PQRS* is a square, all sides are the same length as *QR*. The **perimeter** of a square is equal to four times the length of one side.

$$\textbf{Area of Square} = (\textbf{side})^2$$

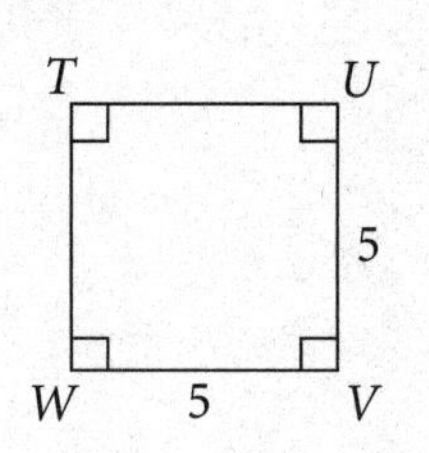

The square above, with sides of length 5, has an area of $5^2 = 25$.

92. Interior Angles of a Polygon

The **sum of the measures of the interior angles of a polygon = (n – 2) × 180**, where n is the number of sides.

$$\textbf{Sum of the Angles} = (n - 2) \times 180$$

The eight angles of an octagon, for example, add up to $(8 - 2) \times 180 = 1{,}080$.

CIRCLES

93. Circumference of a Circle

$$\textbf{Circumference} = 2\pi r$$

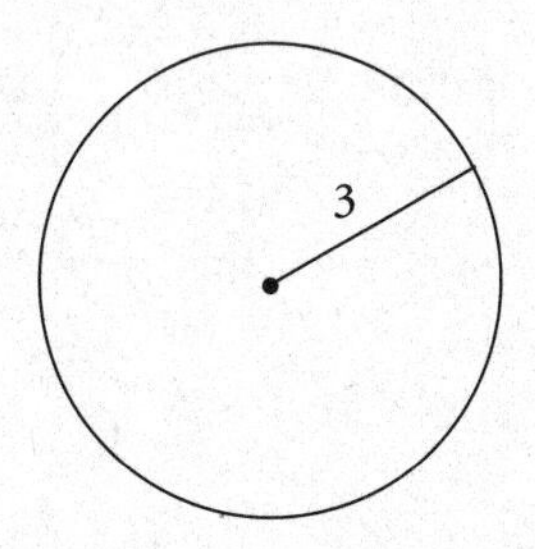

In the circle above, the radius has a length of 3, so the circumference is $2\pi(3) = 6\pi$.

94. Length of an Arc

An arc is a piece of the circumference. If n is the degree measure of the arc's central angle, then the formula is

$$\textbf{Length of an Arc} = \left(\frac{n}{360}\right)(2\pi r)$$

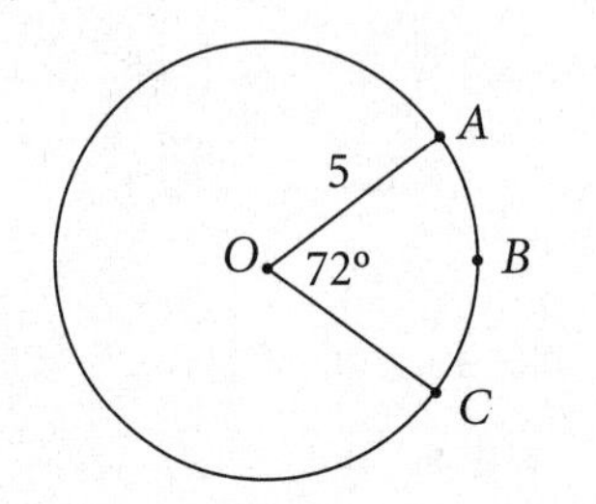

In the figure above, the radius has a length of 5, and the measure of the central angle is 72 degrees. The arc length is $\frac{72}{360}$, or $\frac{1}{5}$, of the circumference:

$$\left(\frac{72}{360}\right)(2\pi)(5) = \left(\frac{1}{5}\right)(10\pi) = 2\pi$$

95. Area of a Circle

$$\textbf{Area of a Circle} = \pi r^2$$

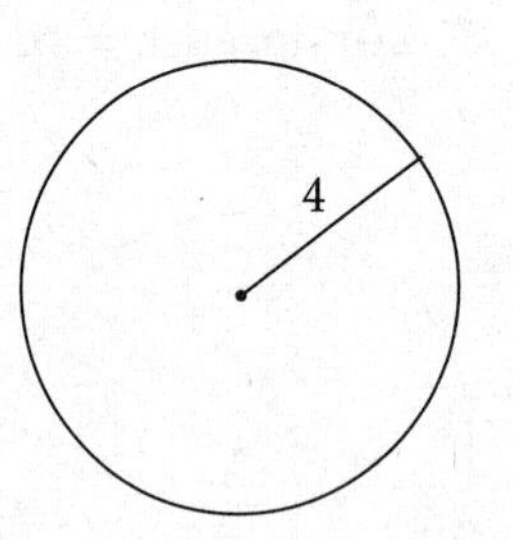

The area of the circle is $\pi(4)^2 = 16\pi$.

96. Area of a Sector

A sector is a piece of the area of a circle. If n is the degree measure of the sector's central angle, then the formula is

$$\textbf{Area of a Sector} = \left(\frac{n}{360}\right)(\pi r^2)$$

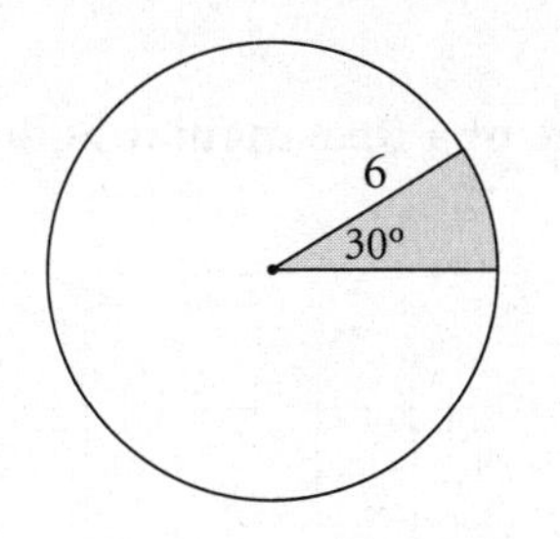

In the figure above, the radius has a length of 6, and the measure of the sector's central angle is 30 degrees. The sector has $\frac{30}{360}$, or $\frac{1}{12}$, of the area of the circle:

$$\left(\frac{30}{360}\right)(\pi)(6^2)=\left(\frac{1}{12}\right)(36\pi)=3\pi$$

97. Tangency

When a line is tangent to a circle, the radius of the circle is perpendicular to the line at the point of contact.

SOLIDS

98. Surface Area of a Rectangular Solid

The surface of a rectangular solid consists of three pairs of identical faces. To find the surface area, find the area of each face and add them up. If the length is l, the width is w, and the height is h, the formula is

$$\textbf{Surface Area} = 2lw + 2wh + 2lh$$

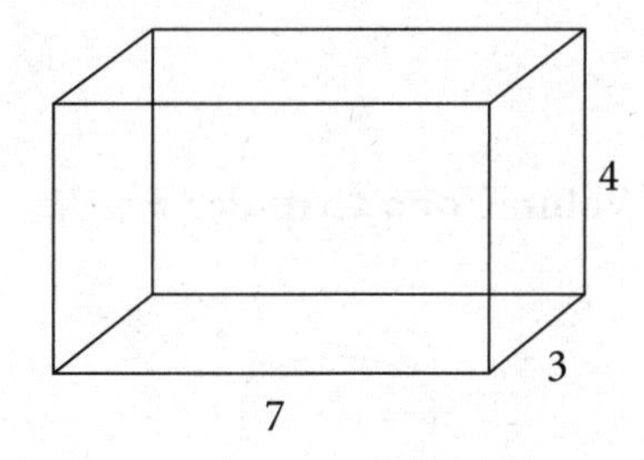

The surface area of the box above is: $(2 \times 7 \times 3) + (2 \times 3 \times 4) + (2 \times 7 \times 4) = 42 + 24 + 56 = 122$

99. Volume of a Rectangular Solid

Volume of a Rectangular Solid = lwh

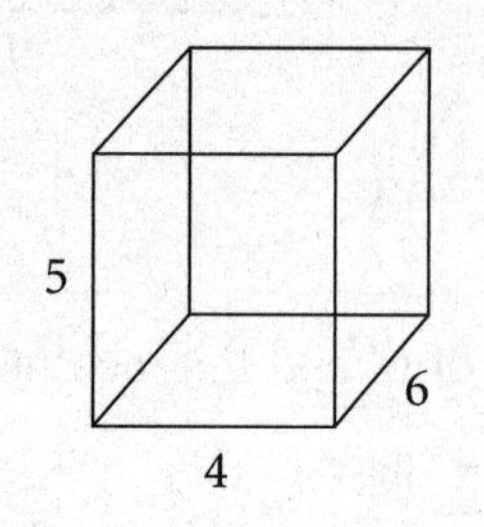

The volume of a 4-by-5-by-6 box is

$$4 \times 5 \times 6 = 120$$

A cube is a rectangular solid with length, width, and height all equal. If s is the length of an edge of a cube, the volume formula is

Volume of a Cube = s^3

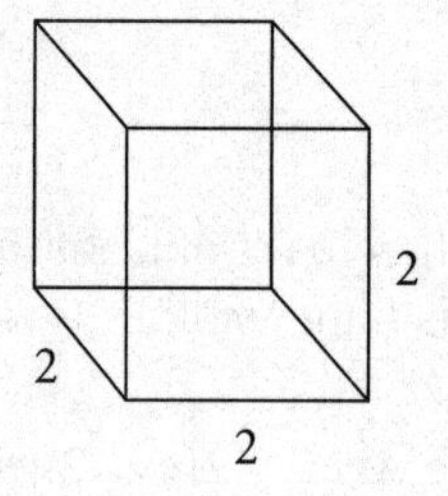

The volume of this cube is $2^3 = 8$.

100. Volume of a Cylinder

Volume of a Cylinder = $\pi r^2 h$

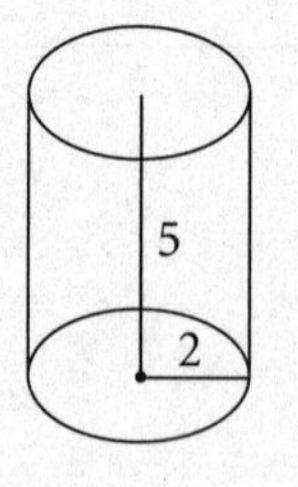

In the cylinder above, $r = 2$, $h = 5$, so

$$\text{Volume} = \pi(2^2)(5) = 20\pi$$

NOTES

NOTES

NOTES

NOTES